Reflexive Cartography

Modern Cartography Series
Volume 6

Reflexive Cartography

A New Perspective on Mapping

Emanuela Casti

D.R. Fraser Taylor, Series Editor

AMSTERDAM • BOSTON • HEIDELBERG • LONDON • NEW YORK • OXFORD
PARIS • SAN DIEGO • SAN FRANCISCO • SINGAPORE • SYDNEY • TOKYO

Elsevier
Radarweg 29, PO Box 211, 1000 AE Amsterdam, Netherlands
The Boulevard, Langford Lane, Kidlington, Oxford OX5 1GB, UK
225 Wyman Street, Waltham, MA 02451, USA

ISBN: 978-0-12-803509-2

ISSN: 1363-0814

British Library Cataloguing in Publication Data
A catalogue record for this book is available from the British Library

Library of Congress Cataloging-in-Publication Data
A catalog record for this book is available from the Library of Congress

For information on all Elsevier publications
visit our website at http://store.elsevier.com/

Contents

The images contained in this book are visible in full color at the URL: booksite.elsevier.com/9780128035092.

Preface

Reflexive cartography introduces, as the subtitle of the book suggests, a new and very important perspective in mapping in both theoretical and applied terms. The book is the sixth volume of the Modern Cartography Series and represents a valuable contribution, not only to cartographic theory and practice, but also to critical social science thinking. Thinking on reflexivity goes far beyond cartography and this book draws attention to the need for scholars from all disciplines to critically examine both their theories and practice. There is a need to escape from the rigidity of many theoretical constructs and embrace a greater degree of transdisciplinary pluralism. There is also a related need to develop more innovative methodological approaches. This book makes a contribution to both of these challenges in cartography.

Cartography increasingly diverged from geography after World War II, in which cartographic technology played a significant role. The strong technical approach to cartography continued after the War and the gap between cartography and geography widened, helped in no small measure by the fact that the so-called "quantitative revolution" in geography in the 1950s undervalued the role of the map. Until that time, cartography was a subsection of geography in the International Geographical Union, but in 1959 the cartographers established their own international organization, called the International Cartographic Association, and the growing divergence between the two disciplines became a formal split. This book makes a strong case for a greater degree of reintegration of the two disciplines and especially argues that cartography must escape the rigidities of a purely technical topographic approach and concentrate on "mapping," in every sense of that word, including a social sense of territory.

Reflexivity is an integral part of the growing field of critical cartography to which this book makes a valuable contribution. As the author points out, cartography has an important role to play in establishing a social view of the world and in linking the local scale of "inhabited space" with the emerging realities of our increasingly globalized world.

The advent of location-based and remotely sensed computer technologies offers opportunities for new cartographic explorations of a complex world.

The black and white metrics of an increasingly reductionist approach must be replaced by a consideration of what the author calls "CHORA." She also rightly points out the importance of learning from past experience and illustrates this by a number of interesting examples, including those from colonial cartography in the African context.

The original version of this book was published in Italian and, as a result, the ideas it contains were not as widely considered as they might have been. This translated volume has also been supplemented by the introduction of new material. Capturing the nuances of complex concepts from the original Italian in English has been a challenge ably met by Dr Davide Del Bello, who translated the book. Prof. Emanuela Casti has developed her ideas over a number of years and she mentions in the acknowledgments that she has been ably assisted by her team at the *Diathesis* Cartographic Lab of the University of Bergamo which provided "a human context shaped by women." I congratulate Prof. Casti on the publication of this outstanding book and I am pleased to have played a small role in bringing it to an English-speaking audience.

D.R. Fraser Taylor, FRSC
Chancellor's Distinguished Research Professor
Carleton University, Ottawa, Canada
June 2015

Introduction: Cartography's Building Site

In the 1980s, studies in critical cartography brought to the fore new modes of representation as an alternative to the Western tradition and promoted a rereading of European Conquest maps themselves. Since then, cartography took on the unprecedented role of establishing the arena for a "geographical decolonization;" that is, a reinterpretation of the meaning of the encounter between the colonizers and the colonized. The next step was to envisage a *counter mapping*, to advocate a cartography conceived as a tool for upholding the rights of local communities against the ruling hegemony, as a current of opposition, criticism, and "counter-project" aimed to redress the asymmetries of power.[1]

At the same time, many researchers—most notably American geographers and anthropologists, but also European scholars—engaged in a program of "research/action" that combines interpretative study with the production of new maps. Such a program relies on reflexivity, because it implies the researcher's involvement both in the study and in the effective solution of socially relevant issues, such as the role digital technologies play in empowerment, or the potential cultural assimilation brought about by cartographic tools.[2]

Reflexivity is, therefore, an analytical feature of critical cartography* that applies to the present book in two forms: both as a research perspective and as the cartographic segment it proceeds from. From the point of view of research, cartography is seen as dithering between interpretation and the construction of a map. Reflexivity offers a set of tools that enables us not only to understand what comes to us from the past but also to rethink what we do and

---

[1] D. Wood, *The Power of Maps*, Guilford Press, New York, 1992; J. Crampton, J. Krygier, "An introduction to critical cartography," in: *ACME. An International E-Journal for Critical Geographies*, 2005, 4 (1), pp. 11–33.

[2] On the thorny issue of mapping among indigenous societies, see among others: D.R.F. Taylor, T. Lauriault, eds., *Developments in the Theory and Practice of Cybercartography. Applications and Indigenous Mapping*, Elsevier, Amsterdam, 2014; for issues relating to complex societies, see the recent volume: D. Sui, S. Elwood, and M. Goodchild, eds., *Crowdsourcing Geographic Knowledge, Volunteered Geographic Information (VGI) in Theory and Practice*, Springer, Berlin, 2013.

thus provide suggestions for future improvements. In the field of the sciences, reflexivity matters because it lays down practical goals to be achieved by making researchers pause to observe themselves and take stock of what they do or have done. After all, their actions ultimately set up the scene they are expected to act upon.[3] Complexity is known to expose uncertainties, doubts, risks, and value conflicts: all these may be addressed only once they have been allowed to the surface and made the object of reflection. Research often relies unconditionally on theoretical principles that tend to deny complexity. By dismissing such rigid principles, reflexive practice enables researchers to rewire their thought patterns and see things differently, to embrace pluralism and encourage methodological innovation.

The second feature, to do with the cartographic field from which reflection originates, is equally important. By favoring competing worldviews that enhance a self-reflexive approach to what is handed down, the present book envisions Euclidean cartography as a sort of prelude from which researchers may derive new metrics able to convey the spatiality of the contemporary world. That is in line with Habermas's view, whereby we need to be able to discuss principles that come to us from the past before we even sit down to identify the principles on which new communicative action is to be based.[4]

To avoid raising false expectations, we need to clarify that this book is not meant to address the theoretical and practical question of laying down principles or issuing prescriptions on how to build an epiphanic cartography. Rather, it provides an outline of current cartographic experimentation by throwing light on a complex and chequered scenario. That does not prevent us from envisaging a kind of cartography able to convey a social sense of territory, to be achieved by leaving topographic metrics behind and adopting what is here called a "chorographic" cartography, able to boost self-reflexivity in the very process of its creation. Cartography is thereby made to take up a challenge we could phrase in these terms: since digital technology offers unprecedented possibilities and overcomes some of the limitations inherent in cartography, could cartography then envisage a world with the features that globalization imprinted on it?

Globalization has been shown to mark a major shift, forcing geography to give up its traditional epistemic assumptions and seek alternative grounding in new categories of analysis. Insomuch as it involves the spatialization of social

---

[3] Among the studies that record a growing interest in this critical approach, see: T. May, B. Perry, *Social Research and Reflexivity. Content, Consequence, and Context*, Sage, London and Thousand Oaks, 2010; in cartography: M. Dodge, R. Kitchin, C. Perkins, eds., *Rethinking Maps: New Frontiers in Cartographic Theory*, Routledge, London, 2011; Id., *The Map Reader: Theories of Mapping Practice and Cartographic Representation*, John Wiley & Sons, Chichester, 2011.

[4] J. Habermas, *Theory of Communicative Action Volume 2: Liveworld and System: A Critique of Functionalist Reason*, Beacon Press, Boston, 1987, pp. 121–126, cit. p.124.

phenomena, globalization is an inherently geographic process, whereby social space encompasses global dimensions that call into question the models we have employed so far to make sense of spatial phenomena. The new emerging model is the network, which restores movement as an essential factor in the relationship between humans and their environment.[5] We envision a world in which local/global scales and their interactivity become synonymous with community/society. The local scale would convey a subjective dimension of inhabiting a place, expressed in a sense of belonging. The global scale would be instead expected to constitute an unprecedented social "unit."[6]

If, as Jacques Lévy claimed in his metaphor, we need to "change lenses" to observe this new world—given that its primary representation, cartography, was by definition a distorting lens—then we will need to change our glasses in order to examine it.[7] Hence, faced with the emergence of new spatial categories, cartography needs to rethink itself and revive what in recent times seemed bound to obsolescence, namely its inevitable link with geography. The field of cartography today is set to provide a new spatiality, a world no longer made up of lands, seas, continents, states, but of human beings, communities that metamorphose such features from physical data into inhabited space. Hence the need to reestablish ties with geography, whose statute it is to analyze that world. It is essential that cartographic reflection should not focus exclusively on a physical rendering of the world, as new as that might seem, but that it should raise questions involving the rendering of its social significance, possibly by looking at areas traditionally quite alien to its field, such as the language of technical and visual arts.[8]

Such perspectives are obviously hazardous, because analysts of cartography cannot possibly master the vast and complex universe of visual communication with multidisciplinary expertise. They may, however, adopt a method that, anchored to their own skills would allow for outside forays and prevent entrenchment within self-contained certainties. The Renaissance interplay between technique, science, and art is the projective domain one needs to embrace in order to engage

---

[5] One of the first researchers to draw attention to this issue was: G. Dematteis, *Progetto implicito. Il contributo della geografia umana alla scienza del territorio*, F. Angeli, Milan, 1995.

[6] In his book *The Society of Individuals* Norbert Elias raised a crucial issue about the nature of the link between autonomous individuals in a global society: does a global "we" exist? If, as research seems to suggest, it does exist, then the global "we" cannot be but "societal," since what is at stake is the identity of communities struggling to safeguard themselves against the diversity that threatens them. In fact, we cannot speak of a world community since there does not exist a different or hostile Other to be resisted. (N. Elias, *The Society of Individuals*, Blackwell, Oxford, 1939).

[7] J. Lévy, "Introduction. Un événement géographique," in: J. Lévy, ed., *L'invention du monde: une geographie de la mondialisation*, Presses de Sciences Po, Paris, 2008, p.11–16.

[8] As recently advocated: W. Cartwright, G. Gartner, A. Lehn, eds., *Cartography and Art*, Springer, Berlin-Heidelberg, 2009.

in a reflection on contemporary cartography. That entails restraining one's claims and conceding that, while knowledge may have been managed systematically during the Renaissance, nowadays it would seem unthinkable even to imply that such management even exists. Rather, scholars are expected to place their analysis within liminal spaces, areas that straddle a number of disciplines. And they are expected to do so "within their own discipline," drawing from various areas in order to ready their exploratory gear. In short, what is needed is a heuristic approach, capable of holding together the outcomes of cartographic and of geographic theories, the artistic hybridizations envisaged by historical cartography, and the possibilities offered by digital technology.[9] The work I present should be seen as an attempt to pursue such line of inquiry.

By taking on a semiotic perspective, which showed the communicative potential of maps, the present book proposes a reflection on the possible adaptation of cartography to a societal view of the world. I challenge topography-based metrics and call for a topology of places with a view to proposing the adoption of new metrics based on digital technologies that show new modes of interaction between cartographers and recipients. On the one hand, within the field of territorial management, participatory maps have been contributing with increasing success to decision-making at the negotiating table and are well-placed to promote governance between the actors involved. In Europe, notably, but also in Africa, within the realm of environmental cooperation, participatory maps feature prominently as tools used in territorial planning to achieve multiple objectives: first, to produce a diagnostics report; secondly, to advocate proactive solutions; finally, to mirror the process of reflection which involves multiple actors who collectively formulate a spatial line of reasoning. Since this type of cartography is a tool for discussing the points of view of both institutional and local actors, it is clear that its relevance is political. And it becomes inevitable to think about which languages to use, which perspectives to embrace, and which interactive potentials to tap. The advent of Google Map/Earth and the development of Geoweb 2.0 brought about a true paradigm shift in the social use of maps, to the point that maps now provide the main ground for spatial indexing of knowledge and information. Maps are no longer mere representations of territory, but configure themselves as the preferred interface for accessing hybrid, and especially urban, spaces. On a wide range of media and "smart" mobile platforms (phones, tablets, online computers, interactive screens…) maps provide direct access and the ability to move across different areas of the "smart city."

Therefore, the call is directed to the Web and to the innovative aspects it injects into cartography. Consider especially the innovative thrust of Geographical

[9] Such "interdisciplinary" approach is all the more necessary when we acknowledge cartography's ontological crisis: R. Kitchin, M. Dodge, *Rethinking maps, op. cit.*

Information Systems (GIS), which, even without the support of a networked environment, explode some of the assumptions behind the creation, transmission and production of a map's meaning. The enormous amount of data a GIS can handle, as it parses an unlimited number of attributes for each geographical phenomenon; the ability to process and to render spatial relationships otherwise hard to detect; the ability to integrate different data at different scales, and coming from disparate sources; the ability, finally, to design ever-new representations by manipulating the same data sets (thanks to the split between the archival function of information, entrusted to the database, and its iconic role, achieved by its output) leave no doubt. By virtue of its immense potential, GIS is the type of innovation that has overtaken maps and gone far beyond its semantics.[10] Having said that, I must admit that the true paradigm shift occurs when GIS technology integrates with the Web. That is where a threshold leap occurs, for once a GIS is made to interact with the Web, final products are no longer possible. The nature of WebGIS shines forth in its endless refashioning, in the dynamism of a cartographic construct that may never be said to be complete or finalized. WebGIS enables anyone to make or unmake maps: their products are never concluded and indefinitely subject to change. That actually marks the breaking point with traditional maps, the most intriguing and compelling aspect of WebGIS, which lies at the core of my research, for we cannot reasonably presume that such a distinctive outcome would fail to engender a special type of semiosis. In other words, even though WebGIS technology may be said to derive from conventional cartography, I submit that it needs to be rethought as a thoroughly new tool, by focusing on its communicative rather than on its technical side. It is a tool that recovers the semiotic models of cartography with a view to formulating its own.

The book is divided into chapters: the first places our research within the landscape of cartographic semiosis, seen as a hermeneutic interpretive approach juxtaposed to the approaches that came before it; the second deals with the debatable quality of topographic maps and their communicative implications as exemplified in colonial maps; and the third evokes other ways of mapping, in use prior to Euclidean models, and shows how landscape featured prominently in them in terms of social construction. These three chapters form the first part of the book, followed by a second section devoted to contemporary

---

[10] Not surprisingly, in reference to GIS, some speak of GIP (Geographic Information Processing) to emphasize less the technical setup which turns GIS into a software and hardware tool than its extraordinary and multiple capabilities to support social processes and projects. Fraser Taylor, one of the pioneers of geomatics, placed the issue at the center of his research interests and publications, and continues to advance it in his capacity as director of the Geomatics and Cartographic Research Centre at Carleton University in Ottawa: https://gcrc.carleton.ca/confluence/display/GCRCWEB/About+GCRC.

cartography and its experiments. The fourth chapter addresses participatory mapping technologies able to capture the topological dimension of places; the fifth proposes a cartography of landscape based on the representation of its iconic values; finally, the sixth surveys ongoing experiments in several workshops aimed at rendering network-like spatiality and its social implications.

Each chapter provides a piece in the mosaic that describes our underlying goal as mentioned in the title: a shift from topography-based maps, centered on the topos, to chorographic maps, based on the chora. Our line of inquiry is open-ended: it presents no final, or even provisional, findings, choosing instead to echo Popper's memorable appeal to cherish the transience of what is achieved by remembering that "research has no end."

Acknowledgments

This is the space customarily devoted to acknowledgments and to the recognition of intellectual debts toward those who contributed to the final shape of the book. I am going to use it, instead, to mention the place where the book was conceived and to take stock of the professional and personal environment that surrounds it.

It all began in the *Diathesis* Cartographic Lab of the University of Bergamo, although the book was actually started long before the laboratory was granted official recognition. Both the book and the laboratory acted autopoietically, imposing their own presences. Articles published over time would provide the framework for the book chapters; the addition of computers, software, printers, plotters, and large screens would in time sanction the existence of a space dedicated to experimentation and research.

As for the people involved in the laboratory, they are many and various. One invariable feature is their gender: the women who now are the mainstay of our workshop (Federica Burini and Alessandra Ghisalberti); the women who pursued other paths (Chiara Brambilla, Francesca Cristina Cappennani, Michela Della Chiesa, Federica Fassi); and the women who came to us by their own independent routes (Annarita Lamberti, Sara Belotti). It is a human context shaped by women, albeit not programmatically. Its features are a passion for research but also for arduous challenges, combined with commitment and determination to pursue one's goals.

The outside world, with which I established a close network of contacts, exchanges, and collaborations at various levels, was instead mostly characterized by the presence of men. The most important: Jacques Lévy, mentor and driving force behind cartographic experimentation; Giorgio Mangani who responded both professionally and amicably to my project; Oliver Lompo and Andrea Masturzo who maintained close and intense ties in the course of their long-distance training and research.

The international release of my book was made possible thanks to the support of men and women to whom I am indebted for this English-language edition: Prof. Fraser Taylor, who shares my interests and intellectual pursuits, was the first to encourage publication of this book in English; his appreciation and crucial assistance were greatly appreciated and call for unreserved gratitude; John Fedor, at Elsevier, who promptly embraced the project and put it into practice; Marisa LaFleur, who took over the painstaking process of editing with passion, patience, and dedication: Davide Del Bello, whose meticulous translation endeavored to overcome linguistic hurdles.

All this would have been impossible without the unconditional support of my family - Carlo primarily - which in the meantime grew larger: Claudia and Niccolò joined the original members while Federica, Alessandra and Francesca gained on the field the right to affective inclusion.

I count myself very fortunate.

PART 1

CHAPTER 1

Cartographic Interpretation Between Continuity and Renewal: On the Trail of *Chora*

I approach deep problems like cold baths: quickly into them and quickly out again.

F. Nietzsche, *The Gay Science*, 1882

This chapter surveys a number of research trends from the last few decades and zeroes in on the semiotic approach as a potential ground for advocating new cartographic modes. By shifting our focus of interest from maps as territory mediators *to maps as* operators *capable of eliciting action, this approach unveils the specific contexts that need to be addressed in order to take control of the communicative outcomes of maps. Two cartographic phenomena are submitted as especially crucial:* self-reference *and* iconization. *The former marks a map's propensity to be accepted by virtue of its mere existence and to influence communication quite independently of the cartographer's intentions; the latter relies on those self-referential outcomes to present highly conjectural facts as if they were truths. Maps provide a model that replaces territory, yet fails to represent it. Through iconization, maps put knowledge of the material world aside in order to assert another dimension that they shaped and established. In the modern period, such replacement occurred via topographic maps, a translation of Cartesian logic, which posited territory as* topos, *that is, territory in its superficially abstract sense. As it sets out to abandon Cartesian logic in order to restore the* chora *– which in fact enhances the cultural aspects of the area and an individual's relation to the place where he or she lives – cartographic semiosis becomes a privileged scenario for identifying areas on which to act.*

CONTENTS

SOCIETY AND CARTOGRAPHY

Over the last few decades, studies in cartography have subsumed a number of approaches, points of view, and theoretical considerations aimed at recovering the problematic nature of maps* and their social role. Attention was paid

* Asterisks refer to the Compass/Glossary section where concepts and definitions are explained. In the specific case of *map*, I devoted most of Chapter 1 to a discussion of the concept because it is vital to the development of my argument.

Reflexive Cartography. ISSN 1363-0814, http://dx.doi.org/10.1016/B978-0-12-803509-2.00001-X

both to the set of tools used to interpret the history of cartography[1] and to the new possibilities offered by Geographic Information Systems (GIS).[2] Scholars agree that the symbolic apparatus used to represent the world derives from the values on which a given society is based, values according to which societal knowledge will be organized. Nor is that all. It has also become evident that each society produces particular views of its territory, according to the specific relationship established with it and the practices it is invested with.

By now, we are quite removed from studies of cartographic history imbued with scientific positivism. Such studies based their analysis either on the technical features of maps or on the self-evidence of what maps represented, thus sanctioning and strengthening the maps' alleged or claimed objectivity.[3] When the notion that maps mirrored reality was finally rejected, maps started to be considered as tokens of the intellectual appropriation humans pursue as they endeavor to master the world. All that enabled scholars to recover a dual cartographic perspective: the notion of maps as social products that show us the ways in which a given society builds its own items of territorial knowledge and the idea of maps as means of communication, whereby these knowledge items are circulated. Maps thus function as symbolic *operators* able to affect territorial agents directly. As for the first aspect, it was finally understood that maps are an entirely special type of representation, able to generate a territorial image that stands as a truthful, unquestioned and wholly authoritative final product.[4] Secondly, once the self-referential working of maps was highlighted, maps could be seen as a means of communication able to supply their interpreters with strategies for the production, use and mediatization* of territory.

---

[1] This was done by explaining the role of maps within the social group that produced them, with respect to the period of history they belong to and the political project they uphold. Among the many contributions on this issue see: (for the Italian context) G. Mangani, *Cartografia morale*, Panini, Modena, 2006; (for the international context) the third volume of *The History of Cartography*, a series edited by J.B. Harley and D. Woodward, University of Chicago Press, Chicago, 2007.

[2] For a brief overview of the Italian context, see: E. Casti, J. Lévy, eds., *Le sfide cartografiche: movimento, partecipazione, rischio*, Il lavoro editoriale/Università, Ancona, 2010. For the international context: P.A. Longley, M. Goodchild, D.J. Maguire, and D.W. Rhind, eds., *Geographic Information Systems and Science*, John Wiley and Sons, New York, 2011.

[3] Among others: A. Robinson, *The Look of Maps: An Examination of Cartographic Design*, University of Wisconsin Press, Madison, 1952.

[4] Especially over the unprecedented social role maps play, despite the fact that they derive from manipulation of various information sources and refer to the context of their production (cosmological notions of the cartographer, institutional interests, the technique used, the conventions in use…). Similar arguments mark a well-established trend within cartography studies. Further details and extensive explanations may be found in the volumes edited by J.B. Harley and D. Woodward (v. 1 and v. 2, book 1 and 2); then by D. Woodward and G.M. Lewis (v. 2, book 3); and finally by D. Woodward (v. 3, book 2) of: *The History of Cartography*, University of Chicago Press, Chicago, 1987.

In this new perspective, it would be clearly anachronistic to assess cartography in terms of geographic distortion or mathematical projection. From the point of view of information, features of a map that may seem "extravagant" cannot possibly be written off as purely cosmetic, superfluous or accidental. On the contrary, they need to be taken as evidence of a specific worldview. Similarly, it makes sense to deny once and for all the claim that the "scientific quality" of maps is vouchsafed by their degree of accuracy. Even in that case we would be witnessing an attempt to manipulate reality in order to convey a very partial view of it. To be sure, a rich and complex panorama opened up over the last few decades. Within it, the study of maps was problematized; maps were inseparably tied to a set of methodological procedures and critical assessments with which each scholar of cartographic theory, history of cartography, historical cartography but even of participatory or interactive cartography must comply.[5]

Here I will explore, albeit in broad outlines, the stages through which this cartographic structure was built. My aim is to show its most innovative features and to give relevance to a path that, having been clearly marked, must now be consciously adopted as one's epistemological framework. At the same time, I intend to illustrate the crucial role geographical studies have in this area, because they have promoted awareness of the problematic nature of maps seen as meta-geographical discourse, but they have also disclosed their multifaceted action as self-referential tools and illustrated the key role cartographic semiosis* continues to play.

THE ROLE OF THEORY IN CARTOGRAPHIC INTERPRETATION

To start with, we need to clarify the assumption that underlies critical studies on cartography. Every interpretation relies on a hypothesis. No cartographic analysis may be considered neutral, for each relies on a hypothesis whereby the bits of information obtained from the map are placed within a precise frame of reference, which affects their meaning. This must be asserted to clear up irrelevant doubts as to the usefulness of embracing a hypothesis in the first place. On the contrary, it should be stated that a theoretical approach is still and ever present in any interpretation, for the simple reason that interpretive activities produce knowledge, which is in fact a hypothesis. Hypotheses are nothing but answers to questions or solutions to issues, which eventually fall under the scrutiny of the scientific community of researchers. And the notion that knowledge may be derived from an unwarranted and purely contemplative activity is glaringly removed from fact. Rather, it is the answer to a need, and may be understood

[5] For a general overview of recent developments in cartographic studies, see: P.I. Azócar Fernández, M.F. Buchroithner, *Paradigms in Cartography: An Epistemological Review of the 20th and 21st centuries*, Springer, New York, Dordrecht, London, 2014.

only in the light of a human interest which justifies its relevance.[6] In our case, therefore, reliance on theory makes sense only when theory is perceived as a tool for clarifying the learning and communicative outcomes of maps. And maps are to be seen not so much as means of recording reality but as instances of a mediatization that intervenes to shape it, as operators able to alter it.

Of course, hypotheses gain relevance in accordance with to their ability to enhance the level of inquiry. Insofar as issues are implicit, for instance ingrained in shared beliefs, hypotheses are undoubtedly prevented from assuming the explicit form they need to undergo validation. In that situation, the identification of issues is therefore a measure which increases scientific awareness, to the point that it elicits and sets in motion new processes of discovery.[7] Scientific knowledge, then, is *always* a theoretical knowledge which *always* presupposes an issue. The latter, in turn, may be either explicit or implicit, that is it may be grasped according to a variable that runs the whole gamut from perfect presupposition to full explicitation, passing through intermediate forms. In particular, an implicit hypothesis corresponds to an unspoken issue, which is in itself something quite close to subjective knowledge. Also, such issue may well be *preventing* a hypothesis therein raised from taking on an explicit form. The final explanation one provides will appear thus severed from its generative substrate and will produce a set of conditions that prevent real appreciation.

So, how to determine the value of a hypothesis, either in itself or in relation to others? How to determine the fairness and competitiveness of an answer? How to justify it without knowing the realm of understanding such hypothesis applies to? Ultimately, an unuttered decision remains an unstated intention and marks the researcher's final denial of responsibility. However, to make a hypothesis explicit is not always compulsory. In some cases, information culled from interpretation is formulated only linguistically, that is, it is rendered in the form of implicit theories, hardly recognizable as scientific points of view about the world of experience. Such information will thus prove unfit to produce a communicative flow. Explication of a hypothesis is, therefore, the first prerequisite for activating a scientific exchange. For, besides providing

---

[6] As stated by Jürgen Habermas, human interests may be of three types: technical, practical, critical. The first call for knowledge, which extends our technical prowess; the second require interpretive knowledge, which makes it possible to steer action within common traditions; the last elicit analytical knowledge capable of freeing our understanding from its reliance on the powers that be. J. Habermas, "Knowledge and Interest," in: Id., *Theory and practice,* Beacon Press, Boston, 1974, pp. 7–10.

[7] Popperian epistemology is well known for assuming as the cornerstone of scientific evolution the idea of a partial and provisional truth, which may be refuted by empirical proof (C. Popper, *The Logic of Scientific Discovery,* Hutchinson and Co., London and New York, 1959). Although criticized by many, including Feyerabend himself, Popper's approach is widely adopted on implicit terms. See for instance: J. Preston, *Feyerabend: Philosophy, Science and Society,* Polity Press, Cambridge, 1997.

a test for validating a hypothesis, the assessment and verification procedures within a community of researchers create a forum of exchange and, therefore, an area of shared growth.

While I am aware of these issues, in this chapter I chose to analyze, among others, a number of theoretical approaches that have not yet reached full explicitation. My choice depends on the fact that such approaches highlighted key issues, which point to a line of research devoted to promoting theoretical explicitation through interpretation.

Before endeavoring to do that, however, I feel we need to consider, albeit briefly, what is meant by "cartographic" interpretation. If, as discussed previously, interpretative activity belongs to cognition and, as such, it bears features presumably shared by all the sciences, what is still loosely defined is the specificity of our object of study: the object to which interpretation is applied, namely, cartography. The meaning of "cartography" calls for some explanation, not so much to comply with terminological rigor, but rather because the term harbors a fundamental ambiguity, to do with a momentous change of perspective in interpretation.

The term "cartography" is a late 19th-century neologism, coined to denote the science that studies and produces maps. Over time it has taken on a number of meanings, and was used to identify: 1) a *corpus* of records that share common features – scaled-down images of the world, rendered on a plane using techniques and languages symbolically encoded in various forms (from *globes* to *road maps*, from *topographies* to *thematic maps*, to *atlases* etc.); 2) the highly implicit theory whereby the complexity of the environment is reduced and the world is intellectually appropriated.[8] These multiple senses record our ambiguous and sweeping use of the term, also related to our unmindful assumption of the meaning of "geographic map" from which it is derived. A map is commonly defined as "the planar drawing of the earth or of one of its regions," which shifts the issue from what a map is, as a technical object, to what it represents. But when the world being represented is taken into account, the map is usually seen as a sheet or a medium of representation. This ambiguity, which refers to content in order to explain the object and to the object in order to explain the content, conceals the problematic nature of maps, derived from the fact that they are a complex medium of communication. Recent studies in fact demonstrated that maps are powerful instances of mediatization, able to intervene in communication in quite autonomous terms. As we saw, the word

[8] See the entry for *cartography* in: J. Lévy, M. Lussault, eds., *Dictionnaire de la Géographie et de l'espace des sociétés*, Editions Belin, Paris, 2013 (first edition, 2003).

"cartography" incorporates mediatization and opens it up to a vast array of interpretations which address both the process of map construction and that of map communication. As they do so, they *shift focus from the features of reality that maps reproduce to what maps communicate about the meaning of territory.* It is precisely the ambiguity of the term "cartography" that discloses the elaborate nature of geographic maps and makes it possible for us to envisage new spaces for reflection.[9]

It should be stated forthwith that cartographic interpretation as a cognitive activity is closely linked to the interpretation of geography. This must be stated less to claim a specific domain than to lay emphasis on the fact that cartographic interpretation is a *meta-geography*, since what maps visualize is territory. We need to start from such self-evident fact in order to understand the new interpretive perspective. It should first be noted that territory cannot naively be assumed as an objective fact. Rather, it ought to be understood as the outcome of a process whereby natural space embodies anthropological values, which are collected in the survey and the processing of natural phenomena carried out by territorial actors.[10] Because such values are later recorded on a map through an act of interpretation on the part of the cartographer, they enter a second level of interpretation which in turn produces a meta-geography. In short, to analyze a map means to refer to many cognitive activities that involve the presence of multiple interpreters with specific roles: the territorial actor, who transforms shapeless space into something ordered and communicable, that is territory; the cartographer, who recognizes that order and interprets it with a view to presenting it in a symbolically coded language; the recipient, who draws instructions for action through a new interpretation of that order. Ultimately, geography and maps are tied in a symbiotic relationship. And that is the prerequisite we must bear in mind if we wish to reflect on cartographic interpretation, especially in its final phase, namely the one activated by the recipient, who has a vested interest in understanding and mastering the cartographic message.

With this in mind, I shall now consider the various interpretive approaches that, albeit in different degrees, all converged to define the critical perspective of interpretation we are going to embrace: the *object-based* approach; the

---

[9] In fact, once maps were released from their role as mere instruments for recording territory and injected into the functioning of complex communication systems, their propensity for setting up relationships with other types of descriptive typologies and for fashioning mashups with other representations became evident. That is the base on which new forms of communication may be worked out. Please refer to Chapter 4 of the present study for information in that regard.

[10] We are here thinking of the theoretical approach borrowed from studies on complexity conducted by Claude Raffestin and formalized by Angelo Turco. Such an approach sees geography as the territorial form of social action. C. Raffestin, *Per una geografia del potere*, Unicopli, Milan, 1981; A. Turco, *Verso una teoria geografica della complessità*, Unicopli, Milan, 1988.

deconstructivist approach; and the *semiotic* approach. Although related and in many cases overlapping, these three approaches followed a chronological curve of development. In an initial phase, cartographic interpretation distanced itself from a "positivist" approach and pinpointed aspects that eventually fostered and implemented the set of considerations that inform the second phase.[11] The second phase consisted in raising once again the thorny issues maps present when placed within a social context. As we raised the bar of research, we came across the need to seek new answers. These could in turn derive solely from theoretical advances initiated by the third approach, the semiotic one.

THE OBJECT-BASED PERSPECTIVE

The object-based perspective has the merit of having freed maps from a positivist approach. By rejecting accuracy and relevance to reality as the sole criteria for interpretation, the object-based perspective drew attention to the role of maps as documentary sources. And the relevance of maps then lies not so much in querying the self-evident items of information they provide but in recovering the social issues from which they originate. This change of course, which started in the first half of the last century, paved the way to the study of the value and significance of maps as records of the relationship between humans and their environment. The first timid attempts to show maps as practical tools in any kind of social endeavor date back to that period. The use of maps in the realms of education, politics, administration, defense, religion or science turned them into valuable documentary sources which demanded attention. Maps thus began to be probed as social tokens, although most scholars continued to focus on their structural features. What was now taken into account was the building process of a map and its distinctive features: medium type; graphics technique; motives of its existence; customer base; the cognitive and expressive skills of its author; its market circulation and, in some cases its value as a prototype or model for future cartographic productions. That in turn fostered the interest of antiquarians and collectors in studies related to the history of cartography.[12] Focused as they were on the technical features of a given map in relation to its authors (designers, engravers, printers, publishers, merchants, and sellers) and on the world of print publishing – from which most of the production issued – these studies responded to the demand for a commercial estimate of maps by art dealers, who were driven by the rarity or inherent interest of

---

[11] One of the most influential exponents of the positivist approach was Arthur Robinson, who was interested in reproducing what he called "map effectiveness" able to render reality objectively (A.H. Robinson *et al.*, *Elements of cartography* (sixth edition), Wiley, New York, 1995).

[12] For a bird's eye view of antiquarian interests in cartography: M. Harvey, *The Island of Lost Maps. A True Story of Cartographic Crime*, Random House, New York, 1999. On the lasting appeal of such endeavor in Italy, fostered by extensive research, see: D. Woodward, *Maps as Prints in the Italian Renaissance. Makers, Distributors & Consumers*, The British Library, London, 1996.

the document themselves. World-famous researchers, united by a common stock of special skills in the history of cartography, operated in this area without in fact contributing, except in a few isolated cases, to a critical assessment of maps. Even those who probed the various phases of construction of maps and, therefore, the process whereby maps were manufactured, produced, and circulated – who did provide valuable insight for enhancing our knowledge of unknown aspects of the everyday life, the environment, the arts and crafts of ancient societies – ultimately failed to draw attention to the clusters of information maps put forth.[13]

Around the same time, some scholars focused on other aspects of maps which, on the contrary, could have led to momentous breakthroughs in the development of critical discourse on cartography. In Italy, we should remember Roberto Almagià, who viewed maps as invaluable records not to be tied down to a metric rendering of reality. He stressed the importance of content, to which maps refer, as found not only in information we now call "referential" – to be discussed later – but also in social items of information, be they symbolic or performative. Thus redeemed from the kind of analysis devoted to itemizing superficial data, maps disclosed their potential as documents capable of recording the territorial practice a given society implements at a given time in its history.[14] To achieve his goal, the Italian scholar provided a critical overview that included images of territory until then considered as "geographic maps," namely administrative maps. Such maps had in fact been neglected by historians of cartography, because they failed to meet positivist criteria of "measure" and quantitative recording of phenomena. Previously they had been dismissed as naive drawings of territory. Since they failed to provide information on the reduction scale being used, on the kind of projection employed to avoid distortion, or even on their authorship, these maps had been excluded from the cartographic genre.[15] By contrast,

[13] My remarks are not meant to censure these studies by playing down their value. Rather, I wish to keep them at the margins of the line of inquiry I am going to pursue.

[14] Suffice it to recall here the painstaking analysis Almagià undertook on a number of key documents, including the "Map of the Verona territory, otherwise called Almagià." His work made it possible to date this map and reconstruct its social context. See: R. Almagià, "Un'antica carta topografica del territorio veronese," *Rendiconti della Regia Accademia Nazionale dei Lincei*, XXXII, 1923, issue 5–6, pp. 61–84. Almagià's work inaugurated a fruitful trend of studies on historical maps, including those of Eugenia Bevilacqua on Venetian cartography: E. Bevilacqua, "Geografi e cartografi," in: *Storia della cultura veneta* V. 3/11, Neri Pozza, Vicenza, 1980, pp. 355–374.

[15] Such maps are mostly owed to the work of unknown surveyors or land measurers (*perticatori*) employed in public or private institutions that were involved in cartographic science as a form of land survey. Eminent – or at least soon-to-be-eminent – cartographers would occasionally contribute, to bequeath us a wealth of administrative records remarkable both for their technical innovation and for their conceptual framework. In this regard, and within the context of Venetian administrative cartography, see: E. Casti, "State, Cartography and Territory in the Venetian and Lombard Renaissance," in: D. Woodward, ed., *The History of Cartography*, vol. 3: *Cartography in the European Renaissance*, University of Chicago Press, Chicago, 2007, pp. 874–908. http://www.press.uchicago.edu/books/HOC/HOC_V3_Pt1/HOC_VOLUME3_Part1_chapter35.pdf

Almagià viewed them as the ultimate expression of the territorial policy whereby modern states put themselves to the test. He claimed they rightly belong to the cartographic genre and included them in collections which, from the second half of the last century, have stood as one of the most remarkable instances of map valorization: the *Monumenta Cartographica*.[16]

These works dealt the first blow against the myth of a "pure" map, presumably superior to other drawings of territory not tied down to the paradigm of geometry. It should not be forgotten that the *Monumenta* marked the very first endeavor to revalue maps by a community of experts on the history of cartography. Such endeavor aimed at replacing 19th-century cartographic collections, making up for the lack of information such collections perpetuated in their cursory written account of maps.[17] Bundled in large formats with a view to reproducing maps for reading and documentary collation, the *Monumenta* were largely instrumental in promoting the genre of cartography and in spreading knowledge of those relics which often paved the way to future regional cartography. The scope and significance of such work is all the more apparent when one considers the dissemination and persistence of those collections over time.[18]

Within the history of cartography, there coexisted at the time several different strains which, precisely by virtue of their multiplicity and their varied outcomes, paved the way to future inquiry into the meaning of cartography and the ways to interpret it. Predictably, the prevailing attitude was to approach maps as detached from their social context and its attendant territorial practices. The activities of promotion and communication which I just mentioned, however, eventually introduced mapping also to the general public. More precisely, we may speak of an object-based perspective on interpretation on

---

[16] Almagià authored the two Italian collections: *Monumenta Italiae Cartographica*, Istituto Geografico Militare, Florence, 1929, anastatic reprint: A. Forni, Sala bolognese, 1980; *Monumenta Cartographica Vaticana*, Biblioteca Apostolica Vaticana, Vatican City, 1944–1955. Later, in collaboration: R. Almagià, M. Destombes, *Monumenta Cartographica vetustoris aevi*, N. Israel, Amsterdam, 1964.

[17] Think for instance of Marinelli's annotated repertoire, one of the first of its kind. In his case, the lack of photographic reproductions, of limited availability at the time, makes perusal difficult. (G. Marinelli, *Saggio di cartografia della regione veneta*, Naratobich, Venice, 1881).

[18] These large collections, which first appeared in the late 19th century, covered many regions including: Kamal, S. Fauat, *Monumenta Cartographica Africae et Aegypti*, Cairo, 1926–1951, reprint: Institut fur Geschichte der Arabisch-Islamischen Wissenschaften an der Johann Wolfgang Goethe Universitat, Francoforte, 1987; C. Armando, A. Teixeira da Mota, *Portugaliae Monumenta Cartographica*, Imprensa National-Casa de Moeda, Lisbon, 1960 (new edition 1988); U. Kazutaka, O. Takeo, M. Nobuo, N. Hiroshi, *Monumenta Cartographica Japonica*, 1972; G.A. Skrivani'c, *Monumenta Cartographica Jugoslaviae*, Istorijski Institut, Beograd, 1974; S. Monchengaldbach, *Monumenta Cartographica Rhenaniae*, Stadtarchiv Monchengladbach, 1984; G. Schilder, K. Stopp, *Monumenta Cartographica Neerlandica*, Uitgeverij Canaletto/Repro, Holland Alphenaan den Rijn, 1986–2013; M. Watelet, *Monumenta Cartographica Walloniae*, Editions Racine, Bruxelles, 1995.

account of the relevance then widely and commonly granted to the "object" as an independent entity and no longer as a mere corollary or marginal support to other sources. Realization of the prominence of maps gave impulse to a deeper understanding of their communicative workings. And the ideological implications which maps as social products necessarily harbored began to be investigated. The studies of John Bryan Harley, well-known for his prominence and his prolific scientific production,[19] first attested to the emergence of a new critical approach towards maps. That in turn brought forth the second phase of interpretation: the one based on deconstruction.

THE DECONSTRUCTIVIST PERSPECTIVE

Although there had been signs portending a change of perspective, the problematization of maps from a deconstructivist viewpoint signaled a sharp break with previous cartographic interpretation, which relied on a progressive layering of research. This new approach did not in fact recover a preexisting state of things but aimed to raise the level of inquiry, opening up new fields of investigation until then ignored. The scientific community was initially loath to accept innovation. Yet in time, new practices were assimilated and went to enhance the community's trove of learning. Ultimately, in the field of cartography the idea that there was only one way to study maps was abandoned and new, separate modes of interpretation came to the fore: those aimed at probing a document as an object, in order to shed light on the implications of its construction;[20] and those that zeroed in on maps as social products and placed them within the wider debate, kindled by the social sciences, over the means of representation.[21] In this new perspective, maps took on the role of instruments for setting up links and interconnections with other disciplines, which in turn put an end to the isolation of cartographers, finally thrust onto the wider landscape of the human sciences.

Such broadening of one's critical horizon inevitably raised new issues: as the discipline of cartography benefited from novel contributions, it also tended to become more fragmented. For the interweave between studies of cartographical history and

---

[19] Harley wrote about 140 articles and essays and contributed to numerous leading monographs, among which I may cite "The Map and the Development of the History of Cartography," in: J.B. Harley, D. Woodward, eds., *The History of Cartography*, The University of Chicago Press, Chicago & London, 1987, vol. 1, pp. 1–42.

[20] I believe cartographic features such as watermarking and heraldry, which were recovered and deemed relevant to cartographic interpretation, should in fact be referred to the competence of experts in the arts and archival systems.

[21] Among the many contributions on this topic, see: J.B. Harley, "Maps, knowledge and power," in: D. Cosgrove, S. Daniels, eds., *The iconography of Landscape. Essays on the symbolic representation, design and use of past environments*, University Press, Cambridge, 1988, pp. 277–312.

other fields diluted the disciplinary cohesion of the former by opening them up to a common ground, shared with other disciplines. In Italy, for instance, studies were scattered over a wide range of fields (including historical science, architecture, urban planning) whose outcomes are often difficult to assess. All that entailed experiencing cartographic interpretation less as an endeavor clearly marked within a discipline than as an activity ultimately dependent on the professional training of the scholar who undertakes it. Thematic areas specific to cartography as a field eventually emerged: studies on historical cartography, and studies on modern maps. And new research areas peculiar to each trend within the discipline or across wider theoretical contexts were established. All that finally resulted in the formulation of multiple critical approaches to cartography. What Harley envisioned when he claimed that maps are too momentous to be entrusted solely to the hands of cartographers was actually taking place.[22]

I venture inside this critical labyrinth in the hope of finding a red thread able to trace a path across an otherwise baffling theoretical landscape. Even in this case, of course, I will limit my remarks to but a few of the scholars whose work provided invaluable landing places in cartography's journey of renewal.[23]

The forerunner of deconstructivist problematization in cartography is, as we anticipated, John Brian Harley. He questioned the communicative outcomes of maps and envisaged the need for a theoretical reassessment paving the way to deconstruction. And to him deconstruction meant the search for different, and at times competing, discourses, able to raise new issues. He took it upon himself to rewrite the history of cartography by placing cartography's meanings, events and outcomes inside much wider social movements.[24] Through deconstruction, Harley severed the link between reality and representation, to claim that the history of cartography must be rooted in social theory rather than in scientific positivism. Thus tapping resources available from a wide range of disciplines, he placed maps inside a movement, the movement of deconstruction, destined to achieve prominence in the 1980s. Drawing inspiration from the deconstruction of literary

---

[22] J.B. Harley, "Deconstructing the Map," in: *Cartographica*, 26-2, 1989, pp. 1–20, reprinted in: T. Barnes, J. Duncan, eds., *Writing Worlds: Discourse, Text and Metaphor in the Representation of Landscape*, Routledge, London and New York, 1992, pp. 231–247.

[23] What needs to be stressed from the start is that these trends never rise in isolation. Rather, they very much rely on the work of scholars involved in cartography and in the development of geographical science. In Italy, for instance, cartography is tied to the names of Claudio Cerreti, Giorgio Mangani, Marica Milanesi, Massimo Quaini, Leonardo Rombai, Paola Sereno, or to the unrivalled epistemological inquiry of Lucio Gambi.

[24] Harley's radicalism and the conviction that animated his position are attested in the debate played out in the magazine *Cartographica* between the years 1980–1982. See for instance: P. Gould, "Une prédisposition à la controverse," in: P. Gould, A. Bailly, *Le pouvoir des cartes. Brian Harley et la cartographie*, Anthropos, Paris, 1995, pp. 53–58.

texts as theorized by Jacques Derrida,[25] Harley set out to achieve three goals: 1) to challenge the epistemological myth (created by cartographers) of a cumulative progress for an objective cartography, devoted to an ever-increasing mimetic imitation of reality; 2) to expose the social import of maps and their role in the consolidation of a world order; 3) to enable cartography to liaise with interdisciplinary studies on representations and on the construction of knowledge.[26] He submitted that strategies of thought of the kind unveiled by Foucault, the notion of metaphor put forth by Derrida, the view of rhetoric as inherent in scientific discourse, as well as the ubiquitous concept of power-knowledge, are traits common to many disciplines and are likely to be highly beneficial to cartographic interpretation. Cartography may in turn contribute to enhancing other areas of study.

Harley did not object to emphasizing the relevance of technology for map production. He refused, however, to reduce the study of cartography to a matter of mere technique. His underlying assumption is that the rules of science adopted by maps are affected by sets of social provisions scholars must be able to detect in the signs maps present. A good deal of the power wielded by maps, he insisted, draws from those provisions, under the impartial disguise of a science that rejects the social dimension of maps while *de facto* sanctioning the existing state of things. Harvey claimed that, in "pure" scientific maps, defined as such by their presumably inherent compliance with standardization and rigorous measurement, science itself becomes a metaphor, because it entails a notion of symbolic realism which is in fact but an affirmation of authority and a political statement interchangeable with any other. He underlined that accuracy and precision in drawing maps are the new talismans of authority, culminating in the creation of digital maps in our time.[27] Harvey's concern clearly has to do with the power maps exert as vehicles of ideology. In all its form, he repeated, knowledge is power. And it becomes crucial to interrogate those who hold such knowledge, those who have access to it, and to investigate how it can be used for good or evil. The best maps, he said, are not rigorous maps, but those that manage to convey an image of authority as if it were impartial. Finally, he held that maps are languages, and did so on the basis of three orders of suggestions: 1) the one drawn from the studies of Jacques Bertin, whose emphasis on semiotics as a perceptual system he however rejects as too narrow for application to the history of cartography; 2) the one coming from Erwin Panofsky's studies on iconology, which enabled scholars to identify two layers of signification in maps: a surface layer and a deeper layer generally associated with the

---

[25] Namely, the search for aporias whereby the tension between logic and rhetoric is disclosed.

[26] J.B. Harley, "Deconstructing the Map...," *op. cit.*, p. 64. Page number refers to the French edition, issued in his honor, in which the article was republished: P. Gould, A. Bailly, *Le pouvoir des cartes. Brian Harley et la cartographie., op. cit.*

[27] J.B. Harley, "Deconstructing the Map...," *op. cit.*, p. 77.

symbolic dimension; 3) the one attuned to the outcomes of the sociology of knowledge, which led him to accept the idea that cartographic knowledge is a social product enmeshed in power interests and thus to shed light on the relationship between cartographic discourse and ideology.[28]

Harvey's relevance, however, lies primarily in the fact that he intuited the crucial nature of the relationship between cartography and geography. He denounced the gap that existed between the two disciplines as inexplicable even though, if one takes territory as the object of representation, maps must necessarily appeal to geography. Harley somehow implied that the representation of territory cannot be reduced to the record of what we see, because landscape,* that is the visual form of territory, is nothing but the simulacrum of a relationship between humans and their environment, whose consistency resides in the social dynamics that animate it. He advocated the use of social theory as a starting point for questioning the implications concealed by cartography.

It is at this juncture, however, that we detect the major shortcoming of Harley's otherwise insightful theory: his apparent neglect of the fact that any sweeping social theory must of necessity have equally sweeping outcomes. Arguably, his line of reasoning is flawed because it fails to acknowledge the unavoidable necessity for a specifically geographic hypothesis, a hypothesis admittedly alert to social issues, yet able to apply the territorial skills to the realm of cartography. In the pages that follow, I will have the opportunity to show how a balanced approach of this kind can in fact have a varied range of unhoped-for outcomes. Presently, I would insist that the innovative thrust of Harley's study lies in his unquestionable insistence on the need to discard the notion of maps as self-evident representations of reality. Rather, maps ought to be reassessed in their unique role of tools for conveying geographical knowledge.[29]

Many scholars followed in Harley's footsteps.[30] Two are worth mentioning here: Christian Jacob and Franco Farinelli, who both dwell on cartographic language, albeit with different goals.[31]

---

[28] J.B. Harley, "Maps, knowledge and power...," *op. cit.*

[29] The cognitive approach also underlies the reflection of: A.M. MacEachren, *How maps work: representation, visualization and design,* Guilford Press, New York, 1995.

[30] Among the followers of this approach we would recall: Denis Wood, Mark Monmonier, Martin Dodge, Rob Kitchin, John Pickles, who extensively demonstrated the ideology inherent in maps (D. Wood, *The power of maps,* Guilford Press, New York, 1992; M. Monmonier, *How to lie with maps,* University of Chicago Press, Chicago, 1996; J. Pickles, *A history of spaces: cartographic reason, mapping and the geo-coded world,* Routledge, London, 2004).

[31] The parts of their work more immediately relevant to our present needs are: C. Jacob, *L'empire des cartes. Approche théorique de la cartographie à travers l'histoire,* Albin Michel, Paris, 1992 (English edition: C. Jacob, *The Sovereign Map. Theoretical Approaches in Cartography throughout History,* University of Chicago Press, Chicago, 2005); F. Farinelli, *I segni del mondo. Immagine cartografica e discorso geografico in età moderna,* La Nuova Italia, Florence, 1992.

Jacob's studies start from the idea that the persuasive power of maps matters because it answers both needs that are quintessentially sociopolitical and needs that are fundamental to individuals as such: it is the instrument of a poetics of space which envisions the world as it could be. In this regard, it should remembered that, from primeval Bronze Age maps engraved in stone to the simulated space of modern computers, maps have always lent themselves to the whims of absolute powers or the ambitions of the media as useful tools for passing off a partisan or self-interested idea as if it were altogether wholesome and impartial. As he examines the visual properties of maps and relates them to the larger framework of their figurative codes, Jacob concludes that maps are less objects than functions, or social mediations. As such, they lend themselves to a plethora of interactive uses: constructions, projects, field work, teaching, cooperation. And even when their spreading is subject to restrictions or monopolies, maps are social objects, power pawns, and strategy tools.[32] Ultimately, Jacob focuses on the complex dialectics of maps, which rests not so much on a generic set of territorial skills but on a socially validated corpus of knowledge: the one provided by geography. As such, Jacob's contribution marks an area of possible rapprochement between two disciplines that have long overlooked each other: that is the history of cartography and present-day geography.

His most telling work in this respect, *L'empire des cartes*, takes a structural and synchronic approach in order to trace the history of cartography as an organic whole and pinpoints the theoretical issues posed by maps, highlighting their trouble areas and their graphical features. Also, he explores the paths and stages of cartography's perception and interpretation over four main domains or realms. In the first, he develops the idea of maps as symbolic mediations between humans and their environment and, at the same time, as means of communication able to trigger interpersonal dynamics of knowledge that are nearly universal, given their visual qualities that allow individuals to communicate. In the second, he traces the linguistic structure of maps and identifies the shaping of a cartographic discourse in grammar and vocabulary. In the third, he expands on his linguistic analysis of maps and, considering the presence of a double communicative plane (both visual and textual), he shows their hypertextual potential and their unique placement within the field of communicative systems. Finally, in the fourth he maintains that maps are graphical devices whose complexity, in order to be understood, requires methodological tools to be found in the interstices between different disciplines. Through Jacob's studies, concepts which had hitherto been solely keen intuitions or faint assumptions finally took shape and substance. Such concepts were indeed shared by many scholars, but still lacked scientific systematization. Jacob's wide-ranging research had momentous outcomes, starting with the acknowledgment of cartography as one single genre, not so much on the basis of its technical and structural features – which had misled quite a few theorists – but

[32] C. Jacob, *L'empire des cartes. Approche théorique de la cartographie…, op. cit.*, p. 458.

on account of its peculiar communicative features, which show maps in their compelling role as tools of geographical communication. Maps are thus placed within a global framework of inquiry: the analysis of their historical phases and variegated localizations (with attendant ideological and graphical solutions) confirms the validity of investigating their nature and their social functions.

For his part, Farinelli followed in Harley's trails and laid the foundations for a critique of geographical knowledge, by giving special emphasis to the ideological implications of maps. Farinelli rereads cartography's evolution in this perspective, maintaining that the changes it undergoes are linked to the political setup of the State. Convinced that maps have in fact been adopted as interpretative models of geography, he feels the need to reflect on their outcomes within geographical epistemology. He interrogates the communicative role of maps to demonstrate that: 1) what maps convey is always subservient to ideology; 2) unless it is subjected to close scrutiny, their message can deeply affect the very idea of territory. Farinelli also claims that maps leave their most treacherous encrustations upon the concept of space. To confirm that, he focuses on geometrical cartography, and shows that "bourgeois" geography rises when the provisions cartographic logic had previously imposed are finally discarded.[33] He argues that cartographic theory of the modern world – i.e., the representation of the world through the rules of 18th-century Euclidean geometry – tends to uphold the status quo and leads geographers to revive the practice bourgeois geography had resisted for a whole century in order to gain ground: the practice of geographical knowledge as issuing from the injunctions of cartography. This process came to an end between the end of the 19th and the onset of the 20th century, when the adoption of a topographical model of space, freed from the strictures of positivist geographers, marked the demise of any critical theory of geographical space. The world was taken as a network of individual phenomena, whose solidity cartographic images sanctioned by endowing them with names and symbols. As a consequence, it was not concepts, but mere representations that marked the epistemological acts of human geography.[34]

THE HERMENEUTIC PERSPECTIVE

With respect to the deconstructionist perspective, the semiotic approach shifts the focus of investigation from maps as *mediators* of territory to maps as *symbolic operators.** The underlying assumption is that maps are complex communicative systems, which internally develop sets of self-referential information. These in turn give substance to maps' representational power. Ultimately, their

[33] On this point: F. Farinelli, "Alle origini della geografia politica borghese," in: C. Raffestin, ed., *Geografia politica: teorie per un progetto sociale,* Unicopli, Milan, 1983, pp. 21–38.
[34] On the role of maps in the construction of geographical knowledge, see: F. Farinelli, *Geografia. Un'introduzione ai modelli del mondo,* Einaudi, Turin, 2003.

power resides in their ability to regulate the complexity of geographical space through a metrics* which is the mainstay and prerequisite of territorial action.

This change of perspective, brought about by an analytical shift from correlations within the cartographic object to its integral role in the process of territorial transformation, clearly means rethinking some tenets on the nature of maps. First of all, maps are no longer seen as one of the many "visual representations of reality," but rather as very particular means of representation, capable of acting within the social dynamics. Secondly, maps are not solely symbolic *"mediations"* of territory. Although they do affect our understanding of territory, they should never be assumed exclusively in their role of interposition between reality and those who interpret it. Rather, they present themselves as a form of *mediatization* which puts forth a most particular world layout, the cartographic one, as an interface between reality and society. In this sense, maps become very powerful metamorphic devices, achieving the map = territory equation not as something objectively definable, but as a potentiality by which and through which the society–space relationship is established. What, however, makes this approach truly "new" is that it endeavors to prove this hypothesis theoretically. Thus going beyond a general outline of the issue, it zeroes in on a possible solution and contributes new lines of argument in support of the mimetic functioning of maps. Eventually, maps are envisioned as powerful mimetization devices organized on multiple levels: 1) they imitate reality, thus offering themselves as tools capable of reproducing it; 2) they trigger a process of concealment whereby they blend in with other representations of territory; 3) they mask the difference between geographical space and cartographical space.

Before taking into account the results such an approach achieves, we should point out that it lies within the scope of those semiological studies that recover the role of textual language and focus on the message images convey. For maps are hypertexts based on a dual system –analogical and digital – from which they draw their iconizing power. More precisely, the semiotic approach attempts to: 1) technically master the semantic layout of maps and their communicative workings; 2) show the ways in which maps steer territorial action. The undisputed forerunner in the first case is Jacques Bertin. Bertin investigated the semiotic functioning of statistical images in their many forms: graphs, grids, maps. He studied the distortions and communicative inconsistencies produced by the visual pairing of colors and shapes and the use of perspective. He provided technical solutions for mastering them and submitted that sign orientation and color-coding draw attention to a specific phenomenon or object represented.[35] At

---

[35] J. Bertin, *Sémiologie graphique,* Mouton & Gauthier-Villars, Paris, La Haye, 1967; Id., *La graphique et le traitement graphique de l'information,* Flammarion, Paris, 1977; Id., "Perception visuelle et transcription cartographique," in: *La cartographie mondiale,* 15, 1979, pp. 17–27.

the same time, Bertin showed how graphical semiology produces self-referential information able to take over the drafter's intention and the information he or she meant to convey. Therefore, although his contribution is limited to the technical aspects of figural language, he problematized maps, placing them within the scope of a broader inquiry that scholars undertook in the 1970s and 1980s to expose the risks of figuratively rendered information and advocate urgent reflection on geographical representations as a whole. This is the line taken by studies on cartographic semiology first promoted by ICA's *Commission on Theoretical Cartography* and later linked to the magazine *Diskussionsbetraege Zur Kartosemiotik Zur Theorie Und Der Kartographie*, brought to life by Alexander Wolodtschenko and Hans Georg Schlichtmann. The latter started from Bertin's studies and tied them to achievements in linguistic semiology – especially by Yuri Lotman, Julien Greimas, and Umberto Eco – in an attempt to formalize specific cartographic theories that respond, above all, to the challenges posed by new Information Technology (IT) systems.

On the second front, namely with regard to the procedures whereby maps enter the dynamics that guide territorial action, cartographic semiosis envisages maps as *symbolic operators* able to prescribe how territory ought to be assessed. To conclude, a map becomes a symbolic operator when seen in its role as an elaborate communicative system, whose distinctive quality lies in the fact that it is a meta-geography. It makes sense, therefore, to address this research model as a legitimate field of study that is both crucial and highly structured. This field is undeniably peculiar and heuristic in its innovation. Nevertheless, it can already produce formalized results that may be submitted for scrutiny to the scientific community. In particular, I will here consider cartographic semiosis, which has come to the attention of researchers over the last few years.[36]

The theory of cartographic semiosis unfolds along two directions: the first sees maps as closely tied to territorial dynamics, which must be taken into account for their interpretation; the second relies on the assumption that semiosis, or the process by which information is produced and transmitted, is activated in the presence of an interpreter, seen in a double role of territorial actor and social communicator. It is therefore an approach closely rooted in the outcomes that geographical research had over the last few years. More precisely: 1) it is rooted in systemic theories able to envisage territory as the outcome of a process; [37] 2) it is grafted onto a geographical processing of elaborate concepts in the field of semiotics and philosophy of

---

[36] The following pages will refer to the line of thought developed in: E. Casti, *Reality as representation, The semiotics of cartography and the generation of meaning*, Bergamo University Press, Bergamo, 2000 (It. ed. *L'ordine del mondo e la sua rappresentazione. Semiosi cartografica e autoreferenza*, Unicopli, Milan, 1998).
[37] I am thinking here specifically of the complex processes mentioned above: C. Raffestin, *Per una geografia del potere..., op. cit.*; A. Turco, *Verso una teoria geografica della complessità..., op. cit.*

language.[38] Acknowledging maps' essential functions, *description* and *conceptualization,* this line of research becomes a tool for disclosing *iconization,** the process whereby maps express highly conjectural facts as if they were truths.

Clearly, this theory is not meant to shed light only on the communicative systems of maps. Rather, it sets out to show its potential as a sophisticated self-referential tool. Its aim is to deconstruct, de-situate and re-encode geographical maps in their theoretical cohesion as powerful mimetization tools.

Maps and the Territorialization Process

It should be noted that this approach conceives territory not as a given, issued in clear-cut form, but as a process tied to the set of social practices that alter natural space, endowing it with human values. Although it straddles multiple lines of operation, territorial action may be broken down into three main categories belonging to *territorialization* as such: *denomination, reification* and *structuring.* First, we need to point out that the term *denomination* covers a complex question, laden with meanings that depend on the type of designator used and, therefore, the different values it may have. In any case, the term tends to be used as "shorthand for descriptions" and also, more or less precisely, a "cluster of concepts." In this perspective, denomination clearly subsumes all the properties of the object in question, whose meaning is condensed to the point of being understandable only via multiple levels of reading: *denotative* and *connotative.* The first level belongs to the referential designator, because its codification took place in order to set up referents and thus refers to a surface meaning that is quite manifest. The second relies on symbolic and performative designators, since it allows for the detection of cultural, technical and "historical" valences which must be examined in depth in order to be understood.[39] The importance of the designator type rests in the fact that it can pinpoint a wide range of values and social systems. Nor is that all there is. What is truly groundbreaking is that, when territorial action is linguistically enforced within a social body, it activates and upholds a semiosis which encompasses various other sub-semioses, related to a single designator or groups of designators. As attention is drawn to the analysis of semiosis, that is the process whereby something functions as a sign for someone who is called upon to interpret it, the communicative role of denomination is enhanced. This makes it possible, in fact, to trace the phases through which

---

[38] See: A. Turco, "Semiotica del territorio: congetture, esplorazioni, progetti," in: *Rivista Geografica Italiana,* 101, 1994, pp. 365–383; Id. *Terra eburnea. Il mito, il luogo, la storia in Africa,* Unicopli, Milan, 1999.

[39] According to A. Turco's classification, referential designators are meant to establish referents and are thus instrumental in securing practices such as orientation and mobility. Symbolic designators, instead, refer to meanings derived from socially produced values. Values of this latter kind also fall under the domain of performative designators which, however, unlike their symbolic counterparts, refer to empirically testable facts (A. Turco, *Verso una teoria geografica della complessità…, op. cit.,* pp. 79–93).

things acquire meaning. And with regard to our own field of inquiry, this in turn leads researchers to highlight the outcomes produced when the means of transmission intervene in the process of communication.

That is especially crucial for societies in which maps are the privileged instruments of territorialization, as largely attested by historical records across the West: maps are adopted as strategic tools for territorial conquest, to meet the hegemonic needs incessantly generated by policies of expansionism. It needs to be stressed, however, that under those circumstances, maps play the key role of *representation devices, thanks to which denomination is enforced.* In other words, maps are taken as representations that humans implement in order to construct the world from a linguistic viewpoint. The names and signs reproduced on maps arrange the world of experience in the form of orderly, communicable items of knowledge.

We need to acknowledge, therefore, the crucial result of such an approach: the symbiosis between denomination and cartography and, consequently, the adoption of names as the cardinal features of cartographic communication. This way, maps are not merely loci of the intellectual appropriation of territory, but become denominative projections, because they convey the meaning designators harbor in themselves. To achieve that, maps pair that meaning to other signs, which take on some of its senses and enhance them in the process of communication.[40]

Maps are indeed the products of denomination, relying on the semiotic dynamics of linguistic codes. Yet the dynamic of semiosis is in itself a semiotic field,* within which the use of various codes triggers yet another semiosis. Ultimately, maps originate from names but show their communicative workings by pairing these names with other signs. With reference to maps it may thus be more appropriate to talk of a "metasemiosis" – that is, a second-level semiosis generated by primary signs and developed according to some of their implications. For the sake of argumentative clarity, I will be using the conventional term *cartographic semiosis,* even though I am aware that it originates in territorial semiosis.

Structurally, maps appear as *sorting systems* that endeavor to master a complex cluster of information by homing in on the most relevant geographical phenomena and placing them in the same sequence in which they are perceived in reality. That translates into using multiple linguistic codes (names, numbers, figures, colors) and multiple schemes: the geometric one referring to the sheet and the symbolic one that subsumes the cluster of codes.[41]

---

[40] On denominative projection see my: *Reality as representation..., op. cit.,* pp. 65–96.
[41] E. Casti, *Reality as representation..., op. cit.,* pp. 35–41.

Already at this first level, maps become hypertexts able to affect communication actively, for the particular nature of hypertextual languages is linked to their communicative outcomes. The information they convey does not equate to the sum of those bits of information derived from each single code. Rather, those bits are enhanced by becoming parts of one single whole.[42] Now, if we recall the data from which we started, namely the importance of the designator, and consider it as a sorting principle that arranges map codes into a hierarchy, then, and only then, will we appraise the true potential of denominative projection. Names are the pivots around which information is communicatively produced.

The Map as a Semiotic Field

From what we have said so far, it transpires that an analysis of maps' structures cannot possibly be the heart of one's research. Once the hypertextual potential of maps has been ascertained, maps must be related to the process of formation and transmission of their contents. It is at this point that the role of the interpreter gains relevance: he or she turns to maps to draw information for achieving specific goals. The presence of interpreters allows maps to appear as in a semiotic field, where signs become actual *sign vehicles*, that is they gain momentum when their meaning is sanctioned and interpreted by someone. As we shift the focus of our discourse onto *sign vehicles* – in our case on the designator paired to another sign – we realize that they elicit three types of relationships to do with: 1) the formation of meaning; 2) the interconnections between signs; and 3) the recipient's interpretative act.[43] We may then start to discern the places where the rules of actual cartographic semiosis abide: in the *semantic domain* meanings are produced via sign encoding; in the *syntactic domain* new meanings originate from the relations signs are made to entertain; in the *pragmatic domain* maps appear as an interpretative cypher and, at the same time, as matrices for social behavior. Cartographic semiosis should then be rightly considered as a complex communicative system, whose momentum lies less in the information it conveys than in the information generated in the process that interpreters set in motion.

It should also be noted that the interpreter acts as a "ferryman" between the two planes along which the semiotic approach moves, the geographical and the linguistic, for he initiates a decoding of both cartographic and territorial language. This statement is consistent with a hermeneutical perspective, which displaces geographical maps in order to implement a re-encoding that proves their momentous contribution to the production and circulation

[42] On the communicative outcomes of the transition from single-structure systems to polystructural languages see: E. Cassirer, *Filosofia delle forme simboliche,* La Nuova Italia, Florence, 1961, pp. 9 and ff.
[43] The approach suggested here was outlined by: C. Morris, *Signs, Language and Behaviour,* Prentice-Hall, Oxford, England, 1946.

of territorial meaning. As I maintained earlier, our research is grounded in the awareness that the interpretation of maps is a stage of territorial action, which prefigures strategies of production, use, and mediatization of territory.

That line of research enables us to shed light on the foundational aspects of cartographic representation. Maps reproduce space according to the principle of analogy, which aims to arrange objects using the same layout, the same network of relationships and the same size in which they are perceived in reality. As such, maps clearly refer to a topological order that in turn refers to a *cognitive spatialization*, in other words a procedure whereby objects are invested with spatial properties in relation to an observer who is in turn "spatialized" with regard to them.[44] This approach reclaimed the whole range of positional features – orientation, point of view, centrality of representation – both from a technical and an ideological viewpoint, giving evidence of their social importance through theoretical and analytical arguments. At the same time, it demonstrated that each designator inserted in a map is endowed with a set of prescriptions which somehow confirm its unique social value. As they neutralize any excess of signification and prescribe directions for interpreting what is depicted and what is excluded, cartographic icons draw attention to some of their features at the expense of others. In topographic mapping, for instance, the encoding-abstracting procedure, which is supposed to enhance the meaningful transmission of representation, in fact undermines the core meaning of territory, since only a limited number of the object's material properties are raised to the rank of social values through the self-referential action of the map itself.[45]

What makes this approach groundbreaking is ultimately the fact that maps are not only important tools for an intellectual appropriation of territory, but also essential instruments for boosting the whole territorialization process. In some social and historical contexts, maps become sorting systems whereby a community relates to the world. Cartography may then be seen as the product of a culture that becomes culture itself: it engages the cognitive heritage of a given society in order to enhance its territorial knowledge; it emerges as an autonomous means of communication; it asserts itself as an innovative interpretation of the world within the control mechanisms of the society that produced it.

The Cartographic Icon*

As we venture deeper into the folds of the semiosis, consideration of the relationship between *maps* and *territory* becomes essential. Relying on the fact that the latter is modeled after the former, we may say that maps are tools for selecting

---

[44] I will discuss the issue of spatialization in detail in Chapter 6 of this book.

[45] Material features take on connotative meanings that are not necessarily those of designators, but rather those conveyed as such by maps (E. Casti, *Reality as representation…*, *op. cit.*, pp. 151–174).

which information is to be highlighted, which is to be sidelined and which is to be concealed altogether. We have already stated that, technically, such modelization is accomplished by pairing names with codes (figure, numbers, colors, or even with reference to their position on the sheet), which convey the designator's distinctive features. We should now add that the semiotic figure that subsumes such pairing is the *icon*. That needs to be stated clearly, because icons play a momentous role in interpretation: they take on the designator's meaning, shape it somehow and feed it into the communicative circuit. I submit that the icons' outcomes are not additive with respect to the meanings that come from the initial codes. Rather, they are consistent with the transformation and implementation brought about by icons, which are self-consistent units endowed with a certain degree of independence from what produced them. It is therefore legitimate to refer to such units as *figures that take charge of the designator and endow it with special worth to determine how it should serve as a model in territorial practice.*

As for designators, so for icons we are in the presence of a double level of communication, *denotative* and *connotative*. The first matches the referential function it performs, that is, its ability to situate designators on maps as they emerge in reality; the second has to do with the ability to propose features that refer to the social environment. In short, once added to the sheet, icons produce a dual action over the designator: 1) they figurativize its localization,* hence strengthening its referential aspects; 2) they establish the relevance of some attributes related to the social context.[46] Cartographic icons are thus at the core of the generative process, whereby information is not only processed and communicated but also effectively produced. Similarly, a map's communicative action is not reduced to the mere transmission of data entered by a cartographer, but also affects the information generated by the icon itself. Since icons depend on the medium in which they are placed – typically, maps – they will on the one hand respond to the communication systems of their own visual representations, namely those unveiled by vision semiotics. On the other hand, they respond to the communication system of hypertexts, which convey information through autopoietic processes.[47] It becomes essential at this point to provide at least a general outline of cartographic communication systems.

[46] It is worth mentioning that the information created by icons derives from the stages of *figurativization*, whereby the values of the designator not only evolve in a communicative sense, but also undergo intensification. The stages of cartographic figurativization are: *spatialization* which uses topography to strengthen referentiality and, therefore, affects denotation; *figuration* which uses signs to take charge of the designator, highlighting its distinctive features; and, finally, *iconization* which, embracing the outcome of spatialization and figuration, invests the designator with social values (E. Casti, *Reality as representation…*, *op. cit.*, pp. 70 and ff.).

[47] On the semiotics of vision, contributions made along the line of the semiological studies inaugurated by Lotman and Greimas later merged with the endeavors of art historians (Panofsky, Arnheim, Gombrich), giving rise to a branch that may be classified as the science of visual communication. I will return to these issues in the course of this discussion.

Communication Systems: Analogical and Digital

It may be useful to remember that, in any instance of communication, information management is subject to the rules of sign encoding, which belong to two systems: *analogical* and *digital*. There is in fact a third system, the *iconic*, but it comes from a meshing of the previous two, and may thus be regarded as a derivative system, though it is in itself able to develop specific and most particular functions we will cover later.[48] Analogical and digital are used either simultaneously or in alternation within all communication processes. More precisely, we may find an analogical system inside tools that work by analogy, namely, according to processes linked to unchanging features of reality (for instance, photography). Digital systems, by contrast, operate over discontinuous scales, pinpointing discrete elements and developing a sign structure that does not necessarily bear a relationship with what it stands for: the relationship is purely arbitrary (e.g., lexical structure, number system, etc.). Analogy thus communicates serially via continuity and relies on an assessment of *difference*: difference of magnitude, of frequency, of distribution and of arrangement. Conversely, digitalization relies on *distinction*, encoded in accordance with the criteria of opposition, identity, contradiction, or paradox. Unlike analogy, digitalization necessarily entails some form of access code in order to be decoded (knowledge of the alphabet, of numbers, etc.). Analogical and digital are then poles of the same communication process, within which one or the other may be found at times to prevail. This matters because they convey the properties of objects in different ways. By appealing either to the criterion of difference or to that of distinction, they may or may not reproduce the identity of what they show. In other words, they may play up either the denotative or the connotative plane of communication.[49]

If we now turn to the investigation of the communication systems used by maps, we come across another important issue that semiotic analysis raises. Contrary to common perception, *maps are not analogical models of reality but models that use both the analogical and the digital systems* in a most particular combination. The analogical system emerges in the very structure of the basemap (intended here as the set of rules that dictate the localization of information on the map). The layout of objects and their relevance with regard to the medium used obviously abide by the rules of reduction, of proportion, and of perspective. Yet such operations are not actually an instance

---

[48] Analog and digital systems may also be found at the somehow primary level of biological information. There is, however, nothing equivalent to that for the iconic level; in this sense, the iconic is a derivative system, and its processing may be considered strictly cultural (A. Wilden, 'Communication', in: *Enciclopedia*, Einaudi, Turin, 1978, v. 3, pp. 601–695; Id., *System and Structure*, Tavistock, New York, 1980).

[49] F. Fileni, *Analogico e digitale. La cultura e la comunicazione*, Gangemi Ed., Rome, 1984, pp. 57–70.

of transformation (in the mathematical sense of the term). In order to understand them, no special access key is needed. Maps aim to portray objects as they are in real life, seen as a *continuum* that is based on physical laws arrived at through differentiation (one object differs from another because it is located at a given point, and because its features are unlike those of other objects).[50] Conversely, digital systems may be identified because, in order to convey the information of a geographical object, they employ different codes (color coding, number coding, or figure coding), which are designed to isolate only some of the features already inherent in the designator's meaning: digital systems, therefore, aim to set up *distinctions*, that is, to stress the distinctive qualities of an object.

We should stress, however, that the relationship between analogical and digital is not oppositional, even though it may at times come across as such.[51] Rather, it could be argued that in maps the analogical system makes up the "context" for the digital one. Although maps are based on an analogical system, they should never be presumed exempt from the implications attached to the digital system they use. In fact, the presence of the two communication systems promotes the creation of a third system mentioned above: the *iconic* system. Being derivative, the iconic system does not convey information based on the features (either distinctive or differential) of the systems that make it up. Rather, it arranges information in a new, markedly cultural perspective. Therefore, it iconizes the world and presents it not as it really is, but as it appears according to a theory: cartographic theory. Thanks to such iconizing process, what is represented is conceptualized and then conveyed on the basis of a dynamics that affords a novel view. The iconic system thus refers to the map's ability to convey the cultural values of territory: its connotative values, even though these are not necessarily the ones that belong there but rather – we should insist – are the ones produced by the map itself.

Cartographic Self-Reference

The claim above gains relevance when linked to the key outcome disclosed by the semiotic approach: self-reference.* As I already remarked elsewhere, self-reference refers to the map's ability both to be taken at face value and to

---

[50] Hence G. Bateson's famous claim, which borrowed one of Korzybski statements: "the map is not the territory" (A. Korzybski, *Une carte n'est pas un territoire. Prolégomènes aux systémes non-aristotéliciens et à la sémantique generale,* L'ECLAT, Paris, 1998 (Or. ed. 1933-1949-1950) to conclude that we "see maps as some sort of effect summating differences, organizing news of differences" (G. Bateson, *Mind and Nature. A Necessary Unity,* Hampton Press, New York, 1979; Bantam Books, 1980, p. 122).

[51] A. Turco, "Analogique et digital en géographie," in: G. Zanetto, ed., *Les langages des représentations géographiques,* Università degli Studi, Dipartimento di Scienze Economiche, Venice, 1987, pp. 123–133.

intervene in the communication quite independently of the cartographer's intentions. By virtue of cartographic self-reference, the names and symbols found on maps no longer reproduce merely empirical data of a physical-natural or anthropogenic nature. Rather, according to their own self-contained logic, they shape other meanings, able to affect the very perception actors themselves may have of the places which fall within their cognitive domain. Such process is triggered by the communication systems themselves, when activated, as well as by the action of icons, which generates a denominative projection. In this perspective, names, colors, numbers, shapes – in short, the whole language of maps – make up their autopoietic system. Ultimately, maps become self-referential because they advertise their ability to affect courses of action over the things they represent. By virtue of self-reference, a map becomes *a sign-based system which speaks for itself once it has been set up, remains relatively independent from all that preceded it, and goes beyond the uses for which it was initially intended.* Its final outcome is its ability to prescribe how territory should be conceived and experienced.

Besides showing an *inner* dimension, on which we have so far focused, cartographical self-reference also exhibits an *outer* dimension, determined by the fact that the understanding of maps is bound to the layers of cartographical records built up over time and to the experience acquired in interpreting them. That will define the existence of any map to come, influence its "perception" and build up the interpreter's "memory."[52]

We may then conclude that a map is a complex system based on compound self-reference. Through a self-contained process, maps arrange and convey cognitive items independently of the goals set by the actors who had originally built them or the actors who habitually have recourse to them. Ultimately, the maps' power of representation rests on their ability to regulate the complexity of geographical space through a metrics which leads to perceiving it as a *cartographic space.* By doing so, maps may direct multiple courses of action based on the "newly created reality."

Self-reference is then the ability of maps to refer back to themselves, as to a system that needs no external aids for defining its identity and instead sets itself apart from territory. One can, in fact, invert the claim that "the map is not the

---

[52] Through memory, interpreters of documents operate within a codification which already occurred and which historical sediments confirmed as legitimate with regard to the attribution of meaning. Likewise, the ways in which signs connect abides by rules of visual perception: the information offered does not equate the sum of information items drawn from individual icons, but is the result of mutual influences which their location exerts over the meaning of others. Therefore, the interpretation of maps occurs through an acknowledgement of the ways in which maps define themselves. See: E. Casti, *Reality as representation…, op. cit.,* pp. 140–144.

territory" and claim the very opposite: the map becomes the territory as such. And it is this paradox that leads to the full development of a map's self-reference potential. Maps do not offer themselves up as territory; if anything, they place themselves at a higher level than territory.

There are many ways to assess the preponderance of the cartographic model over the reality of territory. And they all confirm that the information such model contains *de facto* replaces what direct experience of territory records.[53] Maps may be said to turn their point of weakness into a point of strength. As models, they cannot duplicate reality, but only replace it.

Iconization

In the course of my analysis, I touched upon the possible senses of "iconization." I should only add here that iconization marks the high point in the cartographic production and circulation of meaning. Iconization may be defined as the *communicative process whereby, on the basis of a map's self-referential outcomes, highly conjectural facts are expressed as truths.* Through iconization, maps give us suggestions on how the world works. They do so on the basis of a hypothesis that favors an uncritical acceptance of its propositions. Ultimately, the message conveyed by maps may well replace reality. By promoting unthinking acceptance of its cognitive models as necessarily relevant – indeed as issues that rightly belong to territory – maps can shape social behavior. This is because iconization takes the meaning generated by maps and feeds it into a circuit of communication based on the key functions of the map itself: description and conceptualization. As we look closely at these two functions, we should recall that maps meet two fundamental needs implicit in any intellectual appropriation of reality: firstly, a *description* aimed at rendering features perceived through a first-hand experience of the real world; secondly, a *conceptualization* of the world, which tells us how it "works" in accordance with representational categories derived from interpretation. We would then be in a position to distinguish between maps that favor the communicative mode of description and maps that rely instead on conceptualization, and thus propose a worldview only partially modeled after canons of real-world mimesis. What I wish to stress here is, however, that all this should be linked to the role maps play in the semanticization of the world, that is to their functioning as discursive models of territoriality.* I already noted how the set of communicative procedures activated by icons

---

[53] Think for instance of the self-referential action of maps and the disastrous outcomes it can have, as in the case of the map handed over to commanders of the Italian Army involved in the conquest of Abyssinia (a region of northern Ethiopia) at the end of the 19th century. That map spelled defeat for the Italians in the battle of Adua. Faced with a mismatch between the information given on the map and what came to them from a first-hand experience of territory, commanders decided to rely on the latter. Because of that, the Italian brigades, which were marching separately, failed to recognize the place appointed for their reunion and, having been attacked by the enemy, were annihilated. E. Casti, *Reality as representation…, op. cit.*, pp. 154–174.

to "show" the enunciation of semanticization may be traced to figurativization, the ultimate outcome of which is iconization. Iconization shifts communication from the level of mere description to that of conceptualization, thereby endowing its message with social meaning.[54] This means iconization turns the outcome of figurativization to its own ends, and triggers iconizing processes regardless of whether maps are meant to describe or to conceptualize. As a result, the plausibility attached to description also invests conceptualization. Iconization favors the adoption of maps as hypotheses to be relied upon in order to assess all the information conveyed. It does so by activating a system in which information and concepts are made to circulate and cross-refer endlessly and in various modes along cartography's double plane (description and conceptualization). By appealing to the set of basic provisions inherent in our intellectual appropriation of the world, iconization carves out its own space of existence and overturns the original aims of any informational project.

One might claim at this point that a map is an operator whereby the world is perceived as being of a definite kind and working in a definite way. *Maps are therefore icons in themselves*, in the most compelling sense of tools whereby a metamorphosis of the world is carried out. As representational devices enhanced by their own mimetic reach, maps have greater communicative scope than territory: in fact, through a shift in perspective, they stand in for territory itself. It is by virtue of their effectiveness as mimetic devices, which a semiotic analysis brings to the fore, that maps can potentially compromise the meaning of territory. If, however, their functioning is understood, they can be mastered, iconization can be regulated, and their self-referential outcomes can be used to specific ends.

FROM *TOPOS* TO *CHORA*

Our discussion so far has, for one, provided answers with regard to the role played by the theory of cartographic interpretation. On the other hand, it led us to a new set of questions both internal, to do with interpretation itself, and external, bearing on the social relevance of a semiotic approach or its scope of application. Among the first, the most obvious case is the one of scholars who reflect on cartography: what skills are required of those who study and interpret maps? Among the second, with regard to the lines of research inaugurated by semiology, we should ask ourselves whether maps are in fact able to render the meaning of territory, having been freed from the positivist incrustations which made them look as mere instruments for recording reality. More radically, we may ask whether such semiotic theory has a relevant scope of application or meets specific social expectations.

[54] See note 41.

The first question may be easily dismissed by stating that *expertise*, as a special skill, is indeed crucial for specialist areas such as the one of cartographic interpretation. The possession of skills related to territorial analysis is of strategic importance if one wishes to avoid sweeping prescriptions over the qualities a generic interpreter should have and focuses instead on the qualities expected of a semiotic interpreter. Since a semiotic analysis of cartography operates at a meta-geographical level and activates a second level of interpretation, it would seem the latter can only be understood if the former, that is the one of geography itself, has been mastered in full. Obviously, this does not mean that expertise is strictly a function of disciplinary membership (as in the case of geographers, historians, urban planners, etc.); rather, it means that the scholar must possess suitable instruments supplied by a knowledge of territorial theories as such.

I will be addressing the other queries with examples drawn from specific fields of application in the pages that follow. Still, it may useful for now to outline the main stages of argumentation. First, I will try to show that exclusion of the social relevance of territory does not occur in all the maps ever produced. Rather, it would seem to invest especially maps produced in the period of history that saw the rise of geometric cartography. Cartesian logic,* which favors the visual and material features of territory, is irreparably removed from the idea of landscape intended as an experience of place. And it turned maps into instruments for representing territory as essentially removed from any type of social interpretation. At the time, maps achieved the aim of contextualizing projects: they presented territorial changes merely as material phenomena quite unrelated to symbolical or cultural issues. They trivialized territory, seen as an artifact where material considerations subsume and exhaust the relationships that communities establish with the world. In doing so, maps excluded the transmission of cultural features, which local inhabitants had entrusted to territory. Topographic representation, rooted as it is in Cartesian logic, effectively rests upon a very simple system of signs, which disregards the substance of objects and uses instead analogical distance to record the relationship between them. This representational system renders a neatly circumscribed aspect of territory, which, Augustin Berque reminds us, was called *topos** in classical culture. It makes sense of the material side of geographical phenomena by identifying their size and location. By so doing, however, it erases their social relevance.[55] The *topos* refers, after all, to an abstract view of space, which claims

---

[55] A. Berque, *Les raisons du paysage de la Chine antique aux environnements de synthèse*, Hazan, Paris, 1995; Id., *Écoumène. Introduction à l'étude des milieux humains*, Editions Belin, Paris, 2000, pp. 19–30; Id., *Médiance de milieux en paysages*, Editions Belin, Paris, 2000.

validity in and of itself rather than with reference to the subject who experiences it. When related to a subject, however, territory carries another meaning, which preserves the importance of reference but also highlights the place's cultural function. This alternative conception of territory, expressed by the concept of *chora,** revaluates the relation that subjects entertain with the place they inhabit and the cultural aspects of territorial reality. Nor is that all there is. Berque stresses the political thrust of *chora* (set up against *topos*), which he sees as the sum total of spatial interactions. The *chora*, he argues, is a territorial concept (*Écoumène*) that is necessarily social, for it expresses assets, values, and shared interests that were developed as it were "from the bottom up" and are made manifest through landscape. And landscape contains both a subjective dimension, since it is shaped by a gaze, and a collective thrust, because it condenses and exhibits socially produced values.

Berque's line of research had momentous repercussions, both operative and political. Among the former, I would lay emphasis on a new concept of planning, which sees landscape as a *planning unit*, that is a unifying element between environment and territory.[56] A renewed awareness of the relevance of landscape pertains instead to the second set of developments, which affected international organizations involved in the establishment of new identities. These institutions – notably the European Union – sanctioned the political role of landscape: its importance in creating a sense of belonging and in building an open and mobile citizenship (conveyed mainly by *chora*) against the birthplace-bound notion of the past (expressed mostly by *topos*).

Also, by refusing to set up a clear-cut dichotomy between these two categories and by accentuating instead their deeply symbiotic ties, Berque paved the way to a fruitful line of studies and inquiries. In his survey of the geographic and cartographic debate around the "sense of place" and the loss of a sense of place (as recorded by territory analysts), Giorgio Mangani for instance showed how, in cartography, real territory and depicted territory have been undergoing a very similar process of depletion. To confront this issue, he notes, geographers and landscape architects turned their eyes to the *chora*/*topos* dichotomy in order to set up the critical and conceptual framework they need to intervene in landscape planning and track the creation of new places, which not long ago some would have dismissed as "non places." Having acknowledged that landscape and cartography prescribe the forms

---

[56] R. Gambino, *Conservare, innovare: paesaggio, ambiente, territorio*, UTET, Turin, 1997.

and meanings of places, they now see the first as an expression of *chora* and the second as a topographical expression of *topos*.[57]

To adopt this plausible perspective means to question the assumptions of geometric maps, and especially their notion of space, with a view to finding other metrics capable of recovering *chora*. The theory of cartographic semiosis envisages the abandonment of topographic metrics,* an essential step towards redeeming *chora*. As it shifts focus from what portion of reality the map reproduces to what it conveys with regard to the meaning of territory, cartographic semiosis paves the way to mastering iconization to its own ends. It goes without saying that cartographic semiosis does not aim to erase the map's self-referential outcomes. Rather, it endeavors to find new ways to manage them.

In short, awareness of iconization should be the starting point for further inquiries aimed at a radical rethinking of maps: the goal would be to alter maps' structural features in order to make them suitable for expressing *chora*. Having ruled out a return to the sweeping pronouncements of pretopographic metrics, we now proceed to unravel the semiotic threads that should give us a new metrics. Our aims are: 1) to recover the sense of place expressed by landscape; 2) to give voice to the multiple subjects who shape such sense of place; 3) to engage the pragmatic potentials of Information Technology. In order to do that, in the next chapter we need to step back and look at topographic metrics.

[57] Berque shows how the issue of the progressive geometrization of maps, exposed by cartography historians, finds an equivalent in the current geographic debate over the crisis of the symbolic and narrative senses of territory. The *chora/topos* opposition, originally set up by Plato in the *Timaeus*, rests on the relationship between the physical size of space (the topos) and its cultural identity, its "genius loci" (the *chora*), its narrative dimension. On this point and on the adoption of this pair of concepts in cartographic theory see: G. Mangani, "Intercettare la chora," in: E. Casti, ed., *Cartografia e progettazione territoriale. Dalle carte coloniali alle carte di piano*, UTET, Turin, 2007, pp. 31–41.

CHAPTER 2

The Success of *Topos* in Colonial Cartography: Topographic Metrics

Space itself exists only where surveyors measure it.

D. Kehlmann, *Measuring the World*, 2006

Colonial cartography is the ultimate expression of topos. *As a manifestation of topography in conquered countries, colonial cartography is an ideal field for exposing the power of a metrics that favors the materiality of territory over and against its social value. That power is functional to the protocols of positive science at the time, which often turned to cartography in order to classify and quantify material phenomena and thereby master the complexity of the Elsewhere. Maps serve well to uphold those endeavors of science because their metrics and the abstraction of their language ensure undisputed scientific exactitude. Following the example of French colonialism – which more than any other reflected on its own discursive practices about the Elsewhere and debated over the ensuing terms of misunderstanding between colonizers and colonized – this chapter examines the role maps had both in concealing the holistic meaning of the territory of the Other and in investing it with unduly Western features. A semiotic analysis of the types of maps employed unveils the mechanisms underlying such Western misappropriation. Topographic maps mark the erasure of the symbolic meaning of territory and the ensuing loss of its social value; thematic maps highlight the impossibility of applying taxonomy to manage the complexity of territory; and finally, popularized scientific maps clearly show their role in building and propagating myths about Africa for use and consumption by the West. This close analysis will be used, at the end of the chapter, to address the major theoretical issues of topographic metrics.*

CONTENTS

UNDERSTANDING AND DESCRIBING AFRICA

On the occasion of the 1931 international colonial exhibition in Paris, a general appraisal of the cartographic production made by France in West Africa was presented. Only 5% of the territory was detected through regular topography* (some areas of Senegal, Guinea and Dahomey, now Benin); 20% was mapped through semi-regular surveys (Ivory Coast and the southern part of Upper Volta, now Burkina Faso); 25% from reconnaissance topography (Senegal and Inner Guinea, and southern Mali and Niger); 40% from exploration cartography covering inland areas, while 10% of the latter remained virtually

Reflexive Cartography. ISSN 1363-0814, http://dx.doi.org/10.1016/B978-0-12-803509-2.00002-1

unknown.[1] It should be noted that only regular topography can reference territory exactly, but such mapping poses serious difficulties. Requiring the identification and denomination of each specific point of territory, it entails substantial investment, both technical and intellectual. However, the triumph of positive science at the time made it look like the only possible path to knowledge, which was expected to be extensive and uniform.

A number of technical and theoretical safeguards were put in place to cope with the obstacles that topographic mapping posed. Firstly, initial claims to thoroughness were reduced and the scale factor to which maps were drawn was changed: from 50,000 to 100,000 and later to 200,000. The amount of needed data was thus cut down and the problem of surveying remote areas partly redressed.

But greater endeavors on the part of the French were required to cope with theoretical issues. Colonial territories differed both morphologically and anthropically from the place of origin of cartographers and misunderstanding of the relevance of names given by local populations invalidated cartographic results and stalled progress.[2] The data presented at the International Exhibition in 1931 came from recent surveys, since cartographic information had long remained quite cursory and fragmentary.[3]

It was only in the period after 1923, as commander Édouard de Martonne took over direction of the *Service Géographique* of Western French Africa (AOF), that activity shifted to planned and permanent surveys, in direct contact with the *Service Géographique de l'armée* and with the cooperation of public and private institutions in France and abroad.[4] Even the latter result, however, was achieved

[1] Ministère de la Guerre, Service Géographique de l'Armée, *La carte de l'empire colonial français,* G. Lang imprimeur, Paris, 1931. From a technical perspective, *régulière* topography is the result of terrain triangulation implemented through the plane table (basic measurements; setup of markers; observation and calculation of geodetic chains derived from astronomically exact points). *Exploration* topography translates into a set of sketches, of rapid or itinerant surveys at different scales, rather approximately based on very rare and quite inaccurate astronomical points. *Reconnaissance* topography is also based on rapid surveys, performed along a previously determined direction, which uses astronomical points closer together, thus creating comparatively more uniform maps. *Semi-régulière* topography, finally, is based on general geodetic surveys: information is placed inside a mesh built in accordance with existing roads and trails.
[2] Ample research was undertaken to mediate between the names given by natives and those given by the French in order to produce exact references and at the same time understand the social meaning of territory. Édouard de Martonne is one of the most notable researchers involved. Among his works: "Aspects de la toponymie africaine," Libraire Larose, Paris, 1932, from: *Bulletin du Comité d'Etudes historiques et scientifiques de l'Afrique Occidentale Française,* XIII, 3–4, 1930; Id., "Les noms de lieux d'origine française aux colonies," in: *Revue d'Histoire des colonies,* 1936, pp. 5–55.
[3] On cartographic production prior to that date and for bibliographical details see my "Colonialismo dipinto: la carta della vegetazione in AOF," in: *Terra d'Africa* 2000, Unicopli, Milan, 2000, pp. 15–71.
[4] The activity of the Service, based in Dakar, covers cartographic surveys as well as the research and preservation of cartographic documents relating to AOF; the testing of new data collection methods; and collaborative research in astronomy, geodesy and geophysics (*Gouvernement général de l'AOF, Service Géographique, Inventaire méthodique des cartes et croquis imprimés et manuscrits relatifs à l'Afrique Western existant au Gouvernement Général de l'AOF à Dakar,* Goupil éd., Laval, 1926).

only by limiting the scope of the initial project, which was meant to map territory in order to meet military or administrative needs via a painstaking survey of referentiality. Surveys covered only areas that seemed economically relevant on account of their position (along the coast) or of the resources they harbored (dense population; wealth of materials, mineral resources, arable land, etc.) and that eventually produced what we could call a "gap" topography, which in turn called for integration with thematic cartography. Consequently, areas that were economically relevant (on account of the actual or presumed resources they possessed) or strategically prominent by virtue of their geographic features (accessibility or connectivity) were those diligently and meticulously recorded via thematic maps.

This latter body of records acts as a magnifying glass which shows how Africa was perceived. It enables scholars to assess the colonial enterprise with an eye on its prospected financial and/or political gain. In fact, selection was made not only with regard to the areas but also to the themes proposed, which were recorded on the basis of their quantity, both when referring to natural and to anthropic phenomena. I will return to this homogenizing trend in a while. For now I wish to underline that topographic mapping in Africa over the first decades of the 20th century was restricted with regard to the areas it covered.[5] Thematic mapping, instead, was substantial and recorded both the physical and the anthropic features of territory.[6] Although thematic maps required information that was collated slowly and was largely the result of empirical trials for a good half of the colonial period, the difficulty of implementing topographical surveys and the need to avoid unsustainable delays led to their adoption.[7] This problem was compounded by the fact that, in the period between the two

---

[5] In 1937, scholars recorded the existence of 1,661 maps and sketches covering the area in question. While this obviously attests to the commitment and efforts of the French administration towards cartography, in fact it refers to the whole range of products and not to specifically topographic ones. (E.A. Joucla, *Bibliographie de l'Afrique occidentale française, avec la collaboration des services du gouvernement général de l'Afrique occidentale française et pour le Dahomey de M. Maupoil, administrateur des colonies, Société d'éditions géographiques, maritimes et coloniales,* Paris, 1937).

[6] It is interesting to browse geographical journals of the time, such as *the Annales de Géographie,* the *Bulletin de la Société de Géographie, La Géographie* or those relating to colonial studies *Renseignements coloniaux,* published by the Committee of French Africa, or *Outre Mer,* a generalist journal on colonization. Ever since the late 19th century these journals collected land surveys and travel reports which were conveyed through thematic maps.

[7] Until the First World War (1915–1918) and, to a lesser extent, in its aftermath, cartographers produced largely reconnaissance-based, loose information. The use of *photographic clichés* and the spread of geodesic triangulation between the wars compensated for those shortcomings and made maps more accurate. Thus, production of regular maps often postdated the political and administrative delineation of the territories themselves. It was only in the course of the 20th century, as the colonial enterprise gained awareness and its means and authority grew stronger that production became systematic and theoretically sound. See: A. Maharaux, "Le géographe et le tracé des espaces coloniaux et postcoloniaux," in: M. Bruneau, D. Dory, *Géographies des colonisations, XVè-XXè siècles,* L'Harmattan, Paris, 1994, pp. 349–367.

world wars, knowledge of usable natural resources and the attendant need to monitor the "human capital" became two primary goals of thematic mapping.[8] This means that, although reference and knowledge correspond to two gradual stages in the experience of territory, and thematic maps refer to a more elaborate level of cognitive development, in some cases thematic maps anticipate topography and reference territory themselves.

The interrelation and diversity between the two types of mapping also emerge in the agencies involved in the colonial enterprise, at least when it became fully fledged. At that time areas of competence started to be delineated more clearly: topographic maps and basic documents were drawn by the *Service Géographique de l'AOF*, a government body in charge of that specific function. Conversely, thematization was entrusted to structures which depended on the *Gouvernement général de l'AOF.* The prestige of the latter as a research institute meant that it could maintain at least some form of autonomy from political expectations.[9]

The autonomy of these organizations does not imply that no constraints or ties existed between administrators and researchers. It should not be forgotten that one of the distinctive features of scientific studies in early colonialism was the complex set of ties that existed between a form of political rule and the knowledge constructed and imposed on colonies. The authority of political power and the prestige of scientific discourse were mutually reinforced.[10] Such close alliance seems, however, to have waned since the 1950s, when political and scientific authorities appealed to a progressively different concept of African otherness. The *Gouvernement général* was slow to realize that such alliance between science and colonial rule was changing and that, despite their endurance, institutional ties between colonial administrators and professional researchers were changing because of the different profile that the latter group had taken on. As a consequence, the relationship between researchers and colonial administrators grew

[8] On thematic maps and their ideological implications see: M.A. de Suremain, "Cartographie coloniale et encadrement des populations en Afrique coloniale française, dans la première moitié du XXe siècle," in: *Revue française d'histoire d'Outre-mer,* 324–325, 1999, pp. 29–64.

[9] These were specific institutions, such as the Service géologique (Incorporated in the Service des Travaux et des Mines) , the Service météreologique, or, again, wider-scope organizations such as the Institut Français d'Afrique (IFAN), created in 1939 in Dakar or the Office de la Recherche Scientifique Colonial, a precursor of ORSTOM, based in Paris. The Comité d'Études historiques et scientifiques de l'AOF created in 1915 became IFAN in 1939. In 1982, L'Office pour la Recherche Scientifique et Tecnique Outre-Mer became L'Institut Français de Recherche Scientifique pour le Développement en Coopération, although the original acronym (ORSTOM) was preserved. In 1998 this institution changed its name to IRD (Institut de Recherche pour le Développement).

[10] And yet, throughout the century, positivist principles were ill fitted to the theoretical scope of French geography. I am thinking in particular of Vidal's epistemology and to the cluster of studies it produced. See: V. Berdoulay, *La formation de l'école française de géographie,* Edit. Du C.T.H.S., Paris, 1995, p. 183 and ff.

somewhat strained.[11] By then, research institutions began to gain a more articulate understanding of African difference. This went against the aims of colonizers and their exclusive attempts to impose Western models onto a social context they considered indigent or saw as a "natural" annex to metropolitan France.[12] These were the same institutions that initially refused to take charge of thematic mapping to supplement, integrate or even conclude their own research, but eventually derived a renewed awareness from their work and began to question the relevance of research paradigms they had validated until then.[13] Starting in the 1950s such contradictions were forcefully exposed and challenged, both the use of classifications on the part of colonial administrators and the validity of enumeration itself.[14]

For our purposes, it is therefore important to refer to maps drawn within both contexts and at different times, for these will be able to convey the specific cultural milieu that produced them. I will consider early colonial topographic maps of various types (sketches, reconnaissance maps, relief maps dating back to the period between the start of the 1900s and the 1930s) drawn by colonial administrators. Then I will move to late thematic products (1950–1960), namely the AOF vegetation map, issued by the ORSTOM research institute.

---

[11] On this complex tangle of relationships see my: "Colonialismo dipinto...," *op. cit.*

[12] The notion the French held of their own colonies as a prolongation of metropolitan territory made them especially zealous in supporting all the structures aimed at turning natives from the colonies (schools and universities) into new French citizens and at involving them in public administration, albeit in secondary roles.

[13] André Hauser's position is revelatory in this respect. He cautioned against administrative reductions of the kind found in ethnographic classifications (A. Hauser, "Note bibliographique," in: *Bulletin de l'IFAN*, 1954, p. 220). For his part, Paul Pélissier, in his presentation of the ethno-demographic map of Western Africa, denounced the shortcomings of such map precisely because of the assumptions it embraced. For those assumptions prevent a recovery of all the identity issues which emerge through *métissage* in all their richness and cross-contamination (P. Pélissier, "Notes de présentation," *Carte ethno-démographique de l'Afrique Occidentale*, sheet n.1, IFAN, Dakar, 1952). We should also remember the debate initiated by geographers in 1946 in prestigious journals such as *Annales de Géographie or the Cahiers d'Outre-Mer*, which brought into question the very taxonomic principles research was based on. Special attention in this line should be given to the research conducted by Olivier Soubeyran: "La géographie coloniale. Un élément structurant dans la naissance de l'Ecole française de géographie," in: M. Bruneau and D. Dory, eds., *op. cit.*, pp. 52–90; Id., "La géographie coloniale au risque de la modernité?" in: M. Bruneau and D. Dory, eds., *Géographies des colonisations*, L'Harmattan, Paris, 1994, pp. 193–213; Id., "Alle origini del paradigma possibilista: geografia e colonialismo nella 'battaglia delle Annales'," in: *Terra d'Africa 1995*, Unicopli, Milan, 1995, pp. 59–93; Id., *Imaginaire, science et discipline*, L'Harmattan, Paris, 1997.

[14] Awareness of the colonies changed following a critique of the assumptions, namely positivist assumptions, on which colonialism was based. Such awareness developed over time within the fruitful set of contradictions present in sociology over the land studies conducted by George Balandier. It resulted in a complete overhaul of colonial classifications. See: M.-A. de Suremain, "Cartographie coloniale et encadrement des populations en Afrique coloniale française, dans la première moitié du XX siècle," in: *Revue française d'histoire d'Outre-mer*, t. 86, n. 324–325, 1999, pp. 29–64.

IN SEARCH OF *TOPOS*: TOPOGRAPHIC MAPS

The topographic examples I am going to submit date back to the time of early colonialism, when the mismatch between administrators and researchers was not overt and cartographic rules were not yet firmly prescriptive over the type of information to include. The loose meshes of this topographic deregulation may be said to accommodate both the choices and interpretations of the cartographer-researcher and the obliviousness of the administrator-topographer who supervised surveys. Our first example is a 1903 geodetic survey that betrays the cartographer's intellectual failure in attaching meaning to the social features of Africa. The second is instead a reconnaissance map dated between 1881 and 1885 which strives to recover such features. I shall be comparing the latter map with regular cartography dated 20 years later (1906–1908), after cartographic codification had been adopted in full and those who undertook it were professional cartographers. These had lost their ability to question such issues precisely because codification did not require it, or rather excluded it.

The Birth of *Topos*

We ought to stress first that the production of cartographic maps does not involve only military institutes, but invests other public administration sectors, such as mining and resource valorization. The first document we are going to address confirms this. It is a 1903 sketch map showing the first phase of a topographical survey, the geodetic phase, covering an internal region of the Ivory Coast, where mineral concessions had been granted and needed to be finalized by demarcating territorial boundaries.[15] The *Service des Travaux et des Mines* launched a mission led by a deputy administrator, a certain Mr. Cartron, with a view to defining geodetic points and acquiring trigonometric data to be inserted into a 1:500,000 sketch map. The administrator-topographer was therefore called upon to produce a sufficiently accurate document detailing the "centers" of the many concessions that had already been granted but not yet matched to actual areas of territory.[16] Without this map, the mining company would find it hard to settle disputes or avoid conflict between the parties involved in the mining operation[17] (Fig. 2.1). Referential precision in the survey was therefore vital, even though the procedure used to secure it (triangulation based on finding steady points across territory) turned out to be very hard to implement. The territory to be measured was covered in thick vegetation; no road network existed and, what is especially disconcerting, the landscape was flat, with no

[15] It is a sketch map drawn as part of an administrative record for the *Service des Travaux et des Mines de la Cote d'Ivoire*, preserved in the *Fondo Antico* of the AOF (1903: 146).

[16] Mineral concessions are granted by drawing a circle around the applicant's point of interest within a prescribed radius.

[17] Presumably, the map was used later as the basis for detailed topography work.

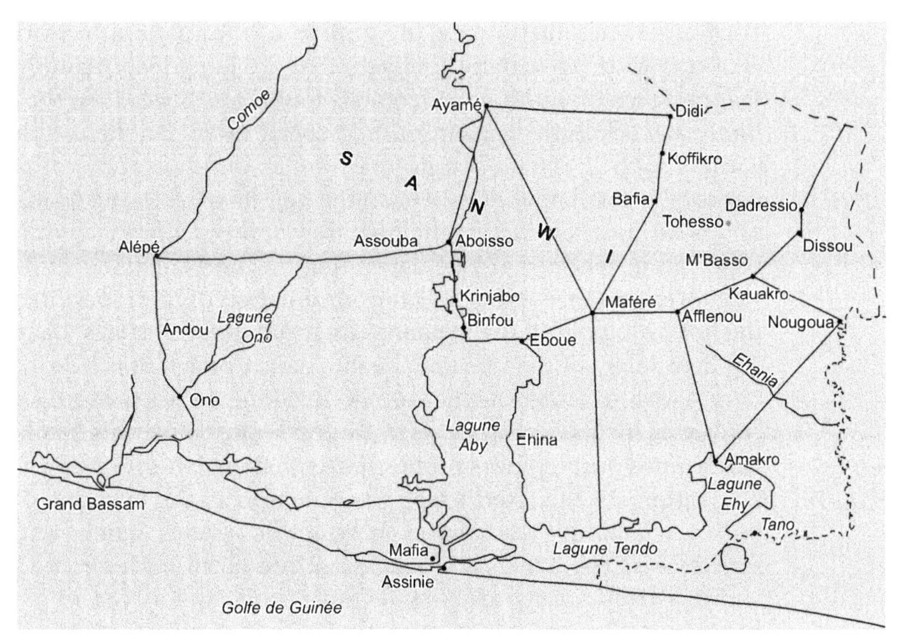

FIGURE 2.1
1903, Service des Travaux et des Mines de la Côte d'Ivoire, *Sketch map of the Ivory region of Sanwi*, 1:500,000.

reliefs allowing a bird's-eye view. Production of the map proceeded slowly in the absence of observation points, which would have been ideal locations for setting up markers, that is, geodetic pyramid bases. For these can be easily detected on the ground and are therefore placed in clearly identifiable sites. Given this environment and these needs, it is inevitable that villages should be seen as ideal geodetic points, because they are the only ones to have all the features listed above. To Europeans, villages provided the only possible grounding for a topographic survey, and the designators that distinguished them – created by local inhabitants – were used as pyramid base markers.

Before we move on to a semiotic analysis of this document, we need to stress that it is particularly relevant to our line of argument, because it enables us to show that a few preliminary assumptions are already fixed by the astronomical survey and necessarily affect all subsequent mapping. I am not claiming here that this is a conscious choice or that geodetic practice in itself precludes the territory's social data. Rather, I submit that such practice sanctions the rise of *topos*, because it establishes an information hierarchy that places reference at the top: steady points and measured lines constitute the information that needs to be secured and safeguarded at all costs. That is how *topos* prevails in cartography: measures trump any other type of information. When the latter conveys anthropic features, as in the case we are going to examine, its meaning is subserved to that metrics and completely absorbed.

Thus, the sketch map marks the outline of the vast region whose expanse is dotted with the names of villages that are joined via trigonometry lines. The presence of anthropic referents should not mislead us, for it does not imply any acknowledgement of their social value. On the contrary, it bears witness to their denial, because they are assumed as nodes of the geodetic network. We will not linger here on the naiveté of choosing villages as steady points without keeping in mind that they never occupy the same site in these regions, but tend to migrate depending on the availability of tillable land. What we would rather stress is the annihilation of their social relevance and the hollowing out of the meanings attached to their names. I am not thinking here solely of the referential value which designators bear. On the contrary, I wish to underline the cultural meaning names possess, as capable of conveying the rules of space vectorization set up by the community members who coined such names and the values that govern their social interaction. The authors of this sketch map are oblivious to the existence of topomorphosis, which, in their capacity as symbolic systems, names activate. It is a process whereby beliefs and values are rooted in territory and reverberate within a given society, affecting its functioning and its system of checks and balances.[18] The case we are analyzing now goes even further: any symbolic value of the designator which is meant to convey the relationship between society and nature is wiped out, because it is made to match geodetic data and given nonterrestrial grounding, quite removed from actual territory.[19] This is undoubtedly a telling instance of pauperization of the cultural value of territory, shown as an empty space upon which colonial expansionism can advance legitimate claims. Evidence of this practice abounds in colonial archival sources. In fact one could arguably claim that Africa's difference and the impossibility of referencing it without alienating it from the context of territory produce the notion of an equivalence between natural and social features. Their equalization enables agents to interpret either using the same taxonomic categories.

Moving Away From *Topos*

As we move on to consider a reconnaissance map, we will have the chance to examine the choices military cartographers make as they try to include in their documents social information about the territory of the Other. In particular, reconnaissance maps, which in the absence of regular topography are the first to be drawn, although they are informationally fragmentary, often include icons which recall cultural features of territory. As I mentioned above,

[18] A. Turco, *Terra eburnea. Il mito, il luogo, la storia in Africa*, Unicopli, Milan, 1999, pp. 128 ff.

[19] Geodetic surveys establish the geographical coordinates of a point on the earth's surface on the basis of a fixed off-earth point that coincides with a star.

the drawing of these maps during the first colonial phase was not aimed to provide an extensive image of territory, but rather to represent regions that were deemed relevant. Consequently, a few areas, such as the one around the river Niger, drew most attention and were mapped in painstaking detail.[20] Significant investments were made between 1880 and 1886 with a view to systematizing knowledge of the region's hydrographic resources.[21] In 1884, a number of exploratory missions were undertaken with the aim to map territory.[22]

The map was issued with these aims in mind. It is the map of a stretch of the Niger River, from Bamako to Timbuktu. It bears no date, but could be dated to the 1881–1885 period, and was drafted by Chief Engineer Ancelle, under the direction of General Faidherbe[23] (Fig. 2.2).

Information came from previous exploratory maps or from rapid surveys and was anchored to astronomical points set quite apart from one another, corresponding to major towns (Bamako, Ségou, and Timbuktu). The map is interesting because it adopts a dual mode of representation as well as various information typologies to describe either the coastal stretch or the inner regions. The river is outlined in its course, with several names placed along it, which refer to cities and villages, but also provide water-supply data, useful for navigation. The rest of the territory is not outlined but scattered with written captions relating to its physical and anthropic features. The double plane of information is here primarily dictated by the map's goal: a precise referential representation of fluvial territory, based on semiological aspects of the map itself; and a written representation addressing general features, both social and physical, of the rest of the area.

First, an analysis of the information level produced by the icons along the river shows that the designators were used as referents for navigation and not

---

[20] In the French colonial project, the Niger River represented the main penetration path, so much so that it was deemed wise to connect it to the coast by setting up a railroad that went from Medina to Bamako and reached the river Senegal.

[21] In 1883, even a collapsible *canonnière* was built. It was a lifeboat (approximately 18 meters long and 3 meters wide) that could carry up to 12 men and 40 tons of food supplies.

[22] See: J. Ancelle, *Les Explorations au Sénégal et dans les contrées voisines depuis l'antiquité jusqu'à nos jours. Précédé d'une notice ethnographique sur notre colonie, par le général Faidherbe, Maison-neuve frères et C. Leclerc*, Paris, 1886, pp. 426–434.

[23] The document, entitled *Carte du Niger dressée d'aprés la carte de Mr De Lannoy, Capitaine du Génie et les renseignements fournis par M.M. le Colonel Borgnis-Desbordes, les Capitaines De Lanneau, Pietri, Valliére et le Dr Bayol* comprises three sheets at a 1:500,000 scale and reproduces the course of the Niger River from Bamako to Timbuktu. It includes the name of its author (Jules-Charles Ancelle, 1850–1941), who took part in several African campaigns, including the one in Morocco. There is also a statement endorsing one of the most enlightened exponents of French colonialism, Louis Léon César Faidherbe, as the person in charge of the document. In his capacity as Colony Governor (1852–1861), Faidherbe saw to the preparation of the first map of Senegal. See: Ed. De Martonne, *Cartographie*, Librairie Emile Larose, Paris, 1927, pp. 28–37.

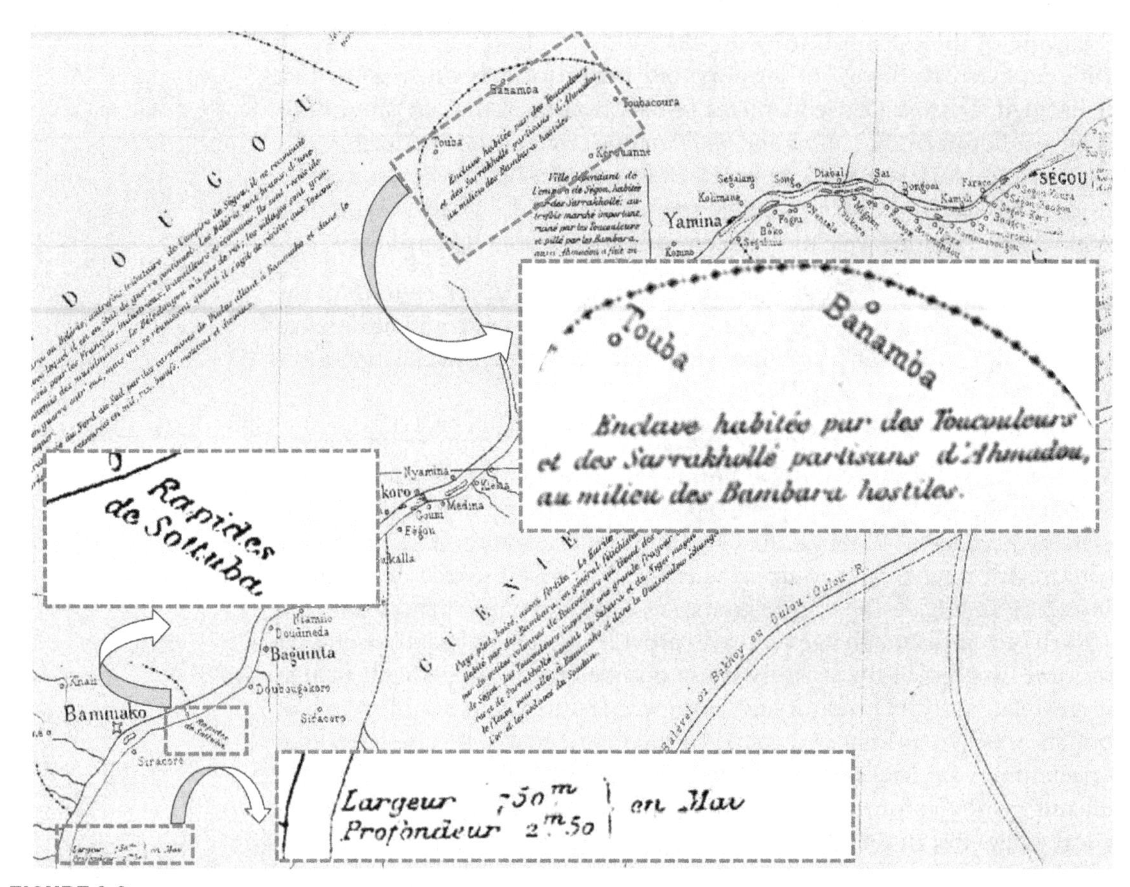

FIGURE 2.2

1881–1885, Jules-Charles Ancelle, *Carte du Niger dressée d'après la carte de Mr De Lannoy, Capitaine du Génie et les renseignements fournis par M.M. le Colonel Borgnis-Desbordes, les Capitaines De Lanneau, Pietri, Vallière et le Dr Bayol* (detail showing the course of the Niger river from Bamako to Ségou, with enlargements by the author).

to convey the layout of riverine territory. It is not by chance that in the lower left corner there should be an indication of water width and depth in a particular month of the year (May), when the course of the river is low and makes navigation harder. Similarly, the mention of the "Sottuba Rapids" responds to the opportunity of warning about another possible obstacle. Settlements are referred to via three types of icons: they rank information, relying both on different typefaces to mark the designator and on surrogates which characterize it. The towns of Bamako and Timbuktu rank at the top and are marked respectively by the symbol of a fortress and by a long text which illustrates its role of market-city. They are followed by towns that are the seats of local powers, famous for historical events or for their role in trade (Sibi, Yamina, Ségou, Sansanding, Tenencou). Lastly, we have the villages, marked by the standard shape of a small abstract circle placed close to the designator. The typeface of their designator, smaller than the one used for towns, and the uniformity of the figural code identify them exclusively on the basis of their localization. The rest of the information in the document regards a few areas whose features are summed up in brief captions.[24] Overall, and despite minor additions, the technique used to render territory along the river Niger is cartographic. In other words, it is based on icons that figurativize the designator. Conversely, regions far from the river are rendered only via short descriptions, arranged in text boxes and superimposed onto the regions they reference. Such descriptions provide social information about local populations, about the setup of territory and the existing hierarchy among basic territorial structures.[25]

This map, ultimately based on a rough and patchy semantics of territory, does not consign information on the social organization to a visual semiotics, but chooses to entrust it to the written text. That way, it shows little confidence or even distrust towards topographic representation. Conversely, the latter seems fit to represent data related to mobility, which the French are more immediately concerned with, as in the case of berth names along the river course. To conclude, although the description of African territory is cursory and conjures the image of an explorer disconcerted in the face of the Elsewhere and striving to find adequate tools of representation, the map records an effort to understand which, regrettably, would soon be lost. Barely a few decades later, following a structural overhaul of colonial administration, new rules for the drawing of maps were fixed and the possibility of a subjective intervention on the part of the cartographer, aimed at making social sense of territory, was permanently denied.

---

[24] For instance: "Enclave habitée par des Toucouleurs et des Sarrakhollé partisans d'Ahmadou au milieu des Bambara hostiles" (Enclave inhabited by the Toucouleurs and the Sarrakhollé partisans of Ahmadou in the middle of the Bambara); "Marché très important Sansandig est indépendant de Ségou. Il a toujours victorieusement repoussé les attaques d'Ahmadou." (A very important market, Sansandig is independent from Ségou. It always victoriously resisted the attacks of Ahmadou.) and again "Population: 100.000 à 120.000 habitants répartis en 200 au 250 villages…" (Population: 100,000 to 120,000 inhabitants spread across 200 or 250 villages…) I am here examining only a small stretch of the river reproduced on the map, namely the one between Bamako and Ségou, because I intend to compare it to a later topographic survey.

[25] We are told that it is a flat, wooded and very fertile land, where "karité (arbre à beurre)" and acacia are the prevailing scents; that the territory is inhabited by the Bambaras and the Peul but also traversed by small groups of Toucouleur who levy taxes on behalf of the Sultan of Ségou. Data regarding other ethnic groups (Mauress and Sarracole) are included, as well as information on the political and military setup of local populations. For an extensive treatment of this issue see the documents analyzed in: E. Casti, "Mythologies africaines dans la cartographie française au tournant du XIXème siècle": in *Cahiers de Géographie du Québec*, Université Laval, Québec, vol. 46, 2001, pp. 429–450.

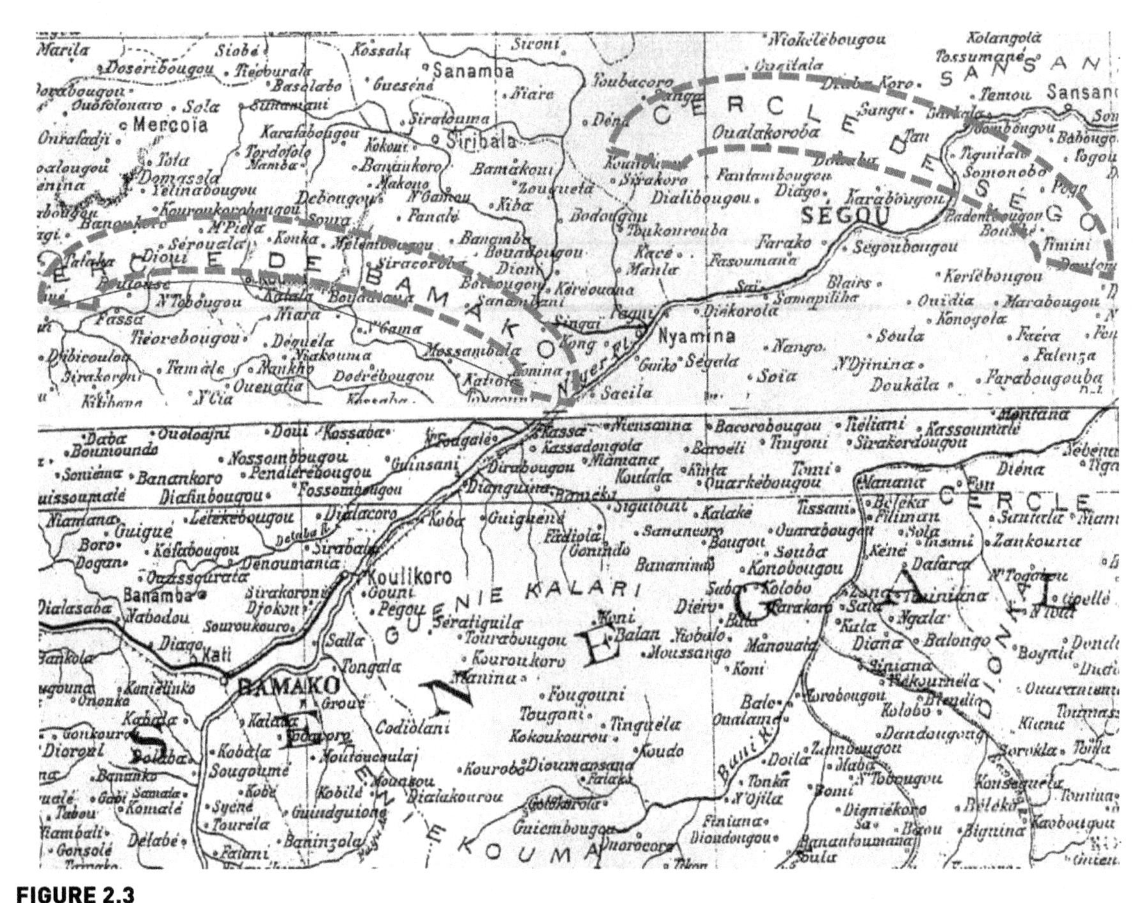

FIGURE 2.3
1906–1908, Service Occidental du Gouvernement de l'AOF, *Carte du Niger* (detail from Bamako to Ségou; highlights are mine).

Delving Into *Topos*

This is the case of the 1:1,500,000 map drawn between 1906 and 1908 by the *Service occidental du Gouvernement de l'AOF* to represent French colonial possessions. It depicts the same territory represented in the map described above (Fig. 2.3).

Even a brief comparison of these two documents shows notable differences in their framework: in the one we are examining now text boxes completely disappeared and the whole territory is sprinkled with names that refer to settlements. A very important market, Sansandig, is independent from Ségou. These names, however, provide no useful information for an understanding of territory, because they are scattered quite haphazardly and without any surrogate for distinguishing them. The icon, consisting of an abstract sign in the form of a small circle beside the name, is flat and generically placed. Even this document conveys a hierarchical order among territorial structures, but ranking is no longer based on the social organization of African territory. Rather, it refers to the political

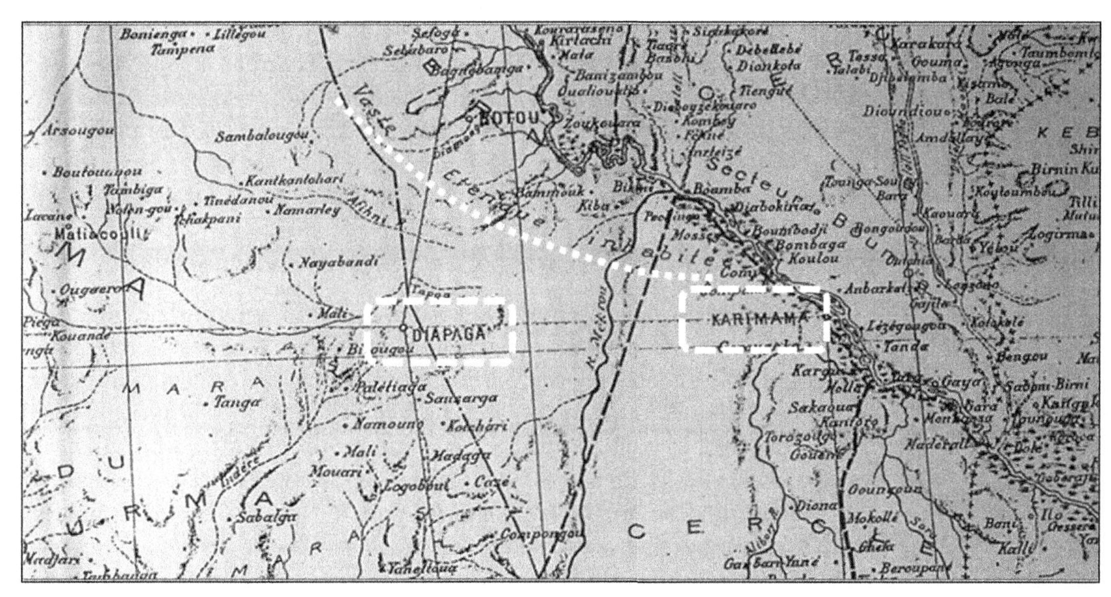

FIGURE 2.4
1906–1908, Service Occidental de Gouvernement de l'AOF, *Carte du Niger* (detail with the W TBR region).

and administrative framework imposed by the French. For instance, the typeface size of the "Haut-Sénegal" caption that lies across most of the map endorses the colony's new political structure and underlines the importance of the "Cercle de Bamako" or of the "Cercle de Ségou," which refer back to their own administrative setup. That entails the annihilation of the historical role played by riparian cities, whose names are cast in a flat and homogenized typeface, in confirmation of the fact that icons play a role both in annihilating the African sense of territory and in bolstering the new meaning imposed by the French. Still along the Niger River, towards the middle of its course and thus in regions now belonging to the States of Benin and Burkina Faso, a number of designators in capital letters (Karimana, Diapaga) assert that those villages were selected as local centers of colonial administration. A few other designators are also present to mark peculiar physical traits or some of the region's natural resources – such as the forest that covers the whole transboundary area of these Countries, defined as a *Vaste étendue inhabitée* – with a view to recording settlement distribution[26] (Fig. 2.4).

The iconizing outcome of this map is obvious. Once represented via typically Western criteria, Africa acquires the prerogatives needed for incorporation into the metropolitan project, whose legitimacy is thereby enhanced.

[26] This forest was catalogued a few decades later and become part of the land registry initiated by the colony to take stock of forestry resources. It is now a protected area named *Réserve de Biosphère Transfrontalière W* (*RBT W*).

The rigorous assumption of topographic norms, documented here, is the result of an overhaul of the French cartographic service, which in the time frame of 20 years between the two documents codified rules and issued strict directives for their enforcement. I should also add that, entrusted to professional topographers, maps achieved a higher level of accuracy as far as reference is concerned, yet abandoned all attempts at making sense of the complexity of African territory. Even those few yet relevant social features that the first document had strived to convey by using text boxes were now neglected.

Thus, even though the latter map shows a much denser semanticization of Africa than the former, no deeper knowledge of the colony is in fact conveyed. The large number of names present on the map (obviously heedless of the density that graphical scale can sustain) and the lack of iconic specifications for them have a negative impact on cartographic communication.[27] To be sure, cartographic semantics in the latter map is prevented from effectively conveying any kind of information, including referential data.[28] Nonetheless, while failing to convey information, excessive name densification does convey the impression that a thorough knowledge of territory has been acquired. The outcomes of topographic innovation seem devastating with regard to cartographic semiosis, because when icons are rigidly encoded, their iconizing power is enhanced. Topographic maps become indisputable tools, for they are purified of any subjective interference and made fit to convey the "objective" meaning of territory.

As for the figure of the interpreter, it should be noted that, while the first map (the reconnaissance map of the Niger River) implies the presence of an explorer-cartographer who eschews a cartographic logic and attempts to convey the meaning of territory, the second finds him or her in the passive role of one who applies rules without asking questions. The explorer is the first person who, albeit with difficulty, endeavors to interpret the social aspect of the Other's place. Conversely, no critical appraisal of what is being done rests with the topographer, except for a rigorous application of Cartesian logic. The latter departs from the concept of place as a socially intended locus and embraces cartography as an

---

[27] In the wide research field of vision semiotics (that is the study of the perceptual functioning of the eye and brain as they process images), an essential monograph is still: R. Arnheim, *Visual Thinking*, University of California Press, Berkeley-Los Angeles, 1969. Arnheim's research played a key role in vision semiotics studies and paved the way to critical analysis in art history and to new experiments in pictorial art, as we shall see later.

[28] Jacques Bertin warns us that while increasing the precision of referential information is a matter of technique, increasing the number of informational variables is a psychological issue, that is, an aspect that touches upon the limitations of visual perception. J. Bertin, "Perception visuelle et transcription cartographique," in: *La cartographie mondiale*, 15, 1979, pp. 17–27; this issue was expanded upon in: Id., "Voir ou lire," in: *Cartes et figures de la terre*, Centre Georges Pompidou, Paris, 1980, pp. 2–8. A few years later, the complex issue of visualization was addressed also by: A. Antle, B. Klickenberg, "Shifting paradigms: from cartographic communication to scientific visualization," in: *Geomatica*, 53, 1999, pp. 149–155.

expressive means, highly instrumental to the rendering of an abstract territory that is detached from any interpretation.[29] Such logic is applied to the map through the criteria of measurement: the concept of nature itself is rendered as an abstract entity, the veracity of which depends on the accuracy of its representation. That consolidates the practice whereby the value of territory – as the outcome and condition of human action – is *de facto* wiped out by the map itself.

THE STRENGTHENING OF *TOPOS*: TAXONOMY AND THEMATISM

We ought now to reflect on the adoption of the quantitative principle as a global and encompassing paradigm in cartographic mapping. That is undoubtedly attested by the frequency of thematic maps built during the whole period, with an exponential growth in late colonialism. In order to master African complexity, Europeans relied on quantity. Suffice it to recall that the term "thematic" is a recent coinage. Until about 50 years ago the phrase "quantitative map" was the usual designation, precisely because it conveyed the quantifiable value of data being represented.[30] Thus, by virtue of their technical features, that is their ability to work as transcriptions of the quantitative properties of space and territory, thematic maps emerge as tools specifically geared to the colonial endeavor, able to reflect both the limitations and the vested interests of their authors.

In the first half of the 20th century, as research was increasingly professionalized – responding both to a scientific interest in African issues and to a demand for knowledge on the part of colonial authorities – Western intellectual categories were extended to all the fields of study, both natural and social. The white man's science broke onto the African scene with unprecedented arrogance, confident in its reasons, and pervasive in its demands. Such craving for performative knowledge opened a semantic gap to be filled, first of all by grafting onto Africa the paradigms of European epistemology. Territory was approached according to a precise cognitive agenda, the Western one, in an attempt to impose order. Once relevance had been established, it became necessary to acquire knowledge by applying methods that met the prerequisites of reliability dictated by the

[29] What we are hinting at here is the adoption of Cartesian logic as a cartographic prerequisite, an issue which will be addressed in the pages that follow.

[30] The adjective "thematic" was first introduced at the beginning of the 1950s in Germany by Nikolaus Creuzburg and was widely adopted in the decade that followed. See: A.H. Robinson, *Early Thematic Mapping in the History of Cartography*, University of Chicago Press, Chicago/London, 1982; E. Clutton, "On the Nature of Thematic Maps and their History," in: *The Map Collector*, 22, 1983, especially pp. 42–43. I would agree with Jacques Lévy as to the irrelevance of the term, since any type of map, insofar as it selects information, may be defined as thematic. (J. Lévy, P. Poncet, E. Tricoire, eds., *La carte enjeu..., op. cit.*). For our present purposes however, and for the sake of clarity, I think it may work as a useful marker for the quantitative features of information.

European world for the making and conveying of science. Needless to say that what is relevant in the European *forma mentis* is not so for the black populations of Africa. Even when knowledge interests do converge, local cognitive patterns may not seem adequate to Europeans. What was ultimately supported and tested was a geography of colonial action based on methods utterly removed from the culture and/or the praxis of African communities. That was when the crucial role of thematic cartography came to the fore.[31] Through maps, in ways that are ever more exacting and far-reaching, there emerged a geography no longer concerned with the cultural issues, established practices, or social interests of African society. In fact the very existence of all these was denied and the Western interpretative model of reality advocated as the only viable option.

By using topographic reference, statistical cartography – as thematic maps are also called in the sciences – was virtually raised up to the level of geographical knowledge even though during the whole century statistical principles were ill-matched to the theoretical scope of geography, especially French geography, which embraced the methods of quantitative mapping belatedly and in problematic ways. At the end of the 19th century, however, geographers seemed to have finally acknowledged the usefulness of graphical statistics, to which due homage was paid. They adopted it as a privileged communicative tool within the discipline, to the point of undermining geographical research itself and, paradoxically, disrupting the analogical principle of topographic maps.

For it should not be forgotten that thematic maps challenge the very analogical reproduction of visible reality: by showing phenomena of reality that are invisible (population flows, networks, moral, economic and political aspects, etc.); by eschewing a conventional description of territory (selection of referential data is subject to the cartographer's will); and by presenting a territorial phenomenon only in one of its aspects (the quantitative one), thematic maps translate into a digital system what was originally intended as the mirror and analogical image of reality.[32] In these maps, information consists of intellectual constructs that cannot be perceived by observing territory topographically. Thus, the first significant difference appears even at the initial level of information production. If we then move on to consider representational norms, such a gap looks even more significant. While topography provides for the reproduction of geometric forms derived from reality, in thematic maps symbols are not determined by the objects' sensible form but by the need to introduce as many variables as possible into represented data: that is the paradigm of

---

[31] On thematic cartography in France at the time and its ideological implications see: G. Palsky, *Des chiffres et des cartes. La cartographie quantitative au XIXe siècle*, CTHS, Paris, 1996.

[32] Recalling the dual communication system present on the map, which I addressed in the previous chapter of this book.

quantity. Therefore, adopting a quantitative paradigm does not mean enabling cartographic data to convey one and only one information set. Having to know and render only one data set entails looking inside that data set for differences. And these will be all the more artificial the more sweeping are one's claims to thoroughness. Each data set is therefore scanned for differences that may be used to break it up, rank it and manipulate it both epistemologically and from the point of view of communication. It is in this line that a project – bound to be very successful – was undertaken: a symbiosis between taxonomy and cartography.

That period coincides with the rise of the great season of taxonomic rules enforced in any applied inquiry, the outcomes of which are depicted cartographically. The creation of a system able to produce and communicate knowledge effectively picks up a challenge the scientific community considered with urgency: that of producing information applicable to social needs. The use of cartography aims at mastering quantity in a communicative sense. A double goal is therefore achieved: 1) proposing a research methodology valid for any disciplinary field; 2) standardizing results, thereby addressing disciplinary fragmentation to some extent and proposing a unity of intent based on the use of the same instrument. For one, maps propose taxonomy as the method fit to master any phenomenon, which turns them into vehicles of virtually absolute knowledge. On the other hand, maps dissolve disciplinary boundaries by creating outcomes that respond to the same underlying logic.

Taxonomic science seems initially set to become the perfect method for a positivist science grounded in quantity. It is based on the detection of regular features and on the creation of differences. It entails collecting heterogeneous data, connecting them by way of analysis, and sorting them according to a classification framework where each botanical, pedological, and zoological unit occupies its own place in relation to an unmistakable descriptive frame. That illusion was bound to be short-lived. Theoretical assumptions differed, inconsistency in experimental protocols emerged, conflict between schools flared up and the various national contexts were mutually affected. That resulted in the paradox whereby each researcher felt entitled to impose his or her own classification, which of necessity frustrated the very objectives taxonomic science had set out to achieve: the creation of an order capable of mastering all phenomena.

The crisis of taxonomy reached its climax when taxonomy started to be taken as a research method capable of mastering any aspect of any phenomenon. When the use of taxonomy began to slide and a shift took place from an exclusively quantity-based approach to a quality-oriented one, the extent of its failure was exposed. Taxonomy is ill-equipped to handle informational complexity unless arbitrarily. And if that were not enough, taxonomy also takes on a markedly cartographic disposition. Faced with facts, one does not limit oneself to establishing classification rules in compliance with given conceptual tenets. Rather, one expects to record them cartographically, because experimental validation is required through analysis at different scale levels. Each ranking consequently brings about a multiplication of classes (so that from a few instances countless subclasses are derived), reaches its own systematization and with it evidence of its success in a cartographic representation. Maps turn out to be therefore much more than technical products. They become tools based at once on taxonomic measure as an empirical science – that is, a conceptual method of systematic empirical trial – and a place where the results achieved through that methodological scheme are assessed. Yet maps, like taxonomies, are based on the detection of regularities and the creation of differences. When, however, they become thematic, and thus fall back upon taxonomic enumeration, the variables they can contain must be greatly simplified and reduced in number. Otherwise both the cartographic message and the classification scheme itself are thwarted. In the pages that follow, as we discuss the vegetation map of French Tropical Africa, we will have the opportunity to note how taxonomy applied to cartography can compromise even the knowledge of territory. All that occurs at the expense of performative designators – that is, those which reference scientifically produced meaning. Once transferred onto the map, they become abstract names, removed from reality and of little use to colonial governance or even less to the creation of new knowledge for local people.

Unfortunately, applied taxonomy is the one exported into Africa. That will be the cause and the origin of an incapacity to understand the Elsewhere and of the failure to produce alternative and innovative knowledge to compete with the products that Western scientific protocols were able to supply. The impression of clarity and intellectual mastery that taxonomic cartography produced was such that it left no space for an appraisal of values endogenous to the Elsewhere, or even for the opportunities of valorization which speculative diversity could offer. Awareness of such opportunities will be achieved only belatedly by European science, once it was eventually called upon to discard the brittle framework of positivism.

In the period under consideration, through taxonomy, the space and territory of Africa were mastered by themes and families of features: geology, geomorphology, hydrology, pedology, climatology, vegetation, use of agricultural soil, population distribution, ethnic, linguistic and religious divisions. In short, it was a world that colonialists extracted from the shapeless complexity that lay before them, with a view to ordering it as a geography by using descriptive categories and interpretative procedures typical of European science. The goal was to ensure the exportability of a model that aimed to assert as scientific, universal and unambiguous what was in fact the outcome of social engineering. The problem is particularly insidious because of those natural features, such as vegetation, which led European observers to believe that they were coming to terms with thoroughly natural data. On the contrary, once intellectually acquired and transmitted as knowledge, this data is no longer perceived in its "natural" inertia: it undergoes a historicizing process, it is assumed and reworked within a social dynamics. So, in the colonial period, vegetation was viewed as a resource of the utmost importance for establishing the conditions of economic and social development and was mapped accordingly.[33]

The point is that vegetation mapping fully exposes the limits of the European taxonomic practice, whatever its theoretical basis, since the presumption to cover both the quantity and the quality of phenomena in fact foils the transmission of useful information. What is even more important is that, by yielding no knowledge, such practice sanctions the inevitability of *topos*. Let us take a closer look at the information produced by the *Carte de la Végétation de l'Afrique Tropicale Occidentale a l'échelle de 1:1,000,000.*

The Orstom Vegetation Map

The vegetation map of French Tropical Africa compiled by Guy Roberty, director of ORSTOM, is a telling example of how the lack of reflection on the theory and methods of one's research can undermine nearly 30 years of work. This map is the result of a land survey that lasted over 20 years (1934–1955) and was sent to press after nearly a decade of cartographic work, between 1962 and 1963.[34] The long time span of the project, in itself a measure of the complexity of the task at hand, covers a period in which the positivist views that had seeped into European culture at the turn of the century gained ascendance. Positivism was not discarded. On the contrary, it was invested with ever more elaborate connotations, which culminated in the adoption of taxonomy as the ordering principle of the natural sciences. We have already highlighted the role of taxonomy and the difficulties faced by the West in mastering complexity through abstract schemes, especially if they relate to territories other than one's own. Now we are going to address the role maps play in thwarting any taxonomic claim.

It is impossible here to discuss the assumptions of a science called upon to create cartographic information, namely biogeography.[35] So, to bring into focus the result of the taxonomy-cartography symbiosis I am going to assess the various construction phases of maps, which include: 1) statement of aims; 2) data

---

[33] See the many studies on vegetation conducted in AOF, summarized in the first part of: J. Tronchain, *Contribution á l'étude de la végétation du Sénégal, Mémoire IFAN*, Larose, Paris, 1940.

[34] The project fell short of its initial goal, which was to cover the entire set of French possessions. The maps published at the chosen scale (1: 1,000,000) only covered the African coast; a few other inland areas were rendered at a larger scale (1: 250-200,000); still others were never mapped.

[35] In this regard, and for an extended treatment of the relationship between taxonomy and cartography, see: E. Casti, "Colonialismo dipinto...," *op. cit.*

collection and attendant preparatory work; 3) cartographic rendering. Perhaps that will enable us to understand how the high standing of the map's author in the field of biogeography, the long African stay of a team of professionals, the competent use of the most advanced cartographic conventions and a graphic rendering implemented by French laboratories to the highest standards, could possibly have led to produce a document that is virtually unusable in practice, and riddled with obvious structural limits. Of these, possibly the most evident and easily verifiable is the presence of an excessive number of symbols (200) in the map. These invalidate the map's primary goal, which was to provide a broad yet clear overview of vegetation distribution and its environmental implications.

The Map's Goal

The map was complemented with some volumes written by Guy Roberty himself, stating the philosophy, methods, conventions, and meanings that the map contains.[36] The aim was to produce a quantification and a qualitative definition of African territories. The perspective we are invited to take on in assessing the map is not to see it as a mere recording of quantitative data but, rather, as an instrument able to trace the territorial dynamics that create these vegetation landscapes. The map's evolving trend is highlighted in the statements of intent itself, which leads our author, as we will see shortly, to make a number of radical choices about the use of taxonomy. His considerations start from the fact that issues of environmental protection, which the tropical areas had to cope with, and the undernourishment of local populations due to the lack of supplies, are sufficient reasons to justify the creation of a "database" capable of setting the scene for planning agricultural interventions.[37] In fact, we must not forget that the map's purpose was to identify the territory's potential for exploitation or, even better, as the French put it, its *mise en valeur*. Knowledge of vegetation seems an obvious choice, because vegetation is the result of physical conditions (climate, soil) one must keep in mind to pursue a valorization of agricultural resources.

As is always the case when attempts are made to pigeon-hole what is different into rigid enumerations, the choice was not devoid of obstacles: the first one was the need to apply tentative names to the various biological types according to the taxonomic principles in use. Now these cannot be ascribed to one line of theory, but vary according to the authors one considers: some draw taxonomic principles from

---

[36] G. Roberty, *Carte de la végétation de l'Afrique occidentale à l'échelle de 1:1,000,000*, ORSTOM, Paris, 1964.

[37] The question of preservation was already being debated in colonial times. With regard to West Africa, France has surveyed and cataloged most of these areas, and applied to them the denomination of *forêts classées*. See: C. Coquery-Vidrovitch, ed., *L'Afrique Occidentale au temps des français, colonisateurs et colonisés* (1860–1960), Éd. La Découverte, Paris, 1992.

the botanical features of landscape with regard to climate; others start with the appearance and composition of the flora, and point out dominant plants; still others combine the various physiognomic data, floristic and climatic, to propose alternative classifications. Despite differences, all do rely on statistics and taxonomic enumeration, based on identifying plant compositions. However, in Roberty's view, this approach had serious flaws, since it did not take sufficient account of those environments undergoing key evolutionary processes, which the African experience qualified as crucial. Roberty exposed the shortcomings of these classifications, based on exclusively theoretical assumptions, because of their inapplicability. These limits convinced him of the need to advance his own methodology for a taxonomy of vegetation. Yet that led him into the trap of those who, failing to apply taxonomic rules already in existence, endeavor to produce their own.

The taxonomy on which Roberty's vegetation map is built is unprecedented and based on the identification of what is called *biogeographic association*: that is, the identification of *vegetative series*[38] and *vertical layers*.[39] By modifying these criteria on the basis of contingent needs, Roberty creates a set of indexes and reference tables that are extremely elaborate and inevitably clash with the simplifying product he had set out to build: a thematic map. Mindful of the difficulty that goes with translating this taxonomy into a graphical view, Roberty made adjustments and tried to design formulas for adapting the former to the latter. He reduced his claims and eventually was able to identify a few "fundamental landscapes." As he did so, he was well aware of the fact that units thus defined take on too generic a value in the map he had planned at a 1:1,000,000 scale. At the same time, however, he acknowledged his powerlessness to remedy the situation and stated "despite all my efforts, this map of west-African vegetation at a 1:1,000,000 scale offers a rough interpretation of observed facts, rather than a full and impartial description. It could not be otherwise in the present state of biogeographic theories and of the knowledge that underlies them."[40] Roberty also gave up his attempt to account for evolutionary factors and entrusted his own opinion on the matter to the written text.

What we are witnessing here is a crisis, due to the impossibility of adapting any taxonomic principle to first-hand experience, unless at the expense of missing the very issues that classification was meant to make up for in the first place, namely, arbitrariness. Yet even Roberty, who held a major institutional position and had ample field experience and firsthand knowledge of African reality, never questioned the basic legitimacy of adopting a taxonomy. He could see

---

[38] Groups defined by the type of environment and by the distinctive plants found in them. To each of these, Roberty assigned a name and a color to be used in the cartographic representation.
[39] Plant distribution based on the vertical development that each plant has in relation to the type of soil and the climate it is subjected to.
[40] G. Roberty, *Carte de la végetation, op. cit.*, pp. 14–15.

the limits of it, but attempted to overcome them, proposing variations that never in fact discarded that founding principle, which initially at least he had tried to master via cartographic mapping.

Preliminary Work

If we look at the phase preceding the drafting of the final map, we can discern Roberty's attempt to render, through special expedients, aspects which the taxonomy/cartography symbiosis invalidates. The same happens in the Thiès map drafted in 1950, under the direction of H. Gaussen,[41] one of the major theorists of the conventions of thematic mapping, with the participation of J. Trochain,[42] an expert on African vegetation in his own right, and of Roberty, creator of the taxonomic classification applied to the map. This is a remarkable instance of collaboration between cartographic, land and scientific skills. As such, it shows how the limits of the product achieved should not be ascribed to professional incompetence or the technical incapacity of the authors, but exclusively to the pursuit of objectives that the cartographic medium cannot possibly meet.

The document features a central map that records vegetation distribution, surrounded by six other smaller figures related to hypsometry, geology, pedology, pluviometry and land use. These are interspersed with long captions that inform on the purpose of the map and clarify the choices and procedures adopted for its implementation, giving directions for interpretation. Theoretical "intricacy" is made up for by great cartographic competence, which renders a large amount of data in a clear and technically flawless design.[43]

Unfortunately, in the final map these limits would become insurmountable, since information variables increased so much that interpreters were forced to select information and focus exclusively on the transmission of quantitative data. This classification attempt was thus emptied of its very meaning, because meaning became impossible to map. It should be noted that the failure of this attempt ultimately rests with its initial goal: that of applying Western protocols to another reality, the African one, whose actual meaning, in this case the role vegetation played in local communities, was never even touched. We might of

---

[41] This is a 1:200,000 scale document published by ORSTOM and currently preserved in its archives and at the Archives Nationales, sect. *Cartes et plans* with the acronym Ge BB 1065, D-28-XX. H. Gaussen gave a lecture on such experiments: "Projets pour diverses cartes du monde au 1:1,000,000e. La carte écologique du tapis végétal" at the International Geographical Congress held in Lisbon in 1949 and published with the ensuing discussion in the Congress Acts (p. 223).

[42] J. Tronchain, *Contribution à l'étude de la végétation du Sénégal*, Mémoire IFAN, Larose, Paris, 1940.

[43] On the importance of collaboration between those who collect the data and those who map them see: F. Vergneault-Belmont, *L'oeil qui pense: méthodes graphiques pour la recherche en sciences humaines*, L'Harmattan, Paris-Montréal-Turin, 1998.

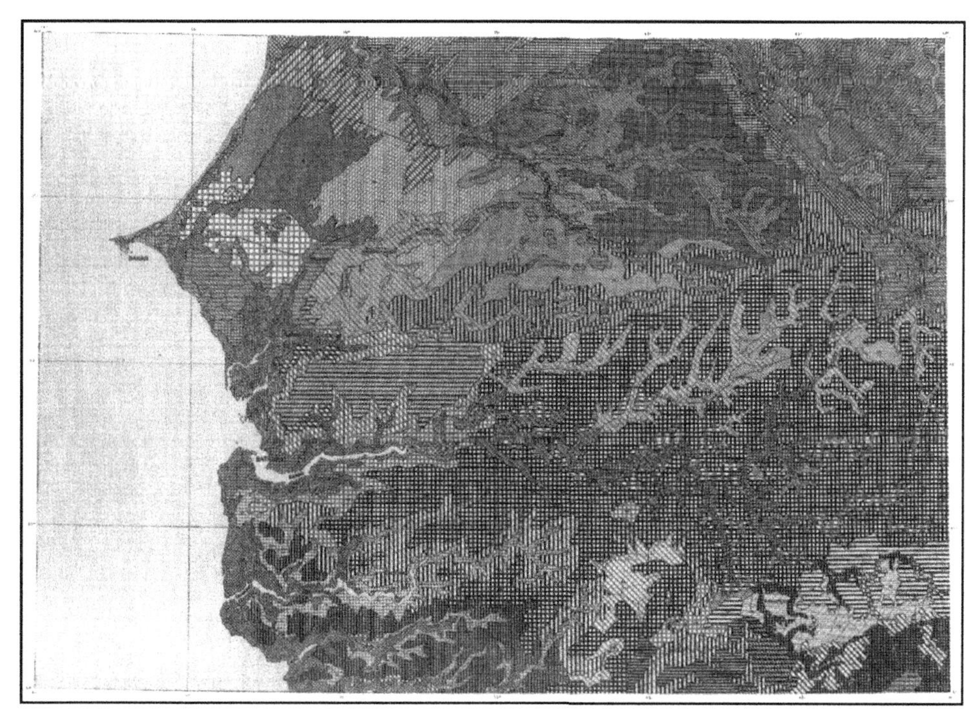

FIGURE 2.5
1962, Guy Roberty, *Carte de la végétation de l'Afrique Tropicale Occidentale*, 1:1,000,000, folio NB 28 Dakar. (See a color reproduction at the address: booksite.elsevier.com/9780128035092.)

course consider the enterprise successful, if any knowledge result had in fact been achieved to justify such a grievous intellectual, financial, and organizational investment. Yet it is not so, as an analysis of the map can easily show.

The Drawing of the Map

Although the design of the *Carte de la Végétation de l'Afrique Tropicale Occidentale* envisaged nine sheets, which would cover the entire region, only three were issued; these covered almost all of Senegal, most of Guinea – up to the regions of Upper Niger – and part of the Sierra Leone coast. The first two sheets reproduced nearly the entire area surveyed (Figs. 2.5 and 2.6), while the third devoted only one corner to cartographic representation, and the rest to a legend and to the conventions adopted (Fig. 2.7).[44]

The latter sheet is the most interesting, as the legend allows us to evaluate the choices – taxonomic and cartographic – that were made, but above all to assess

[44] The analysis that follows is best understood by consulting color reproductions of the images mentioned here at this address: booksite.elsevier.com/9780128035092.

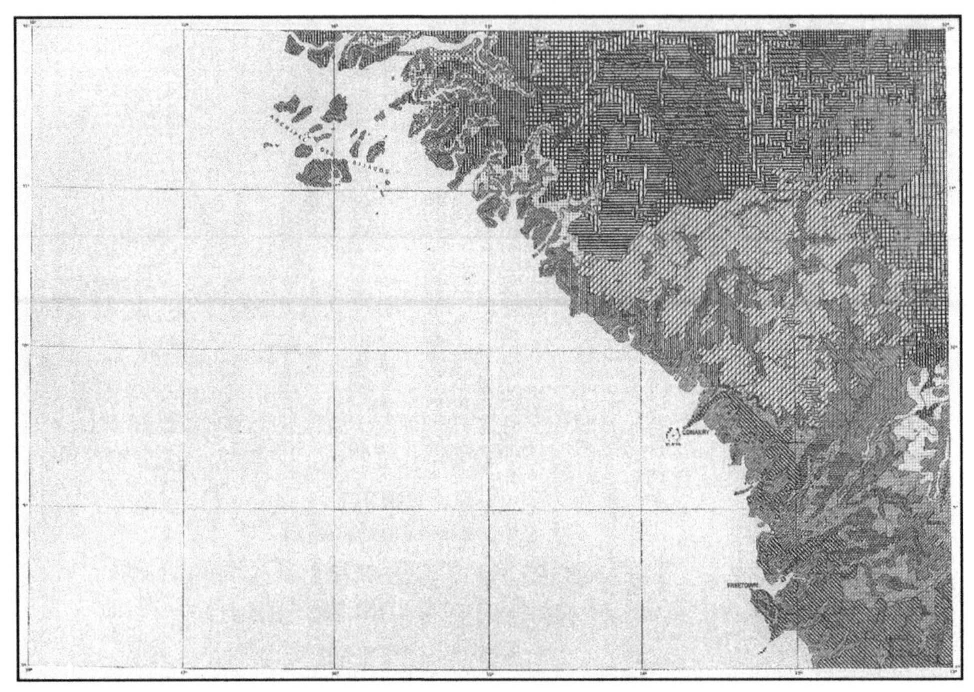

FIGURE 2.6

1962, Guy Roberty, *Carte de la végétation de l'Afrique Tropicale Occidentale*, 1:1,000,000, folio NC 28 Conakry. (See a color reproduction at the address: booksite.elsevier.com/9780128035092.)

its communicative constraints. Even at a cursory glance, it is evident that, being made up of 200 symbols, obtained from color-coded screens, the sheet is difficult to memorize. The alternation of the entire color range in a wide variety of forms (dots, solid lines, dashed lines, and their combination), of trends (vertical, horizontal, oblique and their various combinations), of associations (dots plus lines, dots plus crosses, etc.) and of screens, increases the number of variables. Many of these are within the same symbol, and are expected to refer simultaneously to their original meaning and to their current pairing. This reading method is supposed to identify a precise type of vegetation that, in the drawing, is attributed to a region, marked by a number. Yet the directions given, already quite hard to understand in the legend, become virtually unreadable on the map, even if we were to limit ourselves to deciphering a single symbol. And if we attempt to compare symbols – which operation presumably justifies the creation of a map in the first place – then it all becomes a colossal undertaking.

At this point we can easily conclude that this beautifully colored map, of excellent design, most meticulous and, certainly, very expensive, fails to provide any

FIGURE 2.7

1963, Guy Roberty, *Carte de la végétation de l'Afrique Tropicale Occidentale*, 1:1,000,000, folio NB 28 *Bonthé*. (See a color reproduction at the address: booksite.elsevier.com/9780128035092.)

information at all, because it clashes with overly ambitious taxonomic goals that irrevocably undermine the possibility of cartographic rendering.

We are left to look for the semiotic reasons of this failure by addressing the aspect that more than any other invalidates communication: the use of color with regard to the creation of icons.

Chromatism and the Creation of Cartographic Icons

As a complex ordering system, taxonomy requires a transmission based on easily identifiable intervals and, especially, on a large number of variables. Theoretically, the code that best meets such requirements is color, which ensures an immediate reading, since class progression is made to correspond to that of the color spectrum. Nonetheless, color is ill-fitted to express taxonomically ordered information because its inability to set the order of classes logically (it can only do so by analogy) requires strenuous intellectual commitment on the part of the interpreter. This shortcoming may be understood when related to the map's communication systems.

As I already mentioned, information management in cartography is subject to the rules of sign encoding, which are based on two systems: *analogical* and *digital*. It is easy to understand that while the lexical, numerical and figural codes, which rely on conventional formalism for transmission, may be ascribed to a digital mode of transmissibility, matters are more complex with regard to color codes.[45] Conventions, which do support color, never in fact disrupt the basic associative meaning color has in nature (waters are rendered in blue, vegetation in green, etc.) to the point that color breaks free from the use of a formalism, because its communication draws on the analogical system, which offers its variables as they are perceived in reality. In short, with respect to color, perception and analogical system are shown on the map as facets of the same cognitive/communicative process, interacting in a close, complementary relationship.[46] This is relevant because the relationship between perception and cognitive organization is not linear and gradual, but rather based on a connection, whereby the retina and the brain decide in advance over certain features of the impending phenomenon.[47] And sight, the dynamics of perception, cannot possibly be considered as a purely passive instrument for recording reality.

---

[45] Some codes, such as names and numbers, are abstract conventions that can easily be recognized as digital, though they may originally refer to a continuous system. With regard to numeric codes, R. Thom warns us that the founding aporia of all mathematics is the discrete/continuous opposition, determined by the conventional use of numbers. Such aporia becomes necessary in order to "reconcile our first-hand experience of the continuum with the generativity, necessarily discrete, of operations." In scientific disciplines this opposition has its advantages, because in the course of time it is endowed with "phantasmal" solutions. (R. Thom, *Parabole e catastrofi*, Il Saggiatore, Milan, 1980, p. 150). The problem arises with the figural code, which can in fact reproduce the analog form of what it represents, but – when taxonomic cartography developed – also underwent a process of abstraction driven by Euclidean conventions. As a result, the figural code may be numbered among the codes that use the digital system. On cartographic encoding see: G. de Dainville, *Les langages des géographes*, Pimap, Paris, 1964, p. 330.

[46] This cartography is comparable to the results produced by studies that investigate the perception of color and stress the interdependence between perception and analog communication. Elaborating upon Rudolf Arnheim's thought, Augusto Garau demonstrates that in the perceptual universe there exists a "color continuum" which does not provide for discontinuity within itself, because the range of colors is produced from a mixture of those recognized as basic colors (red, yellow, blue). Such basic colors, present in all derivative ones, determine an associative chain in which no distinctions exist. This insight had already been offered by Arnheim in 1954 (R. Arnheim, *Art and Visual Perception. A Psychology of the Creative Eye*, University of California Press, Berkeley and Los Angeles, 1954 (1974)). A. Garau, *Le armonie del colore. Analisi strutturale dei colori. La teoria delle mescolanze. La trasparenza percettiva*, Hoepli, Milan, 1999.

[47] It thus transpires that, in communication, the color spectrum is a continuum relatable to the analogical system. As such it provides for no symbolic formalism in order to be conveyed, because, as it reflects physical variables, processing takes place continuously on the basis of the perceptual meaning it has. (M. Negrotti, *Cibernetica dei sistemi sociali*, F. Angeli, Milan, 1983, p. 109.)

Conversely, as a processing system, sight draws upon the personal contribution of an individual who belongs to a community and is therefore susceptible to the distorting conventions of culture.[48] In Western culture in particular, sight is closely linked to the idea of truth, and most of our trust is placed in visual evidence even when other senses might be more adequate.[49] In this perspective, to note that cartographic encoding of color responds to the perceptive encoding of reality, means to highlight two key aspects of communication: for one, that color has the persuasive ability to make interpreters accept what they identify as true; for another, that its use must necessarily comply with the dynamics of sight. Color therefore emerges as the communicative code that less than others lends itself to taking on abstract meanings. Rather, it tends to identify itself as a code that refers directly to what individuals process perceptively, on the basis of values entrenched in the society they belong to. In short, from a communicative point of view, color appears as an element unable to sustain important processes of abstraction, anchored as it is to the dynamics of perception directly involved in establishing input rules for the analogical system.[50]

It seems clear that, as color enters taxonomic maps, it is emptied of its analogical implications and used improperly, becoming a code that is subject to no perceptual laws. As color is severed from its analogical range and attached to abstract values, which are the result of mental processing, a change is produced at the level where its meaning is conveyed. From a denotative level – based on the meaning color visually acquires in reality – we switch to a deeper level, the level of connotation, produced by society via the attribution of meanings experimentally constructed. This entails two types of problems: one investing the plane of perception, the other that of comprehension. Regarding the first, as performative meanings are attached to color via abstract conventions, reliance on a legend becomes necessary and the interpreter is required mnemonic skills well beyond the scope of the average human being.[51] As regards the second,

---

[48] The idea that there is a cultural modeling of the "knowledge structure" shared by members of certain communities is proven by the fact that there are rules attached to the ontogenetic development of the human body in relation to its environment. See: A. Wilden, "Comunicazione," in: *Enciclopedia*, Einaudi, Turin, 1978, v. 3, p. 653.

[49] Crucial remarks on the effectiveness of visual communication systems are found in: E. Carpenter, M. McLuhan, eds., *Explorations in communication*, Beacon Press, Boston, 1960.

[50] This does not rule out that feelings, emotions, and sensations may be attached to it. However, these will invariably be ascribed to the field of analogy, that is, they will pertain to perception and cultural processing. The attribution of values to colors related to temperature (hot-cold) and the feelings such pairing entails were investigated by: R. Arnheim, *Art and Visual Perception…*, *op. cit.*, (1974), pp. 369–376.

[51] Arnheim argues that "active selectivity" is a key feature of vision, lost when a subject is relentlessly presented with chromatic stimuli. It would appear that "the retina, informing the brain about color, does not record each of the infinite shades of color using a particular type of message, but is limited to a few basic colors or color ranges, from which all the others are derived" (R. Arnheim, *Visual Thinking*, *op. cit.*, pp. 21–22)

the issue is even more momentous. When understanding of the meaning of color is shifted from denotation to connotation and assigned the task to convey complex concepts and elaborate mental syntheses, it is used as any other digital code which, albeit to a limited extent, is subject to the dynamics of perception. That restricts the color's communicative potential. In short, taxonomy entails, for one, the questioning of the perceptual principle and, therefore, of the analogical meaning of color. On the other hand, paradoxically, taxonomy causes a weakening of the color's communicative potential, on account of the excessive demands its semanticization is subjected to.

All cartographic semiosis is affected by that. At the level of syntax, I already discussed the disastrous effects produced by the inability to connect meanings between legend and map. It should now be added that such failure prevents the creation of any syntactic relationship in the drawing, because the possibility to assign a meaning to each single icon is excluded from the start. In fact, in taxonomic maps icons operate under particular conditions: 1) they are formed exclusively by abstract codes; 2) the information to be conveyed is performative; 3) the quantity of information is particularly large. In meeting these three conditions, the legend plays a critical role, and its presence becomes essential. This information circuit generally poses no issues, except when it must comply with taxonomic criteria and convey particularly complex information. In these cases, icons operate along different paths that remain disjointed: on the map they reference data, while in the legend they serve to connote it. *De facto*, the excessive complexity of the information to be conveyed prevents the legend from working as a reminder or memo to a communication that occurs via the map. Ultimately, in a taxonomic map, the icon's communicative outcome is exhausted in the enumerative form set up by the legend, which refers to its own connotative meaning without integrating itself with the denotative, referential meaning proposed by the drawing.

Semiotic analysis confirms that the cartographic message becomes intelligible only when figurativization* has been completed. Vice versa, in taxonomic maps the gap created between spatialization and figuration (the first in the drawing, the second in the legend) excludes the possibility of ever achieving iconization and with it the establishment of the self-referential mechanisms typical of denominative projection. In other words, failed iconization impairs the process whereby the self-referential mechanism is sustained. The link between icons makes it possible for distinctive features to emerge, differentiating information and arranging it in either conjunctive or disjunctive relationships which lend coherence to the cartographic discourse. The inability to read any one of these relationships on the map, unless by prior and constant reference to the legend, invalidates any cartographic syntax. It may thus be argued that in taxonomic maps, the digital use of color and its combination with other codes prevent the formation of autopoietic mechanisms that could supply interpreters with directions for extricating themselves from the tangle of excessive informational complexity. Interpreters are prevented from understanding the message, because the double circuit of communication (map/legend) disrupts their ability to handle, by their own means, the elaborate cluster of communication that reaches them.

The bleak result encountered in the ORSTOM vegetation map leads us to believe that communication systems are invariably subject to strict semiotic rules, which must be complied with for transmission to be successful. In the case of maps, caution must be even greater, because in their capacity as self-referential tools they can intervene in communication nevertheless. When they do comply with semiotic rules, maps can achieve the goal of actively intervening in communication. Otherwise, if they fail to conform to these norms, they rule out any prescription. That, however, does not exempt their interpreters from considering them as outstanding cognitive tools.

In view of this, the vegetation map, while not producing information, may be assessed in positive terms, because it attests to the commitment of the French government in the field of environmental protection, while also confirming the ability of its research institutions to create knowledge. That may find theoretical justification in the fact that a map can refer to itself in a circular manner, as a self-contained process needing no external input to define its own identity. Maps arrange and convey cognitive items independently of the goals set by the actors who had originally created them or of the will of those who regularly use them. Once again then, colonial cartography, the classifier of African spaces and African identities, wields its power both when it produces information and when, as in the case just examined, it feeds on it parasitically.

Beyond this assessment of the social system for validating scientific results, what concerns us here is considering the outcome to which an analysis of the document leads, namely the inevitability of *topos*. Even when expectations are high and the underlying aim is to recover the social meaning of territory, the presence of competing motives and of reasons linked to the semiotic outcomes restricts such meaning to the level of denotation.

ICONIZATION OF *TOPOS*: MAPS BETWEEN SCIENCE AND POPULARIZATION

To complete my analysis on the persistence of *topos* in colonial cartography, I am going to consider maps published in journals. Although these are designed to be subordinate to the text, they play an important role in conveying a certain notion of Africa. I am thinking here of documents found in the geography journals of the time, such as the *Annales de Géographie*, the *Bulletin de la Société de Géographie, La Géographie* or those relating to colonial studies *Renseignements coloniaux*, published by the Committee of French Africa, or *Outre Mer*, a generalist journal on colonization which gathered land studies and travel reports since the late 19th century. In these essays, physical features, features concerning population, the economy, movement of goods and organization of territory are illustrated by maps, which endorse the scientific nature of the texts. And since these journals are the organ of official communication for knowledge of the Elsewhere, they also provide an authoritative source for the popular press The maps are thus in a position to influence public opinion because their communicative effectiveness is instrumental to the exigencies of colonial propaganda.

In this capacity, these documents played an important role in spreading some myths about Africa, which in the words of Edward Said may be defined as "orientalist," that is built by Europeans for their own sake.[52] In fact, although they deal with a wide range of topics – finding exploration paths, describing physical features, illustrating economic or commercial activities – all these documents unanimously cast Africa from the point of view of those who are about to alter it in order to exploit it. There exist two implicit convictions: that Africa is a rich land and that the people who live there lack the cultural and technical means to take advantage of it. Although these maps are nothing more than the result of cursory reconnaissance, they impose their own authoritative reading as to possible exploitation and economic organization. This not only exposes the imperialist vision that animated colonialism, but reveals that the understanding of Africa at the time was irreparably affected by the insidious assumptions of cartography.

Exploration Between Strategic and Economic Interests

As we turn to maps focused on exploration we notice that, like reconnaissance maps, these refer to financial ventures. The lake region of Faguibine and Télé, near Timbuktu, formed by the branching off of the river Niger, is highly relevant in this respect. This territory is said to be "rich and mysterious," deserving to be known and appreciated, since it has physical features and environmental conditions which can be acted upon to consolidate the French presence in this inner region of Africa. The region is reproduced in a map drawn by two army men: executive Lieutenant Hourst, Niger fleet commander, and Lieutenant R. Bluze, of the Marines, who authored the essay to which the map is attached.[53]

La *Carte de la Région de Tombouctou* (Fig. 2.8), as it is named, is a document attached to a mission report, published in 1895 in the *Bulletin de la Société de Géographie*.[54] It describes the region's hydrographic network and compensates

[52] E. Said, *Orientalism*, Pantheon Books, New York, 1978.

[53] R. Bluze, "La Région de Tombouctou," in: *Bulletin de la Société de Géographie*, n. 7, 1895.

[54] The map is at a scale of 1:500,000 and represents the area around Timbuktu and Goundam. It includes two 1:100,000 inset maps and measures 485 × 572 mm. It is an attachment to the article entitled "La région de Tombouctou" in *Bulletin de la Société de Géographie* (1895: 7, t. 6).

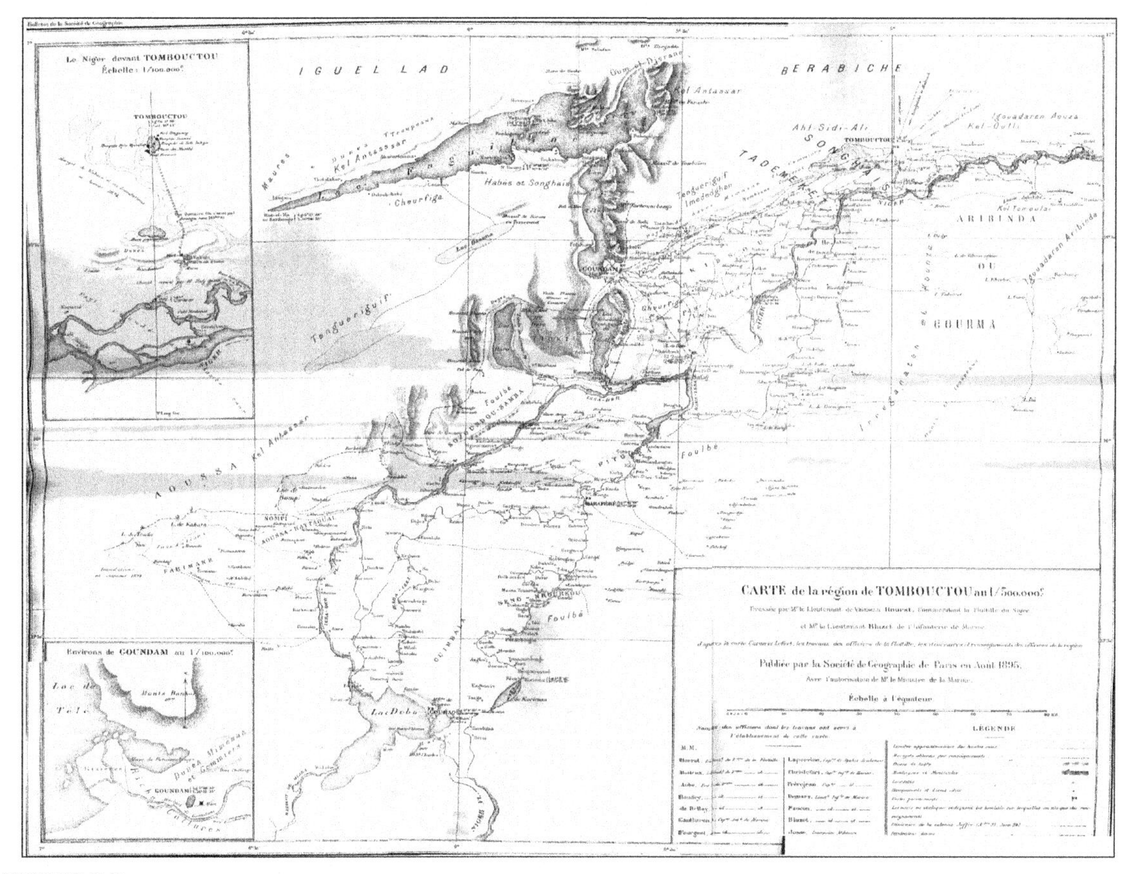

FIGURE 2.8

1895, *Carte de la région de Tombouctou*, 1:500,000, *Bulletin de la Société de Géographie.*

for the absence of topographic mapping by providing an account of its distinctive features. Information is given as to the feasibility of controlling the region militarily, by managing the river, which is both the main communication waterway between Timbuktu and Goundam and a major water supply for farming. Also, the report does not hesitate to point out the region's economic relevance with regard to agriculture and trade, since the river there takes on the features of a wetland which joins the north with the Sahelian south. The kinds of interventions envisaged include military control and the levying of taxes on goods in transit, in anticipation of increased trade towards the West. The document depicts an idyllic view of the region, over which France should exercise its rule in order to increase its influence in the north and secure the livelihood of the four military companies that occupy the country.

Map data, limited to the description of the hydrographic system, are approximate, but the text serves to uphold their reliability. In fact, the text explicitly vouchsafes the accuracy of the survey and even includes a chart of geographical coordinates for key "observation points."[55] Nevertheless, a comparison of this map with one issued only a few years later (1897) at a larger scale (1:200,000) and designed by P. Vuillot, clearly shows the errors the two officers incurred, as regards both physical and anthropic features[56] (Fig. 2.9).

For example, the entire region is dotted with a large number of names drawn in the same typeface, a solution which, while not conveying useful information, gives a reader the impression of being in the presence of a densely populated area. By contrast, the most recent map clearly indicates that the villages – especially those along the banks of the Faguibine lake – are not all permanent settlements, but include many temporary ones, in the form of transhumance camps. By neglecting this feature, the Hourst and Bluze map conveys erroneous information on population density.

Typeface uniformity in names is also clearly meant to highlight the French presence. Although the names of villages are given in double typeface (italic and cursive) – which distinguishes attested names from presumed ones – only the names referred to the four detachments (Gourau, Goundam, Tombouctou and Saraféré) are written in capital letters, paired with the French flag and their geographic coordinates. The goal, which is subordinate to icon diversification, is to rank settlements according to French presence rather than to emphasize their role in the local organization. Not to mention the fact that the use of the same character for all territorial features betrays a kind of failure to distinguish the names of villages from those of regions or those of ethnic groups. We are in the presence of the same expedient already found in the topography we examined previously: that is, covering a mapped surface with names that do not actually provide clear referential data or reliable information on its social framework.

Even in this case, then, the large number of names is not intended to reference territory but rather to lend reliability to the map, by conveying the impression that the region is thoroughly known and subject to French rule. In Bluze's words: "Is the occupation of Tombouctou an increase of our colonial wealth? It most definitely is. But it is above all a necessary consequence of our presence in the Niger valley, to which Tombouctou is the entrance, a door open onto all

[55] The text/map combination then, is deployed to make the map's rather sketchy information convincing. Yet, at the same time, we are deviously urged "to grant the map [...] a limited degree of reliability."

[56] It is a map entitled *Carte topographique de la région des Lacs de Tombouctou*, published by R. Hausermann as a supplement to the *Bulletin du Comité de l'Afrique française* (Nov. 1897), drawn by Paul Vuillot, a member of Comité de l'Afrique française (510 × 642 mm). It was obtained by assembling and comparing information from several explorations and cartographic records, an approach which shows the scarce reliability of the document we are examining.

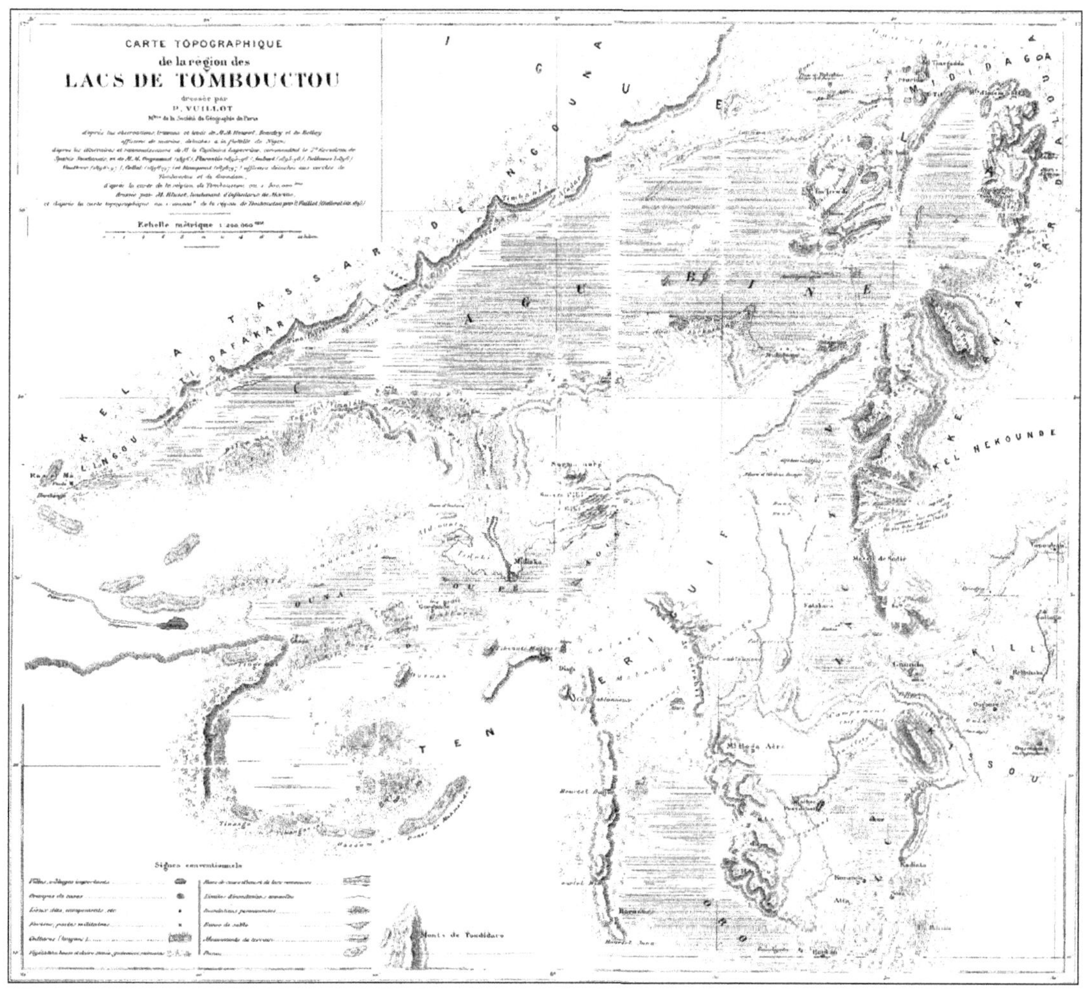

FIGURE 2.9
1897, Paul Vuillot, *Carte topographique de la région des lacs de Tombouctou*, 1:200,000, *Bulletin du Comité de l'Afrique française.*

the countries of the North."[57] It is possibly this vested interest which leads the maps' authors to consider basic features as irrelevant and, as such, susceptible to unconditional valorization.

Interpretation of Physical Features

Similarly, the maps, drawn with the aim to reproduce the physical features of territory, do not envisage information that, on the basis of the conventions adopted in Europe, may record its morphology. Rather, they highlight those

[57] R. Bluze, "La Région de...," *op. cit.*, p. 388.

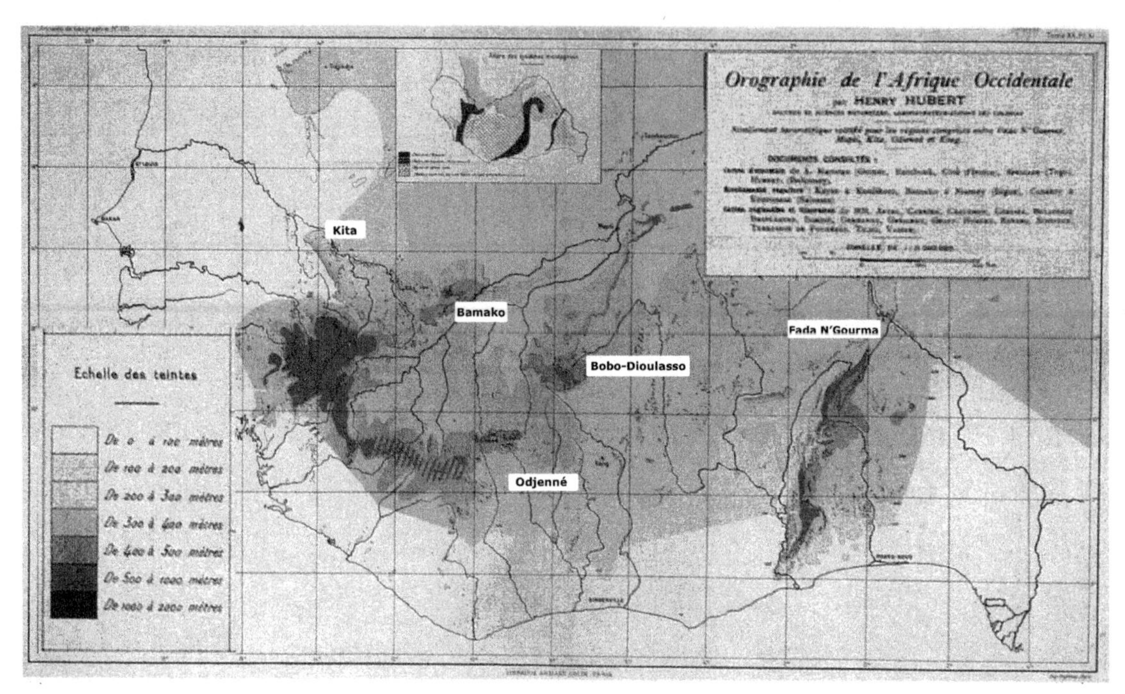

FIGURE 2.10

1911, *Orographie de l'Afrique Occidentale*, 1:5,000,000, *Annales de Géographie* (names are enlarged). (See a color reproduction at the address: booksite.elsevier.com/9780128035092.)

features that legitimize its possible valorization. A good example is the map *Orographie de l'Afrique Occidentale*, designed by Henry Hubert, a naturalist by training and a colonial administrator. The map is attached to a report published in 1911 in the *Annales de Géographie* and describes in very particular terms the range that characterizes tropical West Africa (Fig. 2.10)[58].

It should be noted that the making of this document presents no technical challenges, because its goal of representing a whole vast region may be achieved by using a very small scale (1: 5,000,000), which requires neither a base nor rigorous topographical surveys. Nevertheless, in the report that comes with the map, the survey's technical base, the tools used and the algebraic calculations for estimating the regions to be explored are all shown in detail.[59] And these details, meant to confirm the reliability of cartographic information, seem in

[58] H. Hubert, "Rapport de la Carte de l'Orographie de l'Afrique Occidentale," in: *Annales de Géographie*, 1911, n. 110, t. XX, pp. 155–178. The map is marked with the number IX and measures 345 × 585 mm.
[59] The report states that instrumentation comprised: a compass-alidade Peigné, an aneroid barometer and a hypsometer (H. Hubert, p. 156).

fact redundant in relation to the final product, that is, a small-scale map which allows for no verification of the survey's accuracy.

The claim that the map wishes to submit is that the mountain range poses no obstacle for mobility and that its morphology does not therefore hamper the colonial venture, which is interested in the territory's practicability.[60] Contrary to expectations, the map aims to show that territory can be easily traversed, because it has no considerable or unpassable reliefs. Morphologically, territory is here presented in four sets (a large central plateau, the Atacora chain, going from Niger to the coast, a chain of small massifs and a region dotted with isolated hills) magnified and shown in an inset map in the top right portion of the document. Despite the territory's morphological ruggedness, the objective is here to present the relief only as a mildly uneven elevation. Cartographically, this translates in the use of variously shaded backgrounds of color, which do not refer to any outstanding feature (such as mountain peaks, deep valleys, slopes or insurmountable cliffs, etc.) but are spread out generically. Yet, these color backgrounds are supposed to provide accurate referencing. The map actually presents few designators: those of major centers of the Interior (Bamako, Kita, Odjienné, Bobo-Dioulasso, Fada N'Gourma). All the names of rivers and of other orographic phenomena are missing. It is a telling omission, for it shows that the names of cities are not meant to recover their local relevance but rather to underline their colonial usefulness. The same political reading applies to the text that goes with the map, because it divides the mountain range on the basis of its administrative districts instead of recovering the morphological division the map offers. Hence, the references to the *Cercle de Kita*, the *Cercle de Bamako* and others. These choices (reference tied to the name of cities, division of the mountain range in accordance with administrative exigencies, lack of morphological assessment) may be ascribed to both technical and ideological criteria. Technically, the absence of regular topography rules out the possibility of naming territory accurately and in detail. In addition, cities are the only places recognized as fixed points, on which the rest of the territory should be mapped. Nor should we forget that the absence of roads and the hardships white people have to face to survive in the region make it impossible to carry out a swift altimetry survey of all orographic phenomena. As for ideology, it is obvious that reliance on administrative divisions and, therefore, on a colonially imposed structure produces the impression of a territory that is familiar and easily manageable. The inconsistency of this claim comes abruptly to the fore, however, when the author – compelled to refer to a specific range or a specific orographic system – must forcibly enlist designators that betray their own vagueness by borrowing the names of ethnic groups that inhabit the region (mountains of Mandingo, of Baule, etc.).

---

[60] As clearly stated on p. 177 of the report.

Such inconsistencies are not overtly detrimental to communication, because for a thematic map to work as a corollary to the text, it needs but to be able to pinpoint the selected phenomenon. Thematic maps may, however, rule out the possibility of testing the reliability of what they show, since the information produced by cartographic self-reference makes up for their shortcomings. Specifically, the self-referential information conveyed by maps is the one produced in the connection between the designators and other graphical signs present, for instance colors, shapes, numbers. In short it is the icon taken as a whole. Iconic junction produces the information omitted because of the lack of designators: colors work as surrogates for the names of mountains and offer indications of their altimetry (whose inaccuracy is masked by color shading); figures (black lines) stand in for rivers, etc. Moreover, the absence of names for mountain ranges reinforces the myth that authors wanted to convey: that their altimetry is irrelevant for the purposes of assessing practicability.

It is then clear that the message to be conveyed is supported by the map, for only the map can uphold the unwarranted conviction that it is quite legitimate to disregard the mountain range for a number of reasons: ultimately because it poses no obstacle to Europeans. Reassured of the minor difficulties involved, the French may thus undertake the project of connecting the interior to the coast via a road network.

Wealthy and Metropolitan Africa

A document, covering the same region now considered and built to evaluate the economic and territorial aspects of French Sudan, presents complex information that is worth examining. It is the *Carte économique des pays français du Niger* (1900), drawn by Émile Baillaud and published in the *Bulletin de la Société de Géographie*, to accompany a report entitled "Les territoires français du Niger, leur valeur économique"[61] (Fig. 2.11).

It is the result of the inspection carried out by E. and G. Baillaud Binger, in the years 1898–99, by order of the Minister of the Colonies at the time. The goal was to take stock of the economic and agricultural status of the land and of trade relations with English colonies of Central Africa, with a view to bolstering their economic value.[62] The valorization scheme is based on three actions: 1) entering the trade circuit already in existence; 2) changing the guiding principles of this circuit by building a railway system; 3) expanding cultivation into the fertile regions of the interior. The study keeps track of the various

[61] E. Baillaud, "Les territoires français du Niger, leur valeur économique," in: *Bulletin de la Société de Géographie*, 1900, pp. 2–24.

[62] The map, at a 1:5,000,000 scale, was also published in the journal *Géographie*, t. II, Pl. 2 and measures 354 × 561 mm.

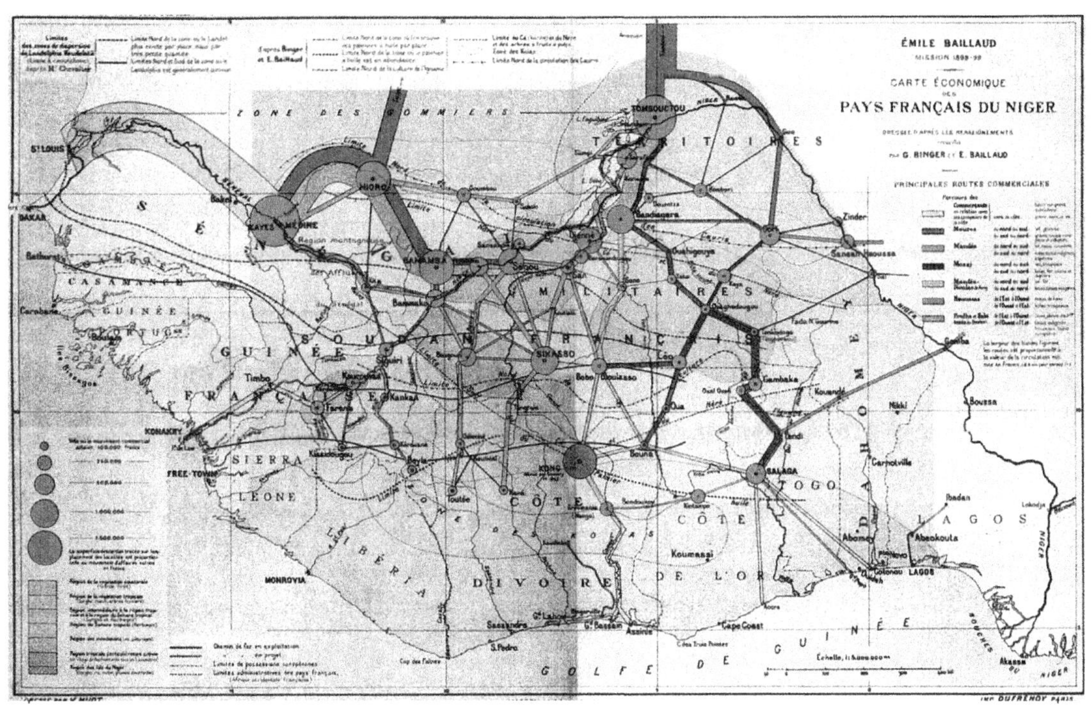

FIGURE 2.11

1900, *Carte économique des pays français du Niger*, 1: 5,000,000, *Bulletin de la Société de Géographie*. (See a color reproduction at the address: booksite.elsevier.com/9780128035092.)

aspects of territory, without neglecting the political situation and the relations between ethnic groups. The map envisages trade both from a qualitative and quantitative point of view, producing information on the agricultural sector by identifying the regions to be exploited; on economic features related to trade routes; on the movement of goods, their direction, their quantity, their quality, and their value. It also specifies the ethnic groups involved in the exchange. So, for instance, products coming from the northern territories thanks to the Mauri (especially salt and rubber) reach Timbuktu and Nioro and are distributed south by the Mandingo and the Mossi who, in turn, carry other goods from the south to the north: cola, spices, textiles and wheat.

The report insists on the fact that the French could intervene in this trade circuit with European products, especially textiles. It also discusses the role that the town of Kita could play when linked to the coast by a new railway system (soon to reach Kayes) and later by navigation on the Senegal River. The key role of the Haoussas and the Peuls along the east-west trade line (herds, ivory, textiles and cola) is mentioned, as is the possibility of using their skills to increase trade with the coast. As far as trade is concerned, the region is represented as wealthy and dynamic; in other words of primary colonial interest. Information on the navigability of the Niger is provided and the fact that safety may be ensured by involving local populations is mentioned.

To highlight its agricultural potential, the region is divided into latitudinal bands that determine the climatic limits within which various crops may be set up: palm oil, yam tree, Shea butter, néré tree and *Landolphia* (rubber liana). Emphasis is also placed on the fertility of the soil in the region between Djenne and Timbuktu, thanks to the Niger's periodic floods which, in captions, are compared to the Nile's. By contrast, the positive assessment of soil fertility and the possibility of expansion into the lake region – much celebrated in the document we already analyzed (Fig. 2.8) – were confuted, since that region would presumably be tillable only in small areas, being already largely occupied by local populations. Different shades of green were used to identify types of spontaneous vegetation (the forest or the savanna) and flood areas or particularly fertile regions were highlighted, and their crops specified. The report advocated a trade increase for the city of Kong, destroyed by the *Almany* Samory Touré in 1897 and called for an extension of trade to the Ivory Coast (Assinie), thereby creating a double link with the sea.

This information was reiterated and expanded in a table that divided indigenous trade exports from imports intended for local markets, with different products, quantities and economic values. The communicative effectiveness of these data is entrusted to the map, which shows the reach and range of influence of trade by arranging data across the territory. The size of symbols used for cities is proportional to their turnover, estimated in francs, while that of trade routes is proportional to the value of traded goods. No information is provided, however, as to how these data were obtained. This omission gives the idea that the nodes and networks of circulation rely on the same commercial logic enforced in metropolitan areas. And that in turns rules out the possibility of recovering the specificity and quantification of trade for this region. When Europeans came upon the stage, there existed a centuries-old framework of West African trade routes, organized into a hierarchical and practical set of relationships governed by the great sub-Saharan and Sahelian markets, the expression of an Arab-Berber religious, technical and architectural culture. Other markets of intermediate rank radiated from these large markets, circulating goods to the Gulf of Guinea, but also to Sudan: they were distribution nodes for products on a regional scale. Finally, third-rank markets were determined by obligatory paths for the supply of food and water, for the recruitment of porters, the purchase of camels and pack animals. The latter ones eventually became nodes of exchange and trade that ensured widespread distribution over a large area reaching Africa's central regions. We should not forget that the organization of mobility is rather complex. Trans-Saharan trade requires a strict organization and is tied to camels; a consortium of merchants is at the origin of each caravan, which travels for two or three months on a season-dependent schedule. Sudanese trade, on the contrary, is very flexible: traffic adapts to the local conditions or the incursions of raiders; it moves by river and, when possible, by land. In the north, it is ensured by animals (oxen, camels and donkeys) capable of transporting large quantities of goods; in the south, in the region affected by trypanosomiasis, animals are replaced by free men or, more often, by slaves. Nothing is comparable to the state of things in Europe.

Nonetheless, the map entertains the idea of an Africa somehow conforming to metropolitan Europe. It does so by downplaying distinctive features and by avoiding the issue of arbitrariness in the criteria used to obtain data. The European public, captured by the myth of an Africa full of exploitable resources and still unvalued products, is an easy prey to the almost explicit illusion the document produced: that of Africa as a territory to be easily incorporated into the European circuit of trade. In short, the map insinuates the idea not only of an Africa rich in raw materials, but also of a land suited for commercial expansion and ripe for assimilation into the international trade circuit.

I will conclude by stating that the maps reproduced in journals bear witness to a hetero-centered view of the black continent, favored and supported by the communicative mechanisms inherent in the medium used. The creation of an opportunistic view of Africa was accomplished via icons: as designators were used in a referential role, there followed a loss of their social meaning. At the same time, the reading of African territories with Western criteria invested them with improper values. Cartographic iconization completes the work, favoring the consolidation of a differentialist Elsewhere that requires no interpretative key to be understood except for a comparison with its Western model. Thus, "Orientalist" mythologies created by cartography are not merely misleading interpretations, but also instruments used to exclude a reality, an African culture that could be recovered through territorial knowledge.

SEMIOSIS AND TOPOGRAPHIC METRICS

Exclusion of the social meaning of territory does not occur equally and indifferently in all maps but especially in geometric maps, that is, those that adopt a precise metrics: topography. We have already noted that Cartesian logic, used in this type of map, favors the material features of territory, irrevocably moving away from the idea of landscape as an experience of the place. More specifically, geometric representation ensures its effectiveness through a very simple sign system, which maintains not the substance of objects but their relationships, their links, by an analogical preservation of distance. It is essential, therefore, that we pause to consider the meaning of *topographic metrics*.

Although it is full of implications that have profoundly affected geographical analysis, the term *metrics* is unmistakable and may be said to refer to the method of measuring and treating distance. The adjective *topographic* instead extends the meaning of metrics in a cartographic sense, retaining that of *topography*, from which it is derived.[63] I will consider at least two senses of the latter term, which refer respectively to the encoding of signs and to the spatial concept that underlies it. The headword topography: 1) identifies a cartographic genre which uses an abstract and coded language, drawn from Cartesian logic; 2) refers to a system of survey and mapping, designed to represent a particular and precise concept of space, Euclidean space.[64] In summary, topographic metrics is *the encoding of signs shaped on Cartesian principles and the preservation of the properties envisaged by Euclidean space* (contiguity, continuity, uniformity) brought together into a measuring system for distance that is not concerned with the representation of the quality of objects but with standardizing it, preserving their relationships, their sizes.*

By way of application, topographic metrics may be defined as the sets of technical knowledge that contributed to the measurement of space and localization of geographical phenomena: locating mountains, rivers, settlements; measuring and demarcating agricultural parcels; detecting the presence of features functional to the construction of buildings and roads; designing the reach of cities and other. At the same time it encoded signs in relation to the representation of an abstract territory, divorced from any social interpretation. Following its widespread adoption – from the 18th century onwards – topography sanctioned a univocal and indisputable view of the world. The process did not obviously take place of a sudden, but is tied to the history of geography as a whole. Measuring the world and taking such data as an unquestionable truth is a project grounded in antiquity, when the idea loomed that not only small distances but also large extensions (like the earth radius or meridian arcs) could be measured.[65] But it is towards the end of the 17th century that maps combined the theoretical resources of geometry and the measurement devices of geodesy (specifically triangulation), which allow exact calculations of plane lengths; in this context, altitude was included within these two dimensions. In fact, one of the major transformations brought about by topographic metrics was the annihilation

---

[63] See the entries edited by J. Lévy in: J. Lévy, M. Lussault, eds., *Dictionnaire de la Géographie...*, *op. cit.*

[64] We will have occasion to observe this logic in section "Symbolization and Cartesian Abstraction," p. 79 of this chapter.

[65] On the intellectual mastery of the world through measurement and on the relations of topographic metrics to other sciences, see: K. Alder, *The Measure of All Things. The seven-year odyssey and hidden error that transformed the world*, The Free Press, London, 2002.

of the third dimension and the flattening of the world. The view inherent in such metrics, presenting a world seen indiscriminately from above – and, therefore, according to multiple observation points scattered perpendicularly to each point of the earth – entailed a flattening of the third dimension that, while still recorded, was made irrelevant for the meaning of the territory. I will return to this issue in the next chapter when, to highlight the importance of such exclusion, I will analyze maps produced before this transformation, and show how a perspectivally rendered survey can iconize territory, restoring a sense of landscape. I should now underline, however, that even accurate zenithal techniques that followed over time (aerial photography, teledetection) changed but marginally the image of the earth that, albeit at multiple scales, reproduced the numberless measurements made on the ground by way of the same abstraction.

I am going to try to uncover the semiotic nodes on which this metrics is based to know it in greater depth and, therefore, better assess its communicative results. To do that I will take into consideration: the double figure of the interpreter, that is the cartographer and the recipient and the axiomatic assumptions of its semantics, summarized in Fig. 2.12. Finally I will focus on its communicative results.

The Interpreter: Cartographer/Recipient

It should be remembered that cartographic communication features two interpreters: the cartographer, who builds the document, and the recipient, who uses the information coming from the map. These two figures are clearly distinguished in their communicative action. The cartographer is the one who activates a sign pragmatics in order to communicate accurate and targeted information. He or she must possess certain features: possess knowledge or at

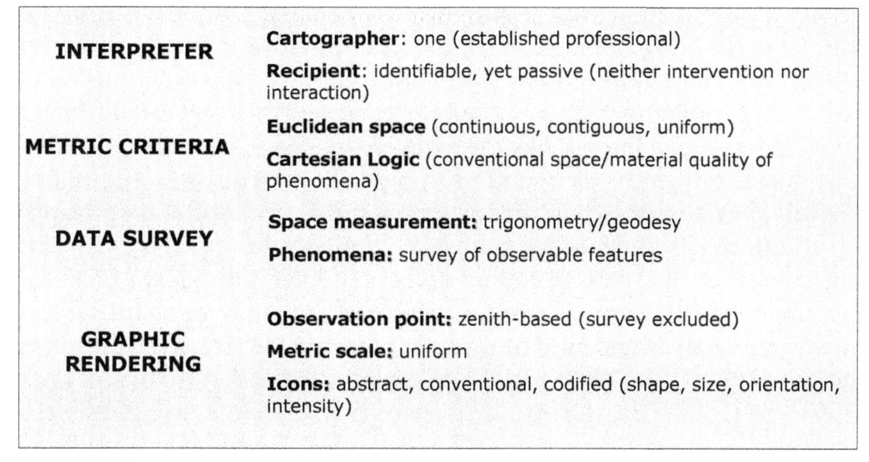

INTERPRETER	**Cartographer**: one (established professional) **Recipient**: identifiable, yet passive (neither intervention nor interaction)
METRIC CRITERIA	**Euclidean space** (continuous, contiguous, uniform) **Cartesian Logic** (conventional space/material quality of phenomena)
DATA SURVEY	**Space measurement:** trigonometry/geodesy **Phenomena:** survey of observable features
GRAPHIC RENDERING	**Observation point:** zenith-based (survey excluded) **Metric scale:** uniform **Icons:** abstract, conventional, codified (shape, size, orientation, intensity)

FIGURE 2.12

The semiotic nodes of topography.

least a mental representation of territory; have mastery of technical equipment; be familiar with the conventions in use; have a purpose to build a map; have a recipient belonging to a community or a group of individuals. The role of cartographers is defined when society acknowledges their ability to transmit cartographic information. Conversely, the recipient is the one who also activates a pragmatics of the sign but does so with a view to receiving information. This second figure requires no specific skills. However, he or she too must be familiar with cartographic standards in use, because as models, maps always fall within a general system of conventions. In addition, recipients must set themselves or already possess a goal for using a map – if solely the goal of gaining information.

Although both actors are able to make choices, the former moves along the plane of construction of cartographic information, the latter along the one of its reception. These individual prerogatives do not rule out the need for compliance with social norms. In fact, just as the encoding of cartographic language is the expression of a given society, at a given historical time, so interpreters are affected by the social norms that compel them to express themselves and to interpret maps through established conventions. Therefore, while preserving their status as subjects capable of autonomous action, interpreters matter because they belong to the same society and are members of a specific culture.

The one aspect to be noted is that – with the birth of topographic cartography – the role of the interpreter underwent a major change. In the case of recipients, their interpretative possibilities were reduced, since messages conveyed via rigid attributions of meaning claim to be unquestionable and do not require – at least not specifically – critical intervention on the part of observers. Recipients were thus placed in the position of having to accept passively what the map showed: that signaled the triumph of cartographic self-reference. As for cartographers, their role, both technical and social, was weakened. First, cartographers lost any chance to express themselves subjectively within the rigid grid imposed by cartographic language, whereas previously, as we saw, although called upon to comply with current conventions, they were still allowed margins for free interpretations. In addition, the creation of institutions responsible for cartographic mapping – driven by heightened interest on the part of nation states – and the subsequent inclusion of cartographers in corporate bodies, debased their role by reducing it to that of technicians who could transpose onto a map what had already been established elsewhere. Their social role was transformed: from that of interpreters able to draw the world to that of surveyors, expert in technical methods of surveying and rendering. Of course, they continued to apply the norms which regulated the representation of territory, but they no longer governed them.

Topography thus favors the presence of a single interpreter, the State, which sets rules for construction and for language codification: having become a territorial actor, the State needs a suitable form of cartography to wield its political, administrative and military control. It is well known that mapping always reflects the intention of a political power to take control of a territory. Here, however, we should stress the special role played by the State as the holder of a planning policy that invests maps with new social tasks. In fact, modern cartography, born in the Renaissance, closely traced the evolution of the concept of the modern State, to the point that when nation States included among their foundations the control of a well-defined territory in its extension, maps became instruments of power – of the Powers that be – that favored topographic metrics, for various reasons coterminous with the ideologies on which this political entity is based. In short, the aim of creating an accurate, extensive and homogeneous representation of territory surely comes from inescapable practical exigencies, such as producing information about movement, defining boundaries, planning infrastructure design. It is also based on an ideological concept that sees maps as ideal tools for securing the intellectual appropriation of territory for political purposes. More precisely, with the development of nation States, there arose the idea that knowledge of territory should be conceived as a socially relevant enterprise, able to promote the shaping of a national consciousness. The success of this project comes from the use of scientific instruments that are supposedly based in objectivity. Topography stands out among them, since the measurement and quantification of phenomena are considered scientifically accurate criteria. Thus, during the 19th century, topography not only sanctioned political and administrative control, but also lent itself to promoting the shaping of a national consciousness and inaugurated widespread cartographic campaigns.

While at the end of the Napoleonic wars, that is the beginning of the 19th century, only France possessed a general State map, based on astronomical readings, and surveyed through a vast triangulation network; by the last decade of the 19th century all European States, with the exception of Greece and Turkey, undertook or carried on the preparation of topographic surveys, based on the new norms of astronomical reference. To achieve this, governments hastened to include in their national programs the creation of institutions devoted to the implementation of cartography. Large-scale maps were designed to render territory with great precision, and from these smaller-scale maps were derived. That spelled the demise of previous maps and highlighted the need to organize campaigns for producing maps measured on the ground, according to geometrical and mathematical systems of surveying and rendering.

In Italy, for instance, in the second half of the 19th century, the State undertook cartographic mapping for its whole national territory to uphold its own legitimate authority. For at the time of its unification (1861), one of the issues Italy had to face was the acquisition of quantitative and qualitative knowledge about the territory under its national sovereignty. To this end, in 1872, the *Istituto Geografico Militare* (Military Geographic Institute or IGM)[66] was set up, with the intent to establish a technical referent responding to ministerial coordination for implementing a nation-wide cartography. There were of course other public and private institutions involved in the production of maps, but they all played subordinate roles. Clear-cut areas of competence were defined between the Ministry, IGM and other institutions. And while overlaps and cross-referencing did occur, the roles each of them played were unequivocal. In detail: official cartography was co-produced by the Ministry and the IGM; the former played the role of the client, who orders, plans and finances the operation, playing a leading role. The survey itself and all topographic rendering were instead a prerogative of the IGM, which was entrusted with translating the Ministry's directives into action.[67] This latter ensured what the new systems of surveying and cartographic mapping require: highly specialized technical skills but, above all, rules encoded into a set of norms able to standardize the process. It needs to be stressed that even this presumably neutral approach is in fact closely tied to vested ideological interests. The State takes upon itself the role of legislator and guarantor of the ways and conditions for constructing the image of the territory under its sovereignty: it acts as the sole interpreter, laying down the principles on which the world must be ordered. In short, systems of representation reveal, even in their technical aspects, an underlying linkage between the aims and the realizations of power. Through techniques, understood here as accuracy in representation, an order of the world is unveiled, promoted and finally translated into principles aimed also at upholding a social order.[68]

---

[66] The Institute was initially called *Istituto Topografico Militare* but its name was changed in 1882. On the venture of mapping Italy's territory, see: A. Mori, *La cartografia ufficiale in Italia e l'Istituto Geografico Militare*, Stab. Poligr. Per l'Amm. Della Guerra, Rome, 1912. This institute was to become the privileged referent in colonial surveys, because it possessed the technical expertise needed to carry out the project. On IGM's production see: *I.G.M. L'Istituto Geografico Militare in Africa Orientale 1885–1937*, IGM, Florence, 1939.

[67] There is a cartographic production linked to various private bodies such as the Società Geografica Italiana (Italian Geographic Society), cartographic institutions (Istituto Geografico G. Cora, Istituto Cartografico L. Rolla, Studio Cartografico G. Giardi, etc.) and to a number of publishers (Paravia e Vallardi, Istituto Italiano di Arti Grafiche, Istituto Geografico De Agostini, Touring Club Italiano). It must be remembered, however, that within this intricate panorama most cartography is still derivative, and thus production from the institutes mentioned above comes in fact from maps produced by IGM. See: E. Casti, "Elementi per una definizione della figura del 'cartografo coloniale' all'interno del progetto DISCI," in: *Geostorie*, XIII, 3, 2005, pp. 203–229.

[68] As demonstrated by: F. Farinelli, *I segni del mondo. Immagine cartografica e discorso geografico in età moderna*, Academia Universa Press, Acqui Terme, 2009, p. 60.

Topographic Semantics: Euclidean Space and Cartesian Symbolization

As we come to terms with the semantic repercussions of topographic metrics, we realize it signals a veritable cartographic revolution. Adoption of Euclidean space and a codification-abstraction largely based on Cartesian premises paves the way to action on two levels: 1) territorial phenomena are placed into a conventional space; 2) such phenomena are made to correspond to abstract signs that refer to a cluster of properties selected according to the logic of matter. Let us consider these two aspects.

Euclidean Space

The adoption of Euclidean space as a cartographic foundation entails compliance with certain prerequisites: 1) *continuity*, i.e., that space be understood as a perfectly seamless continuum; 2) *contiguity*, i.e., that proximity and, therefore, the effects of contact or distance between phenomena be represented in a uniform manner; 3) finally, that such measurement and treatment of distance be *uniform*.[69] The concept of spatiality that underlies Euclidean space is absolute, and refers to a purely conventional mental construct. Conversely, geographical space cannot be but relative, since its content is determined by the presence of humans who establish its relevance. Thus, the Euclidean tenet leaves irretrievably behind any opportunity to represent the social substance of geographical phenomena. Determination to reduce the world to a Euclidean space and to treat it cartographically meant addressing two operations that substantiate its graphical transposition: land *data acquisition* and its cartographic *rendering*. The former was addressed by applying trigonometry rules, drawn from Euclid's theorem, whereby distances are inferred algebraically rather than measured on the ground. Once the magnitude of one angle and one side of the right triangle has been established, the rest of the missing data can be inferred. As I reiterated in my illustrative discussion of colonial cartography, while the determination of land measurements posed practical problems, triangulation presented no issues from the point of view of theory: reliance on geodetic points guarantees positioning, on which subsequent measurements can be based. By contrast, cartographic rendering demanded the use of geometric and mathematical reducers – metric scales, projections, contour lines – able to prospect the world as a model based on a dual system of communication: analog and digital. We have already mentioned that any modeling is based on particular implementation norms and that each model – processing a given amount of data – takes shape through the classification of objects and their abstraction from the context in which they exist. The procedures

[69] J. Lévy, "Euclidien" entry, in: J. Lévy, M. Lussault, eds., *Dictionnaire de la Géographie…, op. cit.*

involved in the topographic translation of Euclidean space drew from both usable communication systems: the analog and the digital. This boosted their reliability, which however relied less on a geometric-mathematical level than on an iconic one. As a matter of fact, the gaps, approximations, and errors that such reducers introduced were deemed irrelevant with respect to the rendering of a rigorously measured world. Thus, reduction to scale according to an analog principle between linear distances verifiable on the ground and those present on the map was achieved by means of a mathematical and graphical relationship and taken as a discrete criterion, which ultimately pertains to the digital system.

I do not intend to assess here the semiotic outcomes that the use of this double system of communication involves. Rather, I wish to remark that the adoption of scale as a principle of metrical accuracy was thwarted. The relationship between distances is never constant in topographic maps – because of distortions produced when moving from the curved geoid surface to a plane – and deviate quite significantly from the one given in the legend. This nullifies the topographic claim as to a presumed correspondence between linear distances in reality and those on a map, and leads us to declare that the premises adopted are insufficient to ensure accurate results. By contrast, scale performs other functions, such as determining the map's degree of information – so much so that the wealth of detail depends on the type of scale used: selection is not only about the number of objects represented, but also their degree of detail, their shape and attributes. Scale intervenes on both miniaturization* – and, therefore, on the creation of icons – and on their number. Moreover, scale has repercussions on the representation of the relationship between the objects, and influences the choice of the model whereby reality is to be rendered. The choice of scale is therefore conditioned by the intentions of those who govern the map. It would be an error to think that it depends exclusively on the extent of the territory to be represented, for the same area could well be portrayed on more maps. Rather, scale depends on the use to which the map is intended. Ultimately scale plays many important roles, but these do not include meeting the accuracy prerequisites that the Euclidean system establishes.

The other reducer, projection, that is, the geometric-mathematical system whereby the spherical surface of the globe or a part of it is transposed onto a plane – i.e., the sheet – consists basically of the use of a grid (lattice or mesh) formed from lines corresponding to meridians and parallels, necessary to determine the exact position of individual points of the surface represented. Yet, as I said, the impossibility of representing a spherical surface like the earth on a flat surface like that of the map leads to using methods that allow for transposition with some degrees of approximation. Projection is an ambiguous transposition system, because, while it endeavors to respect real-world measurements, it actually generates a deviation, an unavoidable error. At the same time, by stressing accuracy as a geometric prerequisite, projection conveys the idea that, as a building system, it is capable of ensuring accuracy. Much has been written about the ideological implications of this issue. However, if we focus on the technical side, although projections do in fact remedy some of the abuses caused by plane representation of a curved space, they entail inevitable distortions that affect the measures of territory, whether they are related to the distances, surfaces, or incidence angles between meridians and parallels.[70] Even *conformal projection* (based on an internal variable scale) does not solve the issue but rather amplifies it, because it fails to comply with another principle of Euclidean space, i.e., uniformity, which guarantees standardization in measurement and in the treatment of distance.

Finally, as regards altitude, topography adopts contour lines which, although derived from trigonometric survey, favor the referential, or analog, system. These are rendered graphically by means of digital signs that make information communicable through very exacting mental processing. When I taught at the Department of Geography, University of Padua, one of the hoops students were put through was the reading of *tablets* (1:25,000 IGM map). The exercise involved no extensive studies or any particular geographical knowledge, except for training in map reading. Students were expected to demonstrate their ability to resist the map's informational denial, contrasting its abstraction with mental processes able to recover the third dimension and, therefore, terrain morphology. In short, they were supposed to see contour lines in volumetric terms: circles or flat lines had to be imagined as overlapping, thereby obtaining information on the relief trend. The efforts students made corresponded neither to their level of preparation nor to their geographical knowledge, but solely to their ability to match the digital system to the analog one, thus bridging abstraction via imagination. More than a knowledge test, it was a test of skill: the skill of imagining the world's third dimension within a representation system that, by turning it into a plane, had virtually erased it.

In conclusion, the use of topographic reducers aimed to render Euclidean space falls short of the claim to represent the world in metrically accurate terms. Worse still, reducers completely annihilate the world's social meaning. Awareness of this issue leads us to discard the constraint of Euclidean space as the only possible criterion, and to advocate the search for other concepts of space.

---

[70] Only globes can preserve with mathematical accuracy the lengths, angles and areas corresponding to reality, through equidistance, orthomorphism and equivalence. In the case of maps, instead, any expedient can thoroughly fulfil but one of the last two conditions (orthomorphism or equivalence), because equidistance is confined to set limits and in certain directions. It follows that distortions are not of equal magnitude within a single map (they normally grow from the center toward the edge).

Symbolization and Cartesian Abstraction

As we move on to analyze the rules of abstraction to which the symbolic structure is subjected to adapt it to the Cartesian worldview, the constraints imposed by the latter view become evident. In "Cartesian space," the rules that govern the relationships between objects are independent of the objects' nature. Objects in their relations actually do not affect the construction of space. For Descartes, it is the geometric features of a unique reality called "extended substance" that provide the foundation for spatiality.[71] Objects must therefore be identified by rational methods and their representation is the sum of a cluster of properties selected according to the logic of matter. A whole series of misunderstandings, in fact already widely reported, arises in topography from this initial assumption, including the one that has to do with the use of symbols. On this issue, Farinelli argues that topography adopts "a concept of symbol that has nothing in common, except the name, with the authentic one."[72] Such "symbolization," which sets up strict norms on both the type of information to select and its graphic rendering – abstract and geometric – has momentous semiotic repercussions. The symbol is reduced to a sign and communication limited to a surface level, which narrows the scope for an understanding of the world capable of appreciating the many facets of the whole, typical of symbols.[73]

It should also be pointed out, still in the semiotic perspective, that the inclusion of signs on the sheet is the result, translated graphically, of a series of mental operations carried out by the surveyor, as a member of a given society, to communicate a name. The graphic product is therefore derived from culturally shared operations of selection, simplification, classification, synthesis, and finally, in some cases, of symbolization. The entire procedure is strictly encoded according to rules that do not involve interactions between geographic objects and their relationships are independent of their nature. In short, the Cartesian worldview is taken in topography as the grounding principle for a symbolic representation of the world: a set of technical criteria that tend to abstract the sign – even the smallest and most insignificant sign – in order to endow it with a univocal meaning in terms of quantity.

This, however, foreshadows consequences even on the technical level. For the encoded set that substantiates information (figures, colors, numbers, and

---

[71] It is a relative space the substance of which depends on the presence of the objects within it, albeit beyond the limited "content/container" metaphor. See the entries "Cartésien" and "Espace" in: J. Lévy, M. Lussault, eds., *Dictionnaire de la Géographie…, op. cit.*

[72] F. Farinelli, *I segni del mondo, op. cit.*, p. 17.

[73] For a discussion of the issues inherent in symbolization, Barthes's work is still essential. See for instance: R. Barthes, *Empire of Signs,* Hill and Wang, New York, 1983.

names) does not imply the same type of abstraction. In the case of numbers and names, topographical signs follow the conventions adopted in the language of origin, numerical or lexical; as regards figures and colors, however, it becomes necessary to produce a figuration based on maximum abstraction.[74] With regard to figures, action is taken: 1) on *form*, by simplifying the features of the object to be represented in a geometric perspective; 2) on *size*, by establishing a hierarchy of magnitude which matches a quantitative aspect of the object represented; 3) on figure *orientation*, to show its development; 4) on *intensity*, i.e., the size of points, lines and hatch lines, used to indicate, even in this case, a hierarchy of importance.[75] Nor are names exempted from these last two options: via typeface, orientation, and size names also partake of the object's hierarchical ranking. As for color, the process of abstraction is more complex, as we saw previously, and involves both arbitrariness – and thus the lack of analog references – and conventionality, already in use in ancient maps. In any case, color deviates irreparably from analogy with reality: blue used to mark water is conventionally borrowed from previous cartography; green is indifferently matched to either vegetation or the plains in relation to map detail; black is attributed to artifacts in wholly conventional terms. Finally, variation in color intensity corresponds to changes in certain quantitative data: depth, altitude, density, etc. In short, derivation of the map's symbolic structure from Cartesian criteria spells the death of the figurative sign and the adoption of the abstract sign. From that moment, icons combine figurative and numerical signs referred to the object's size. Having lost their analogical tie with reality, chromatic signs are used to establish a purely conventional linkage. Such modeling is implemented according to rules of sign encoding that come from the double-communication system – analog and digital – which has iconizing outcomes: analogical signs which undergo a process of abstraction validate the substance of a world based on its materiality. Yet, it is not so much by abandoning abstract signs or recovering figural signs that such semiotic outcomes can be resisted. The path to purse lies elsewhere: in the abandonment of signs and the recovery of symbols in their most authentic sense.

---

[74] With regard to types of signs, G. Anceschi argues that those used in cartography cover a very wide spectrum, ranging from fiction (corographic cartography, shaded cartography, color cartography), hatching (Euclidean cartographic base), to the iconic (thematic maps). In this regard see: G. Anceschi, *L'oggetto della raffigurazione...*, *op. cit.*, pp. 45–51. However, if we consider the contribution of J. Bertin, who worked in a milieu very close to that of linguistic semiotics, we identify three categories of representations: the diagram, the lattice or graph (réseau) and the map; in the discussion of symbolization issues, the latter is the only reference to the figurative vs. abstract question. On this question see: J. Bertin, *Sémiologie graphique*, Mouton & Gauthier-Villars, Paris-La Haye, 1967. The progressive scale of figural abstraction permeates the research of: R. Arnheim, *Visual Thinking*, *op. cit.*, p. 173.

[75] For instance, the placement of the "wedge tip," a sign that marks the elevation trend of terrain.

The Communicative Outcomes of Topographic Metrics

I submitted that topographic metrics coincided with the birth of a sign system that interprets what it represents through abstraction. In other words, as Arnheim warns, abstraction is not a shorthand representation that invites the viewer to fill in missing realistic detail.[76] Abstraction is not incompleteness. If it were meant as such, it would require the interpreters to make decisions on their own about the nature of what they observe, whereas the sign sets itself up as self-contained and undivided to the point of acquiring autonomous relevance with regard to interpretation. Each icon advances a hypothesis about the distinctive features of an object, the univocal meaning of which comes from the fact that it derives from the same convention used by the interpreter. The main goal thereby achieved is to ensure unreflecting acceptance of the cartographic message. The possibility of juxtaposing one's own perceptual experience to the object represented is ruled out and what is envisaged is accepted uncritically. While abstraction has this negative effect, it still allows what the figurative sign (i.e., the sign that refers to the object's shape) excludes. And by moving away from analogical representation, we in fact gain the ability to convey concepts more easily. As illustrated by Lotman, the encoding of semiotic meaning triggers a process of abstract generalization, in which simpler intellectual mechanisms generate more complex forms of knowledge.[77] On the one hand, complexity is mastered more effectively; on the other, as the abstract sign advances its own meaning, it excludes the possibility/need for verification with the reality represented.

The result is a self-contained system of signs which nevertheless still needs a passkey in order to be communicated: it needs a *legend*. The latter is the interface needed to cope with the transformation that the sign has undergone in its transition from an analog system to a digital one. In particular, a legend makes up for the transformation which signs were subjected to in the transition from a continuous system to a discontinuous one. Through the legend, every sign acquires a univocal and self-contained meaning. Yet, most importantly, if such meaning refers only to the visible features of objects (their shape, size, quantity), it relies exclusively on denotation. The final outcome is that the encoding and abstraction of cartographic language bring about the loss of the object's connotative and cultural meaning, while attention to its material features enhances its denotative, or referential aspect.

Possibly, the most startling result of the encoding-abstraction process is the fact that, while it marks an increase in knowledge, at the same time it entails a

---

[76] R. Arnheim, *Visual Thinking*, *op. cit.*, pp. 153–158.

[77] J.M. Lotman, *La semiosfera*, Marsilio, Venice, 1985, p. 105.

substantial loss of meaning. The paradox lies in this: when cartographic language undergoes the process of codification-abstraction (formal, graphic, iconic), supposed to boost meaningful transmission, deeper meaning is undermined, because selection is aimed at enhancing only a few, limited material features of the object.

That is why the irruption of topographic metrics into cartography sanctions the loss of the map's ability to capture the symbolic essence of the world, present in previous cartography. What is lost in the understanding of the world as a symbolic gesture that survived from the Middle Ages throughout the modern period, progressively enriched by what empirical testing is able to deliver.

In the next chapter, I will look at some instances of the latter type of cartography to assess in greater detail what such exclusion entailed. For that is where the loss of the social meaning of territory has its origin.

CHAPTER 3

Landscape as a Cartographic Icon

He who measures the earth and the heavenly bodies and describes all things, wanting to possess sure memory of the world, is at once a madman and is banished.

S. Brant, La nef des Folz du Monde, 1497

The goal of this chapter is to highlight how, before the introduction of topographic metrics, the kind of space maps referred to, and the symbolic structure geographical phenomena were supposed to reflect, corresponded to the notion of a socially constructed world. By looking at some of the maps drawn in the modern period, when studies on the representation of landscape proliferated, I will try to shed light on the relations between the two forms representing territory – landscape and map – in order to make sense of the different roles that they play in communication. I will rely on the essential functions of maps – to describe *or to* conceptualize *– and investigate what from now could be called an* iconic junction, *that is, the link, identified by the categories of* conjunction *and* disjunction, *which comes to exist between landscape figuration and cartographic figuration. This chapter aims not only to reflect on abstract categories but also to illustrate the relationship between these two figurative forms, recovering their underlying metrics. This will be done with a view to understanding the nature and outcome of that connection, and with it its inner meshes, over which other forms of representation freed from topographic metrics may be tested.*

CONTENTS

CONNECTIONS, HYBRIDIZATIONS

In front of the Grand Canyon, in the whirlwind of emotions that it arouses in us, it is natural to consider the inadequacy of any representation in rendering the majesty of landscape. What stands out in front of our eyes is hard to reproduce, not only for the exuberance of colors, shapes, distances, but also for the technical difficulty of recreating the mechanism through which the eye endeavors to master such a view. The countless attempts, often inadequate, that painters, photographers, and graphic artists have made to render what our experience gives us of that landscape are a clear demonstration of how real this obstacle is. Yet, today, thanks to studies on the semiotics of the vision, we do have instances of representation that closely approximate our goal and are quite successful in rendering

Reflexive Cartography. ISSN 1363-0814, http://dx.doi.org/10.1016/B978-0-12-803509-2.00003-3

the deeper meaning of what is perceived visually. Starting from the assumption that landscape cannot be seen exclusively as the result of sensory experience but rather of an intellectual processing and turning, therefore, our attention to the rendering of the concept rather than to its perceptual aspects, we wish to iterate that what we see is not solely due to the physiology of the eye, but to the processing such physiology undergoes through the brain.[1]

This latter remark leads us to reflect on the central role of the semiotics of vision in the study of images and suggests that we trace a link between all the systems of representation that render territory figuratively. As genre distinctions are dropped and analogies introduced on visual representations, a question too hastily shelved by geographers is brought back to the fore, i.e., whether the two forms capable of rendering territory *figuratively* – landscape and map – can somehow converge and, if so, how such a convergence is to be understood. After the introduction of topographic metrics, landscape and maps became figurative systems with radically different features. In the past, however, this split was not so clear-cut and marked off, as maps produced in the period that goes from the Middle Ages to the birth of Euclidean cartography (18th century) demonstrate. These exhibit obvious contaminations between the two types of representation, even within the same document. And if that were not enough, by restoring landscape images to maps and creating complex logical and formal hybridizations, the new figurative systems of *cyberspace* question the very boundaries between the two representations, revealing liminal spaces that are replete with productive ambiguity. Such ambiguity is suggested by the fact that the landscape, in its most trivial sense, is the visual form of territory and, in that capacity, it expresses itself through the visual arts (painting, photography, film, graphics). The map, in turn, presents itself as a figurative form of territory, released from its traditional role of trivial tool for recording reality to become a very elaborate means of visual communication, able not only to describe the world but also to conceptualize it, i.e., to state how it works on the basis of a theory.

LANDSCAPE AND MAPS

Our initial assumption is that landscape is the visual form of territory. This form, however, is not taken as a given but as a concept, that is, a mental processing that took shape within Western culture in a particular historical period, the

[1] Bibliography on the subject is vast. See for instance: G. Bettetini, *La simulazione visiva: inganno, finzione, poesia, computer graphics*, Bompiani, Milan, 1991; A. Appiano, *Comunicazione visiva. Apparenza, realtà, rappresentazione*, UTET, Turin, 1993 (reprint 2008); J. Fontanille, *Sémiotique du visible. Des mondes de lumière*, PUF, Paris, 1995; R. Eugeni, *Analisi semiotica dell'immagine*. Pittura, illustrazione, fotografia, ISU, Milan, 1999 (reprint 2004).

Renaissance.[2] As such, landscape cannot be assumed trivially as the recorded image of territory, which remains elusive to the eye in its making and unfolding. Rather, it must be taken as a vision that organizes its own meaning, according to procedures that refer to what Rudolf Arnheim calls "visual thinking" and what the studies of geographers prove to be the communicable dimension of the experience of place.[3] Landscape is the result of culture seen as a social device, paired with a subjective dimension of the body in the perception and enjoyment of place.[4] Eugenio Turri, in this regard, argues that in order to be communicated, landscape must be ordered according to *iconemes*, landscape features that make the idea of landscape communicable. Iconemes, he claims, generate the concept of landscape, since they reflect the iconic quality of place, referring to the coherence of the relationships established around a relevant element or between elements and, thus, between iconemes. Iconemes, therefore, turn out to be signs able to render territoriality in a figurative manner and to target communication. This means that, in its semantic setup, landscape is a mental construct for which the presence of an interpreter proves indispensable.[5]

Even the map presents itself as a visual form of territory, able not only to describe the world but also to iconize it. In other words, if we favor the communicative

---

[2] Although Augustin Berque explained that this is not a Western prerogative, because landscape is one of the pillars of Japanese culture, in order to highlight landscape's relationship with painting I will examine the rise of this concept in the West at the time of the Renaissance. The prevailing view is not that the pictorial veduta arose to express landscape but, conversely, that the concept of landscape was engendered by pictorial elaborations. In this regard: E.H. Gombrich, *Norm and Form. Studies in the Art of the Renaissance,* Phaidon Press, London, 1966, pp. 107–121.

[3] R. Arnheim, *Visual Thinking,* University of California Press, Berkeley-Los Angeles, 1969. As I stated previously, the work of this scholar had a major impact on studies around the semiotics of vision, and gave momentum to critical studies in art history and to pictorial experimentation.

[4] One outstanding contribution among the many works on this subject is Denis Cosgrove's *Social Formation and Symbolic Landscape* (2nd edition with additional introductory chapter), Wisconsin Univ. Press, 1998 (1st ed., 1984). See also: P. Castelnovi, ed., *Il senso del paesaggio,* IRES, Turin, 2000; C. Raffestin, *Dalla nostalgia del territorio al desiderio di paesaggio. Elementi per una teoria del paesaggio,* Alinea editrice, Florence, 2005.

[5] The work of this scholar follows the substantial trend of research that investigates the semiotic aspect of landscape, a trend outlined in: K. Lindström, K. Kull, H. Palang, "Semiotic Study of Landscapes: An overview from semiology to ecosemiotics," in: *Sign Systems Studies,* 39 (2/4), 2011, pp. 12–36. More precisely, this area of study examines ways of interpreting the conceptual substance of landscape from a semiotic perspective, in the wake of research aimed at restoring the role of the subject (Y.F. Tuan, *Topophilia: A Study of Environmental Perception, Attitudes, and Values,* Prentice Hall, New Jersey, 1974; Id. *Space and Place: The Perspective of Experience,* University of Minnesota Press, Minneapolis, London, 2005 (1977); E. Relph, *Place and Placelessness,* Pion, London, 1976). This is part of the so-called "cultural turn" in geography, which introduced a heightened awareness of the role of language and representations, focusing on the communicative aspect of landscape and isolating the "iconeme" as the unit of signification able to give symbolic cohesion to other signs: E. Turri, *Semiologia del paesaggio italiano,* Longanesi, Milan, 1979; Id., *Il paesaggio come teatro. Dal territorio vissuto al territorio rappresentato,* Marsilio, Venice, 1998.

aspects of the map and thus consider it in its capacity as a semiotic field within which signs work as sign vehicles for the interpreter, the map appears as the manifestation of an intellectual appropriation of reality that aims to construct the world linguistically. The cluster of names and signs organizes the world in an orderly frame of knowledge and makes it communicable. Therefore, to take the map as an instance of symbolic mediatization, which plays an active part in communication and develops the communication process autonomously, means to recover both the role of the interpreter and the self-referential nature of the map. Both are crucial in their communicative outcomes, which are made evident through the map's ability to iconize the world.

Landscape and maps appear, therefore, as two forms of communication of territory which converge, because: 1) they have common semiotic links (both are included in the communication); 2) they share common features (an interpreter, a theory); 3) they strive towards the same goal: to advocate a precise worldview; 4) they figurativize territory and rely on visual communication systems. With regard to this last feature, it is clear that landscape and maps can be easily linked to vision and the iconic potential it has. In the case of landscape, communication is expressed through the visual arts, primarily painting, a discipline which in fact gave rise to the concept of landscape in the first place. Landscape may of course be conveyed via other forms – such as a verbal description – which are not considered here, since we aim to focus on the semiotics of vision and to assess whether cross contamination between landscape and map, taken as *figurations*, promotes iconization. Our choice depends on the fact that semiotic studies allow us to draw a link between different views, focusing on the instrument used to express them, to treat them and to process them.[6] Within these studies, in fact, the medium used is crucial because for one it lays down rules by which vision is organized and communicated, while on the other it partakes of communication actively, determining multiple possibilities of figuration. The medium is not taken, then, as an instrument of representation. Rather, it is intended as a mediator of the communicative process which may be analyzed in its transactional concatenation.

The path ahead is, therefore, to consider/take on the "modes" of landscape rendering within the expressive medium that is acknowledged as its own – painting – and then to look for such modes in what appears to us like a map, that is to say a document which, first of all, exhibits its own descriptive intent.

[6] As I recalled earlier, studies on the semiotics of vision within the area of semiotic research inaugurated by Lotman and Greimas later merged with the endeavors of art historians (Panofsky, Arnheim, Gombrich) and communication scientists (Fontanille, Appiano, Goodwin) giving rise to a branch that may be classified as the science of visual communication. The purpose of this subject area is to query communication processes and their self-referential implications.

As a visual form of territory, landscape is a perspectival figuration. Perspective is the linchpin of the theories and speculations of great artists (Leon Battista Alberti, Leonardo da Vinci, Piero della Francesca, Albrecht Dürer), who relied on it to render the spatial image that the eye processes and that we perceive as volume. In art history, this technique occupies a key chapter for studying the birth of the concept of landscape and the relationship that societies established with nature.[7] It is art history that enables us to see perspective - and consequently the theory of proportions - as instruments for interpreting the idea of landscape that was shaped over the centuries, by playing on the relationship between reason and imagination, abstraction and analogy between formalization and perceptual imitation. Thus, recognition of landscape figuration inside maps may reasonably be based on analyzing the use of perspective, which arranges objects hierarchically by acting on their shape and size, and generates in the observer the perception of the "near" or "far," appealing directly to landscape.

In cartography, perspectival view clashes with vertical view, towards which cartography naturally tends, given its propensity for classification and its reliance on the regularity of phenomena. Cartography uniformly removes and abstracts such phenomena in order to fulfill its essential function: description. For the most trivial goal of maps is to render the physiognomy of the world. Even modern cartography, when not yet deprived of the opportunity to render the visual form of territory, rendered it in an analogical reference system, the essential descriptive feature of which are names.[8] Names are undoubtedly the framework to which icons are anchored in their communicative function. Maps may thus be defined as documents that reproduce the world, respecting the relations and distances between objects while also prescribing the recognition of places. We would rather not be tied down by definitions that could well be disproved by the heterogeneous *corpus* of documents archives provide to us. So it should be remembered that, in this context, what matters is to underline the ways in which the descriptive function of cartography manifests itself, with the aim to stress its peculiarity forcefully with respect to any other type of representation. That is above all meant

---

[7] For an overview of the concept of landscape between geography and painting, see above all: D. Cosgrove, S. Daniels, *The Iconography of Landscape: Essays on the Symbolic Representation, Design and Use of Past Environments*, Cambridge University Press, Cambridge, 1988; E.S. Casey, *Representing Place Landscape Painting and Maps*, University of Minnesota Press, Minneapolis-London, 2002. For an account of research on the relationship between art, geography and landscape, see: S. Daniels, "Landscape and Art," in: J.S. Duncan, N.C. Johnson, R.H. Schein, eds., *Companion to Cultural Geography*, Blackwell Companions to Geography, Blackwell, Malden, MA. 2008, pp. 430–446.

[8] I am referring to cartography prior to the adoption of Cartesian logic and the application of Euclidean rules which, we have seen, irreparably foil any chance of rendering landscape.

to refute a number of statements coming from art historians. Attempts to inventory figurative kinds on the basis of their primary function – describing or narrating – have in fact acknowledged that the former belongs properly to the map, perhaps taken superficially as a "meticulous reproduction of reality." That conclusion tends to promote unwarranted comparisons and shows only apparent – and completely misleading – affinities between maps and other representations, ultimately debasing the maps' unique modus operandi.[9] By contrast, here I intend to assert and to stress the particular nature of the descriptive function of maps. We cannot certainly disregard the fact that as they sanction place names and present them in their distribution and their actual spatial relationships, maps irreparably deviate from the function of the landscape, rendered through painting, which proposes itself from the start as a visual expression of the world.[10] This position is consistent with my objective, which is less a generic discussion of the convergence between these two descriptive forms than an examination of their iconic junction,* to be addressed shortly. With this goal in mind, I will shed light on the particular descriptive function of maps first and then move on to consider their conceptualizing role. I believe we ought to focus on the latter because, if the map is seen as *a theory that expresses highly conjectural facts as truths*, we can examine how the concept of landscape becomes part of it and *ask whether landscape may be said to favor or to thwart iconization*. First, let us turn to the technical features that allow us to examine the modes of this intriguing convergence.

PERSPECTIVE AND THE SEMIOTICS OF VISION

One crucial element in the study of perspective is the point of observation,[11] so in order to examine the encounter between landscape and map, I will start with this element and look at the distinctive features of the two figurations.

[9] This is the case of Svetlana Alpers' study on Flemish painting. The researcher argues that in the 17th century such art was descriptive and narrative because reality was rendered with the accuracy and meticulousness of maps. The argument is flawed by her apparent unawareness of the setup of maps and by the direction of her analysis, based on similarities rather than on differences between maps and painting, as keenly shown by Victor I. Stoichita (*L'instauration du tableau*, Drozz, Geneva, 1999, p. 235). Alpers' study is not only oblivious to the semiotic nuances of maps, but seems even unaware of the most elementary technical cartographic norms. (S. Alpers, *The Art of Describing: Dutch Art in the Seventeenth Century*, University of Chicago Press, Chicago, 1983, p. 119 and ff., esp. 119–168).

[10] In the 16th century there existed a hybrid painting, which inserted place names into landscapes, as in the case of the Flemish painter Joachim Patinir, shown at the Prado Museum in Madrid. But those were sporadic examples, dating back to a time when the commixture of painting and cartography was still current.

[11] In this context, I would use this phrase rather than the more commonly used "point of view" to emphasize its technical aspect and – at the same time – the role of the interpreter understood as an observer.

Undoubtedly, with regard to landscape, observation is perspective-based: the world is viewed by an observer who is located at a given point. The latter may belong to the landscape observed and, therefore, be within it; or it may coincide with a relief (a mountain, a hill, a cliff), which allows the eye to pan over what is being observed. It might even be a point so high – an airplane, for instance – as to allow a sweeping view of large tracts of territory.[12] These are all perspective views anyway, because they involve only one point of observation that matches the observer's. The observer's position not only determines the near and far from what is represented but, assuming different distances between objects, it determines the point from which the image unfolds. Perspective is, therefore, a system for representing depth. No matter which one is adopted (natural, central, or vertical), perspective is the system that renders what in art is called a *veduta*.[13]

To assert the importance of perspective in rendering an idea of landscape may seem redundant at this point. It should be noted, however, that perspective was taken as the key element of research in the fields of painting and vision semiotics, and concerted experiments made in those fields so far have had remarkable outcomes. Once identified as one of the main modes[14] of visual communication, perspective is acknowledged as a generative mechanism which can process information in a given direction. The work of David Hockney is enlightening in this sense, especially with regard to the representation of a landscape we mentioned earlier: the Grand Canyon.[15] In his large Grand Canyon canvasses, Hockney recreates the emotions aroused by a direct

[12] Obviously, I will not keep track here of all the implications related to the optical illusions which painting deploys in order to include or exclude the observer. And I should also point out, in this context, that I will not mention the second interpreter involved in representation (the one to whom the work is addressed) but limit myself to the first interpreter (the one who creates it).

[13] The vast bibliography on perspective produced in art history and other disciplines leaves no room for direct citations. However, for a useful and concise overview of the issue, see: H. Damisch, *The origin of perspective*, MIT Press, Cambridge, MA, 1994; and what is more relevant to our present purposes: D. Meneghelli, *Teorie del punto di vista*, La Nuova Italia, Florence, 1998.

[14] The others are color and shape.

[15] The artist, who has long pondered over the role of painting in the contemporary world, claims that it can generate new works only focusing on the semiotic mechanism of vision. He therefore paves the way to visual experimentation, which he endows with the ability to reinvent painting. Rejecting any rule of perspective, which he considers responsible for excluding artistic expressiveness from painting, David Hockney problematizes the point of observation and focuses on the physiology of the eye as a feature that can affect the communication process and hence pictorial expression. As a testing ground, the painter chose the landscape that is at the core of his studies. In 1999 (28 January – 26 April), a one-man exhibition was held at Centre Georges Pompidou in Paris, titled *Espace/Paysage*, and devoted to his Grand Canyon works. See the catalog: Centre Pompidou, David Hockney, *Espace/Paisage*, Paris, 1999. Also in the London Exhibition titled *David Hockney RA: A Bigger Picture* held at the Royal Academy of Arts (21 January – 9 April 2012) landscape is obviously his major source of inspiration (AA.VV., *David Hockney: A Bigger Picture*, Abrams, 2012).

FIGURE 3.1
1998, David Hockney, *A Bigger Grand Canyon*, oil on 96 canvas paintings (207 × 744.2 cm), David Hockney collection. (See a color reproduction at the address: booksite.elsevier.com/9780128035092.)

view of that place, relying on mechanisms through which the eye perceives and processes images[16] (Fig. 3.1).

He experiments with a painting technique that goes back to – and yet at the same time inevitably departs from – that of photography. He breaks down his painting into many pieces, each of which has an autonomous point of observation. When all these are gathered and joined in an image, they reproduce exactly the view processed by the human eye[17] (Fig. 3.2). His goal is twofold: 1) to affirm the need to find technical systems able to return the dynamics of the eye's functioning and, aware of the fact that what the eye processes cannot be seen only as a perceptual event, 2) to advocate a reappraisal of "visual thinking," which is the conceptual outcome of landscape.[18]

Hockney's underlying assumption is that, by relying on the optical experience of distance and using unreal modes of expression, it becomes possible

---

[16] The artist starts from the assumption that when one adopts a geometric perspective – i.e., one assumes perspective as the single vanishing point – one irreparably deviates from the image produced by the processing of the eye. The eye is not fixed but moving and can therefore superimpose within a very limited range multiple perspectives with multiple vanishing points. He also notes how photography itself, failing to provide for the eye's mobility, returns a highly conjectural and abstract image, which inevitably deviates from perception. For this reason, he believes photography is bound to succumb to interactive systems that have irrevocably demolished its presumption of truthfulness.
D. Hockney, *Ma Façon de voir*, Thames & Hudson, London, 1998.
[17] Hockney starts by superimposing partial photographs, and artificially recreates the rapid and subtle eye movements which translate into an overlapping of images. Recently, however, he has used an i-Pad to build a painting via first-hand observation, thus avoiding the intermediation of photographs.
[18] It should be noted that Hockney does not take the easy way out of interpretative freedom for the artist, who departs from reality reproducing abstract forms. Rather, he relies on distance-based perceptual data, rendering reality according to the principles of vision and intervenes on colors and shapes, enhancing them in surrealist terms.

FIGURE 3.2
1986, David Hockney, *Grand Canyon with Ledge, Arizona.* Photographic collage made in October 1982 (113 × 322.5 cm), David Hockney collection. (See a color reproduction at the address: booksite.elsevier.com/9780128035092.)

to represent the intellectual processing of landscape, which is obviously the outcome of perception, but also of the knowledge and the value that society ascribes to it.[19] The final effect is stunning: his paintings allow the outside observer (the recipient of the picture, in this case) to recognize the experience of place. At the same time, they suggest interpretative paths that go well beyond direct perception. The artist shows how, in visual processing, the eye is the ultimate locus of creation, which develops the concept of landscape. That way he underline the precarious position that the communication medium (a map, a picture, a video or other) occupies if taken merely in an object-based dimension. For it is in processing that meanings take on unprecedented directions. He qualifies the medium as a tangible expression of the processing performed by the interpreter-author, who then hands it over to a recipient for new interpretations. In this instance, therefore, the medium is the expression of a visual self-reference able to transform any message. Rephrasing Turri, we could say that the medium stages landscape as theater, where humans dramatize representation and make it possible through the ways in which they create it and express it.

---

[19] In fact, Hockney is convinced that the only painting that can render reality without artificial conditioning is Egyptian painting. Egyptian art shows how the achievement of archetypal forms takes place by means of an abstraction from perspective and through a montage technique: the various parts of a figure in an overall design, as well as the assembly of the various parts of a figure and of single episodes in a more complex scene. Artistic creation is the result of a naturalistic syntax based on a nondissociative view. This model considers the focal aspect not of near vision but of linear vision, typical of distance viewing. It is in that line that the research of the British artist, a naturalized American, attempts to find unprecedented modes of expression.

Indeed, Hockney notes the centrality of vision semiotics and assumes the problematic nature of perspective as an irrefutable starting point. The most significant outcome of his painting is the exclusion of the volume of objects, in the urgent attempt to redeem landscape from perspective. The artist is aware of the importance of volume in the aesthetics of landscape painting, attested in art history ever since the Renaissance. Yet he believes that perspective has irretrievably affected the expressive possibilities of what is meant by landscape today.

Hockney's studies would not be as momentous if his ultimate goal were not to reformulate a "way" of rendering the concept of landscape which, as I noted, developed in Renaissance Italy.[20] When Leonardo da Vinci sought rules for rendering landscape pictorially, his reflections addressed precisely the rendering of volume. He was not interested in geometric perspective, able to express depth according to abstract norms that deviate from what humans perceive. Rather, he strove to codify rules that could render the volume of objects pictorially as it is processed by the mind.[21] Geometric perspective, based on a single vanishing point and theorized by Leon Battista Alberti, could not possibly satisfy Leonardo, because it departed from his idea of nature as a human construct, whose existence and consistency depend on the presence of humans able to interpret it.[22] Alberti sees space as a shell in which people and architectures are placed. Leonardo, by contrast, strives to render landscape pictorially as a concept in which coexisting "values of nature" and "values of culture" reflect the human soul, its experience of the world, and its conception of landscape. While Alberti does not exclude the observer's psychological point of view in painting, Leonardo delves deeper, into the driving forces of the creative process itself. By recovering the volume of objects and not attempting a faithful rendition of what is seen, his studies celebrate the harmony of the universe itself in its interpenetration of nature and culture.[23] Leonardo's view is shared by

---

[20] The importance of Flemish painting in creating the "landscape" genre is unquestionable. However, theories on the possibility of representing landscape originated around the Mediterranean. E.H. Gombrich, *Norm and form…, op. cit.*, p. 110; Id., *The heritage of Apelles. Studies in the Art of the Renaissance,* Phaidon Press, London, 1976.

[21] In his treatise on painting (Codice vaticano urbinate, 1270) – probably edited by one of his pupils in 1651 – Leonardo addressed the issue of perspective via nongeometric expedients. He states that there are three types of perspective: the first has to do with the causes of the shrinking of objects as they move farther from the eye; the second refers to the way in which colors change as they move away from the eye; the third consists in defining the decreasing level of details needed to refine objects as they get farther from the eye. (L. Da Vinci, *Treatise on Painting,* trans. J.F. Rigaud, Dover Publications, New York, 2005).

[22] On Leonardo's interpretation of nature and landscape but also, more generally, on the wider Renaissance perception of that conceptual pair in the European context, see: N. Broc, *La géographie de la renaissance (1420–1620)*, Bibliothèque nationale, Paris, 1980, esp. pp. 211–214.

[23] E.H. Gombrich, *Norm and form…, op. cit*, pp. 111–112. It should be remembered that Leonardo da Vinci himself engaged in cartographic works, without however recognizing these as the locus devoted to the rendering of landscape. On Leonardo's cartographic production, see the contributions on which later bibliography still relies: M. Baratta, *I disegni geografici: Leonardo da Vinci,* La Libreria dello Stato, Rome, 1941; R. Almagià, "Leonardo da Vinci, geografo e cartografo," in: *Scritti geografici 1905–1957,* Rome, 1961, pp. 603–611.

the first systematic treatise on landscape painting which – I must stress – was written by a cartographer, namely by Cristoforo Sorte.

Sorte's education, shaped by great painters, together with his own experience in the visual rendering of landscape, would eventually convince him that landscape must be rendered through volume. But what is truly remarkable is that he reached that certainty after spending most of his life on cartographic works. I think that entitles us to conclude that cartographic rendering of the visual form of territory is what ultimately produces critical reflection on the concept of landscape and on the means used to convey its perceptual dynamics.

Recovering this awareness in cartography is all the more relevant if the issue is related to the changes cartography would undergo over the following centuries. Cartography would depart from the concept of landscape as an experience of place and take on the connotation, borrowed from Cartesian logic, of an expressive medium instrumental to the rendering of territory. This process was not clearly seen at its inception. Originally, it seemed rather an attempt to superimpose two worlds. From the 17th century onwards, however, such attempt was to gain overbearing momentum in the field of cartography. Descartes had claimed and established that perspective "is not a tool that provides us with exact representations: it is a lie."[24] Truth was to be reached, therefore, by freeing oneself from perceptual stimuli and surrendering to rational speculation. What Descartes, but especially his followers, failed to understand was that, by disregarding substance and being (that is, what is real) and founding a whole new paradigm on the inventions of the "cogito," they shifted the issue of truth to a matter of rational constructs and ran the risk of severing reality from reason, to the sole advantage of the latter.[25] Moreover, as we already noted, Descartes' rationalism, faced with the rhetoric of images, sparked off a whole, seemingly uncontrollable current of issues and stimuli in the field of the figurative arts that would develop in the following centuries and invest cartography in full. The principle of rationality, which extends to any feature of the map via the principle of measurement, has no other purpose than to render the concept of nature as an abstract entity, whose measurement ensures accuracy and thus truth. That paves the way to the annihilation of the value of territory as the outcome and the condition of human action.

The very communicative outcomes of the change in the point of observation support this conclusion. We looked at this issue in the previous chapter with regard to different cartographic encoding, which brings about

---

[24] Quoted by: J. Baltrušaitis, *Anamorphic art*, Chadwyck-Healey, London, 1976, (ed. 1978) p. 68.
[25] These reflections were taken up in the important work on simulation phenomena by: G. Bettetini, *La simulazione visiva…*, *op. cit.*, pp. 60–63.

a loss in the value of the symbol to the benefit of the referential value ascribed to the sign. Now, with regard to perspective, it should be noted that cartography shows the world as seen from above, from a perpendicular vantage point capable of rendering territory in a uniform fashion. The underlying assumption is that the exact localization and size of objects would ensure the accuracy and objectivity of representation. We saw how zenithal projection,* which is the most abstract way of rendering the world, was achieved within the Euclidean paradigm, which would ultimately spell the exclusion of humans from what is represented. Zenithal projection discards the human gaze and hence the hierarchization that perspective preserves, producing a loss of its iconic meaning to the advantage of its descriptive sense.

If zenithal projection expects each feature of territory to be rendered from a point of observation perpendicular to it, it follows that such projection implies not only one point of observation, but as many observation points as there are objects being represented. In fact, strictly speaking, the representation of the earth, seen from an airplane or even satellite representation itself, does not actually adopt zenithal projection. That is because it prescribes a single point of observation which, even if located off the earth and extremely far from it, is linked to perspective and, therefore, presupposes the presence of an observer. This type of transformation, however, obtained by translating photographic image into cartographic product, ends up excluding the unique point of observation – which does in fact exist at the beginning of the survey – to the benefit of the zenithal one or, in other words, of the presence of multiple points of observation. Perspective was undoubtedly a loose, imperfect means of rendering territory. Yet it still preserved human presence in the way the experience of landscape conveyed it. Conversely, zenithal projection excluded in the first place hierarchical data of the near and far; secondly, altimetry and, therefore, the height of the objects; thirdly, volume itself, annihilated in the flattening of the point of observation perpendicular to what is being represented. In short, with the exclusion of the observer, territory is rendered as something alien to humans (Fig. 3.3).

The point of observation is thus the core I will focus on in order to distinguish, within the maps I consider, landscape figuration from truly cartographic figuration. The former uses perspective to render not only the depth but also the very shape of objects, and as such it appeals directly to landscape as a visual form of territory. The latter refers to uniform description, enhances their localization and consequently also their referential value. Therefore, we can identify maps based primarily on a landscape+perspective view and maps that are centered on a vertical+cartography view. Here, however, I would emphasize the fact that since they both serve one purpose, which is to act as a reliable figuration of reality, they inevitably fall within the

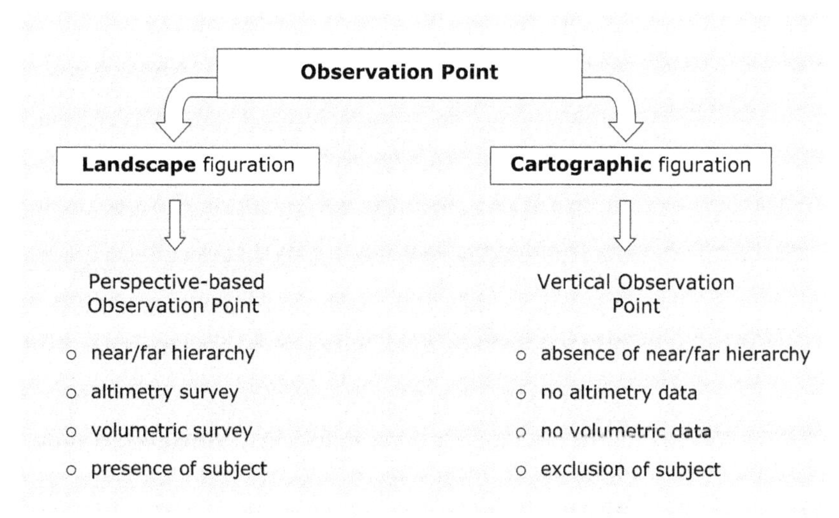

FIGURE 3.3
Point of observation and its communicative outcomes.

project of iconizing the world. There is no difference between perspectival map and vertical map, or between ancient maps and modern maps, in this respect: each tries to present itself as a reliable representation of the world. But it is precisely by adjusting the efficacy of the ways through which this objective is achieved that it will become possible to reflect on the outcome of the iconic junction.

We saw that his mode of representation does not follow a rapid or progressive development. It evolves slowly and in relation to what pictorial studies produce in the representation of landscape. That makes it difficult to adopt zenithal projection which, in the end, will nonetheless prevail. The two visions coexist, are integrated, and work within the same iconizing project over a long period of time. One can then hope to grasp the two different figurations by drawing on cartographic production over a large time span, in which the progression to verticality is not perceived as a persistent and seamless phenomenon but is seen to undergo advances and retreats.

ICONIC RESONANCES

In the historical period between the 16th and 19th centuries, documents were drafted in which the mixture of landscape figuration and cartographic figuration manifested itself in many ways and with varying degrees of explicitness. It shows itself as joint within the same drawing (*conjunction*); in separate documents united by the same subject matter (*disjunction*); or in disjointed figures within the same document (*disjoint conjunction*).[26] We can trace the different roles that such a mixture plays in communication by focusing on the map's essential functions: *describing* or *conceptualizing*. Let us now examine a number of documents to verify the presence of an *iconic junction*, and of its communicative results, by using the categories of *conjunction* and *disjunction*.

We are going to start our inquiry with some maps by Cristoforo Sorte, the leading exponent of a reflection on landscape in cartography, without neglecting to mention his extraordinary contribution. Born in Verona, Italy, around 1510, Cristoforo Sorte lived nearly 90 years, spent mostly in the service of the Republic of Venice with public office at various *Magistrature* and *Uffici*, carrying out the *engineering* profession, or rather what at the time was called the job of cadaster expert, surveyor-cartographer and territory analyst.[27] The production he handed down to us is cartographic, although art and especially painting were a fundamental trait of his training.[28] In his youth he worked in Mantua with Giulio Romano and, much later, in 1578, he was called upon to work on the restoration of the Palazzo Ducale in Venice, after the building was damaged in a fire.[29]

---

[26] *Iconic junction* may be understood as the link between landscape figuration and cartographic figuration. This link may be explicit, implicit, or hybrid, respectively in *conjunction, disjunction*, or *disjoint conjunction*. The link is explicit in *conjunction* when, within the same representation, both the perspectival and the vertical view are present; the link is instead implicit in *disjunction* when the same territory is represented in many distinct maps that use separate figurative systems using different intertextual systems (perspectival and vertical). Finally, the link is hybrid in *disjoint conjunction* when, within the same document, territory is shown through the combination of multiple representations based on different figurative systems.

[27] Cristoforo Sorte, son of Giovanni Antonio, engineer by profession, soon followed his father's business by engaging in the solution of hydraulics problems and gaining remarkable competence in land survey. Summoned to become a member of the Magistrato dei Beni Inculti (Authority for Uncultivated Resources) in Venice, in the capacity of temporary advisor, he used his expertise in the service of the Serenissima for decades, planning a very large number of hydraulic works and territory interventions, which were recorded in surveys, projects, plans, drawings made in connection with various assignments he received from the Senate. These make up a corpus of records held today at the Venice State Archives. See: R. Almagià, *Monumenta Cartographica Italiae*, Florence, 1929, pp. 37–39; Id., "Cristoforo Sorte e i primi rilievi topografici della Venezia Tridentina," in: *Rivista Geografica Italiana*, XXXVII, 1930, pp. 117–122; Id., *Le carte dei territori veneziano, padovano e trevigiano e del Friuli di Cristoforo Sorte*, Istituto di Geografia, Università di Roma, sd (1954), s. B, n. 3.

[28] A wide range of experiences contributed to the shaping of his multifaceted personality: apprenticeship in the workshop of the painter Giulio Romano in Mantua; practice in the survey of land, especially alpine land, acquired with his father in the period when the latter was in the service of Bernard of Cles Cardinal of Trent; and, in the latter city, contacts with the northern European tradition of landscape-cartography.

[29] Having completed the restoration of the Doge's Palace, Sorte was commissioned to compile the great map of the Domini Veneti di Terraferma (Venice Mainland Dominions). The map was to hang publicly in the Sala dei Pregadi inside the Palace, but for safety reasons it was later decided that the size of the map should be reduced and the map kept secret. It was agreed, however, that it should be accompanied by five detailed maps, reproducing the five parts of the Dominion. The five maps, completed in 1595, were kept in the Doge's Palace, inside a purpose-built cabinet, to the end of the 18th century. With the fall of the Republic they were removed and preserved in separate collections.

However, his major commitment was, by his own admission, cartography, which he undertook both in the form of theoretical debates and in the production of maps.[30] Retracing the phases of his work lies beyond our present scope. Suffice it to say that he is numbered among the greatest cartographers of the 16th century, and among those who excelled for their unique rendering of landscape.[31] It is the use of landscape in cartographic representation, as achieved by Sorte, which offers one the most striking examples of *iconic* junction.

Studies conducted so far on the work of this cartographer address issues which, for different reasons, deserve some mention. To begin with, it should be remembered that Sorte's production drew the attention of both geographers and art historians. Discovered and analyzed in the context of geography during the first half of the 20th century, his work was revalued by art historians in the second half of the last century. We owe initial research on his cartographic production to Roberto Almagià, who has extolled Sorte's role since 1929 and gave a comprehensive appraisal of his work in the 1950s, acknowledging its insightful elements.[32] It was, however, art historian Ernst Hans Gombrich who in 1966 claimed that Sorte's *Osservazioni nella pittura* marked the very first treatment of landscape painting.[33] In the same field of interest, Juergen Schulz, also an art historian, compiled a thorough catalog of Sorte's artistic and cartographic output, accompanied by an in-depth study.[34]

It should be noted, however, that the relevance of Sorte's work, acknowledged in both disciplines, is alternately traced either to his painting skills or to his cartographic abilities. Critics seem unable to view the mixture of the two fields as a new aspect, which called for an interpretation based on criteria other than the ones that circulated separately in either one of the two disciplines. Of course, his cartographic technique was bound to arouse interest among geographers, since some of its features recalled systems used in trigonometric survey,

---

[30] On Sorte's output as a writer of tracts see: *Per la magnifica città di Verona, sopra il tratto ultimo del Magnifico Signor Theodoro da Monte,* Venezia, G.A. Rampazzetto, 1574; *Osservazioni nella pittura di M. Christoforo Sorte al Magnif. Et Eccell. Dottore e Cavaliere il Sig. Bartolomeo Vitali,* Girolamo Zenaro, Venezia, 1580; *Modo d'irrigare la campagna di Verona e d'introdur più navigationi per lo corpo del felicissimo Stato di Venezia trovato fino del 1565 da M. Christoforo Sorte primo Perito ordinario dell'Officio de Beni Inculti, con molte altre cose sue in proposito di acque molto giovevoli, et anco di M. Antonio Magro e del Sign. Theodoro da Monte,* Verona, Stamperia Girolamo Discepolo, 1593 (including, in fact, writings dated between 1594 and 1595).

[31] On his cartographic output, in addition to the titles already mentioned, see: L. Pagani, "Cristoforo Sorte, un cartografo veneto del Cinquecento e i suoi inediti topografici del Territorio bergamasco," in: *Atti dell'Ateneo di Scienze Lettere e Arti,* XLI, 1980, pp. 401–425.

[32] See Almagià's contributions, mentioned above.

[33] E.H. Gombrich, *Norm and Form..., op. cit.,* p. 108.

[34] J. Schulz, *La cartografia tra scienza e arte. Carte e cartografi nel Rinascimento italiano,* Panini, Modena, 1990, especially pp. 65–95. The book includes a comprehensive bibliography on Sorte.

which would come into use only two centuries later.[35] Similarly, his pictorial stroke could not escape the attention of art historians, who recognized in Sorte a landscape painter, although in fact his production was exclusively cartographical.[36] In substance, apart from sporadically insightful remarks, few observations were made on Sorte's rendering of landscape, and the importance of his treatises on cartography went largely unnoticed.

No one so far seems to have wondered why a man trained in the arts should find in mapping his privileged expressive medium and write a treatise on the rendering of landscape only after consolidating his cartographic technique. Relying on the notion of iconic junction, I will here endeavor to offer a few interpretative suggestions on this issue as well.

In his *Osservazioni nella pittura* Sorte clearly expounded his reasons for writing a treatise on landscape figuration.[37] Having warned readers that the subject is new and his observations unmethodical, he claims that he feels the need to engage in this reflection for the relevance it has in cartography. So, even though he is not officially a painter and relies almost exclusively on his experience as a cartographer, he embarks on this venture through "those terms and words which are most familiar in Painting, and thus most easily understood by all" ("quelle voci e qu'vocaboli che sono più famigliari nella Pittura, acciochè più facilmente siano anco da tutti intesi"). His first warning concerns the title of his work, which should not mislead and be mistaken as referring exclusively to pictorial art. In fact, he implies that his treatise is aimed at those who wish to try their hand at mapping. In answer to Bartolomeo Vitali, dedicatee and author of the book's preface, Sorte claimed that his conviction came from the fact that "in studies of Chorography I placed all my thoughts and exerted myself in those to no end" ("nelli studi della Corografia posi ogni mio pensiero, e mi sono in quelli di continuo affaticato").[38]

To confirm his geo-cartographical competence, Sorte devotes the first part of his work to clarifying some of his statements on the origin of rivers.[39] Then

---

[35] Sorte's mapping technique, a forerunner to exact, Euclidean cartography, is analyzed in: L. Pagani, "La tecnica cartografica di Cristoforo Sorte," in: *Geografia*, II, 2, 1979, pp. 83–89.

[36] No evidence remains of Sorte's artistic output. His involvement in the restoration of the Doge's Palace covered the design of friezes and slotted mouldings. See: J. Schulz, *La cartografia tra scienza e arte, op. cit.*, p. 66.

[37] *Osservazioni nella pittura di M. Christoforo Sorte al Magnif Et Eccell. Dottore e Cavaliere il Sig. Bartolomeo Vitali*, Girolamo Zenaro, Venezia, 1580.

[38] *Ibidem*, p. 5. All quotes are from the reprint of the treatise edited by Girolamo Corsi da Feltre in 1696, which also bears witness to the enduring success of Sorte's work. A ringing endorsement of Sorte's technical skills comes from Vitali's own preface to the work, where he claims "da un colore, à guisa d'una sottilissima rugiadetta semplicemente disteso si possa lo sterile e il fertile Paese nel vostro disegno conoscere" ("barren and fertile Land may be known through your drawing from color simply laid out and made in the likeness of fine-spun dew").

[39] These were expressed in: *Trattato dell'origine de' fiumi, di me CHRISTOFORO SORTE, All'ill.mo et ecc.mo Signor, il signor Sforza Pallavicino, cap.o Generale dell'illustrissima Signoria di Venezia* (autograph transcript of papers dated between 1561 and 1585), preserved at the Biblioteca Marciana (Cod. cart. Ital. IV, 169-5265).

he goes to the heart of the matter, and addresses the rendering of landscape by involving both color and perspective. As regards the former, he focuses on the watercolor technique,[40] noting its importance for perspectival rendering and arguing that the reproduction of distant objects may be obtained through stumping and the use of a particular color nuance – "azzurrino" (sky blue) – which can render objects loosely and can naturalize them, by surrounding them with haze. Close objects, by contrast, ought to be defined with vigorous and solid colors. For rendering depth, he suggests, these chromatic expedients should be combined with a proper system of perspective, achieved by applying some geometric rules to which he adds the foreshortened view and the one obtained with mirrors.[41] He then dwells at length on technical descriptions, explaining how such perspectives are to be obtained. A link to cartographic rendering is made explicit in Sorte's directions on how to handle map orientation, which challenge the lack of clear astronomical motives.[42] It is, however, the statements he makes on the use of multiple points of observation and, therefore, multiple perspectives within the same document that sound most striking. In this regard, Sorte claims that he draws lowland plains at "plan view" to respect their measurements and distances, while he uses perspective for the rest: "cities, castles and villas with mountains and hills I placed in the map as standing [...] which thing I considered necessary to make sure that the Sites are known" ("cittadi, castella e ville con le montagne e colline ho poste in mappa e in piedi [...] il che ho stimato necessario per far si che si conoscano i Siti").[43] There is no doubt, therefore, that perspective diversification here is not accidental. It shows that Sorte made a conscious choice related to his idea of landscape: not a visual datum but a concept.[44]

Without doubt, Sorte can rightfully be taken as the leading exponent of landscape rendering, not only because he practiced it, but also because he adopted it knowingly: his awareness is sanctioned by cartography.

---

[40] Based on experience in mapping he had acquired when commissioned "by Emperor Ferdinand (Descrizione del Contado di Tirolo e d'altri suoi paesi – Description of the district of Tirol and other possessions of his)" and by the "Repubblica di Venezia nella descrizione di molti luoghi del suo Stato." ("The Republic of Venice for a description of many State sites.") C. Sorte, *Osservazioni nella pittura..., op. cit.*

[41] In this regard, he recalls the teachings he received from Giulio Romano. *Ibidem*, pp. 15–16.

[42] Since orientation, with north at the top, is derived from the flow of rivers from the mountains to the sea and, therefore, in the case of the region of Veneto, from north to south.

[43] *Ibidem*, p. 7.

[44] Commenting on Giotto's landscapes, Sorte maintains that they achieve figural effectiveness when what they paint takes on verisimilitude to reality, not so much because objects are made similar to what nature shows, but by virtue of the meaning that the artist manages to endow them with. *Ibidem*, p. 9.

Iconic Junction in Cristoforo Sorte

To substantiate my claim, I am now going to analyze Sorte's work directly, by addressing two documents built with different objectives in order to reproduce the same area, namely the Treviso area. The first map, which belongs to the set of maps describing the Dominion, designed for display in the Senate hall of the Doge's Palace, is a 1594 document, written for descriptive and political purposes. The second is a 1556 map entitled *Dissegno da adaquar il Trivisan*, built for administrative purposes and linked to a reclamation project to be implemented in the high plains of Treviso by the *Provveditori ai Beni Inculti* (Superintendents for Uncultivated Resources).[45] We will compare the two in order to assess whether their different goals affect the technique with which they were built, or if, conversely, the documents were drawn using the same criteria, that is the ones theoretically supported by Sorte.[46] His particular rendering of landscape relies on the use of perspective: he varies projection within the same document according to the type of territory being represented, using plan view image for flat terrain and bird's-eye-view perspective mountainous areas.[47] Such rendering, however, present in the five maps that were built to be preserved in the Doge's Palace, is absent in other documents compiled for other purposes. So my comparison of the first map, belonging to the first group, to the second, produced for administrative purposes, is meant to suggest the possible meaning of different modes Sorte used to achieve iconization.

The large map reproducing a wide area of the Veneto region allows us to consider the rendering of heterogeneous territory: lowlands and mountains[48] (Fig. 3.4). In this map we shall focus on the representation of the high plains of Treviso,

---

[45] An administrative body responsible for the management of land reclamation and the concessions of water resources to private individuals.

[46] I will not account for the chronological gap between these two documents (the first in 1556, the second in 1594), because it is irrelevant to the purpose in view. I do not in fact set out to assess Sorte's unique artistic style, which could have evolved over time, but rather, to determine when the encounter between landscape and map occurs with maximum intensity.

[47] He complements this with a measurement of the lowlands and calculates the approximate distance between the highest points in the mountains and a number of fixed points of observation he established himself. By doing that he obtains elevation profiles which, once superimposed on each other, yield a volumetric view of mountain landscape.

[48] A large-format drawing (measuring 295 × 160 cm) in ink, watercolor and gouache on multiple sheets of map pasted on canvas. It bears a caption in its bottom right corner: "Io christoforo Sorte ho fatto il presente dissegno, il qual sono uno dillj cinque pezi di tutto il Stado di Terra ferma dilla Ser.ma Sig.a de Venezia, il qual dissegno sono il Padoano, Trevisano, lagune. et parte del Polesene, il qual si puol vedere le distantie da luoco a luoco col compasso sopra la presente scala, et fatto fidelmente quanto ho saputo; levato col bossolo di Venetia il di 10 luglio 1594." (I Christoforo Sorte made the present drawing, which is one of the five parts of all the State of Terra Ferma of the Serenissima Republic of Venice, which drawing includes the Padua area, the Treviso area, the lagoons and a portion of the Polesene. The distances from place to place may be seen by compass at the given scale on this drawing, which I made as faithfully as I could and surveyed with a compass. Venice July 10 1594). Scale 1 mile = 7 mm (Vienna, Osterreichisches Staatsarchiv, Kriegsarchiv, n. B-VII-a-154). Here I shall consider only the frame reproducing the Treviso area.

FIGURE 3.4
1594, Cristoforo Sorte, *Territorio del Padovano, Trevigiano, Lagune e parti del Polesine* (detail).

described by vertical projection, thanks to which the course of the rivers and the distribution of settlements can be easily referenced. No feature is seen in perspective and the cities themselves –Treviso and Venice – are depicted with their urban layout. Treviso is shown as enclosed by walls, and Venice as surrounded by water. Watercourses leave no doubt as to the verticality of the viewpoint: small settlements are marked only by a conventional symbol (a church) while larger ones have slightly more realistic features. Designators referring to the features depicted are many, which shows the map's descriptive intent. By contrast, rendering of mountainous areas is based on perspective, and much space is given to the volume of mountains and the convex shape of valleys. Volumetric rendering in this case leaves no doubt as to the kind of landscape-based figuration: it is approximate and, despite its overall effect, imaginary.

The latter part of the drawing has a clear iconizing aim, which refers to the inviolability of the mountain and to the fact that its territory is uneven and somewhat inaccessible because, as others noted, the landscape is rendered through "rapid flickering signs [that] evoke the vague outlines of the wooded hills, the spikes of rock, the bushes and secluded trees reminiscent of Flemish painting, which Sorte seems to be very familiar with."[49] In fact, what matters here is less artistic

[49] J. Schulz, *La cartografia tra scienza e arte, op. cit.*, p. 84. Some critics see a Flemish influence in Sorte, although it is not corroborated by sources. Schulz, for instance, argues that Sorte's stay in Trent would necessarily have put him in touch with transalpine painters, just as his apprenticeship at Giulio Romano's Mantua workshop would have made him familiar with Flemish artists who worked there.

rendering than the split the document creates between landscape and map. Lowland terrain is expressed via cartographic features (presence of designators, vertical view and, therefore, accurate referential rendition), while mountains are drawn as landscape (perspective view, use of color and volume to render the picturesque, the inaccessible). Thus, a descriptive aim clearly guides the delineation of lowlands, while a markedly conceptualizing intent emerges in the representation of mountains. This map confirms the dual rendering technique put forth by Sorte. At the same time, it reasserts that iconic communication differs with respect to areas. Landscape figuration is convincing but does not show the features of the mountain region: it conveys information at a rhetorical level, typical of images. Conversely, vertical figuration embraces analogical description as indisputable evidence. In this case, even though the document's intent is political and descriptive, landscape is inserted to show rhetorically the iconic meaning of mountains. Nor should this choice be taken as an end in itself. Rather, it should be compared and contrasted with the rendition of lowland areas, conveyed vertically. For, as Sorte claims, by respecting relations and distances the latter highlights the demonstrative value of the description.

Iconic junction is indeed quite distinctive in this document but its basic aim is to take charge of the overall meaning of representation, in order to convey some of its precise and particular features, such as the difference between the two territorial typologies and their hierarchical relationship. The lowlands had higher social impact than the mountains for several reasons: proximity to the lagoon, soil fertility, river and road practicability, not least, administrative dependence on the Serenissima, whose land apparatus did not have to guard its supremacy via unwieldy local bodies such as Mountain Communities.

This first document takes on considerable more relevance when compared to the second, built with administrative purposes and based on the exclusive use of landscape figuration[50] (Fig. 3.5).

In this case, we are dealing with an iconic junction that pervades the whole representation. Here, landscape and cartographic figuration are not juxtaposed but interweave across the whole drawing, giving rise to a junction in which landscape figuration trumps cartographic figuration. The map represents the high plains of Treviso with the lagoon to the south and the Venetian Pre-alps

[50] Watercolor on canvas-reinforced paper, measuring 105.5 × 183.5 cm. No scale given (Archivio di Stato di Venezia, Savi ed esecutori alle acque, Diversi, 5). The document bears no indication of the author. Yet, documentary evidence, in the form of a *commissione* (a letter of appointment) from the Venetian senate to Giovanni Donà, Provveditore sopra i luoghi inculti, (Superintendent of Uncultivated Sites) mentions *maestro Christoforo da Verona,* who was to accompany Donà and "poner in disegno i luoghi et siti visitati." ("make drawings of the places and sites visited.") See: F. Cavazzana Romanelli, E. Casti Moreschi, eds., *Laguna, lidi e fiumi. Esempi di cartografia storica commentata,* Archivio di Stato, Venice, s.d. (1984), pp. 37–44.

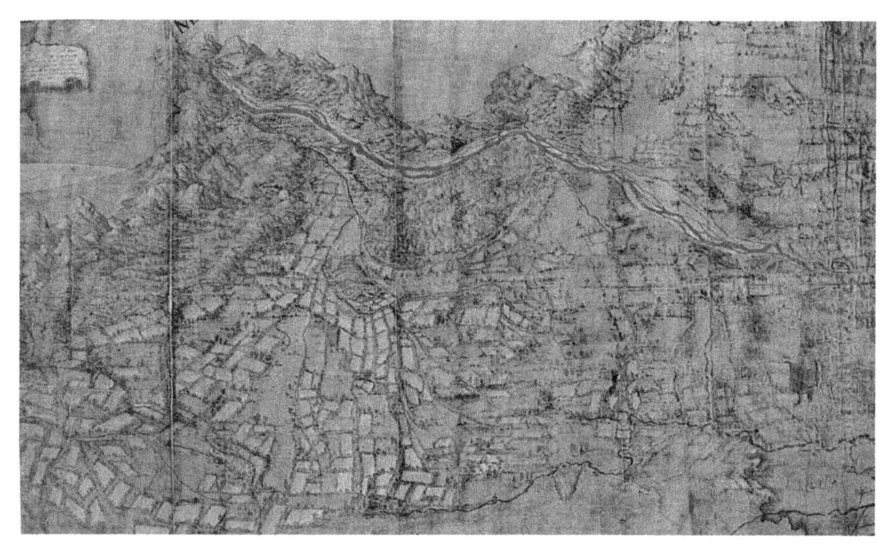

FIGURE 3.5
1556, Cristoforo Sorte, *Disegno da adaquar il Trivisan*. (See a color reproduction at the address: booksite.elsevier.com/9780128035092.)

to the north. It is aimed to illustrate a scheme of territorial intervention based on the use of water for agriculture. Following the drainage of marshlands bordering the lagoon in the middle of the 16th century, the Serenissima started to consider schemes for boosting agricultural output through the creation of an efficient irrigation network. In 1556, to address this need for an expansion of tillable land (and thereby agricultural output), the Venetian Senate established three permanent *Provveditori sopra Beni Inculti* (Superintendents of Uncultivated Resources), members of the body who had commissioned the map. The document shows that Sorte was involved in mainland administrative operations since the early origin of the *Magistrature*. It also gives us an example of how, in the 16th century, cartographers who worked in Venice endeavored to play more roles and to expand their range of representations: cartographers possessed a set of skills that enabled them to draw diverse pictures of territory.

From a technical perspective, the map uses the image of a large bird flying over territory to envisage a point of observation high up, above the southwest area, which corresponds to the lower-left corner of the map. The map itself is meant to push through an irrigation scheme via the competent body. Its underlying claim is that by boosting existing irrigation canals – the *Brentella di Pederobba* and the *Piavesella di Nervesa* – and granting adequate concessions for water access, the naturally arid area of the Treviso upland could be made fertile. Sorte's depiction of the clearly imaginary large fields covering most of the area, surrounded by tall rows of trees, alludes to the land's fertility and is intended to convince readers that the scheme was sound. Similarly, the map represents outposts along the canals with a view to highlighting the importance of waterways, the improvement of which would ensure widespread settlements. Ultimately, the aim was to show not only that the scheme was sound but that it was unquestionably effective. The map does not want to describe but to conceptualize territory. Hence it uses landscape figuration extensively, although the modes of cartographic representation are present as well. More precisely, the map includes orientation markers: an arrow in the upper margin of the sheet points to the north and the initials of winds mark the other points of the compass. Similarly, with regard to denomination, the map presents a few designators in the form of fixed points of reference: Treviso, Montello, Piave, Sile and other settlements along the river. The use of two visual features of landscape, that is shapes and colors, entrusts them with the role of conveying empirical data for knowing territory. The incline of the terrain and the fertile ground, set apart from arid land, help readers to identify the uplands and the beginning of the low plains. The former uses an ocher color, which indicates its aridity and its altitude. By contrast, the lower plain, rich in standing water for lack of an adequate incline to ensure drainage, is characterized by uncultivated land or, anyway, by moist soil (blue-green). That shows that the entire economy of the region was based mainly on agriculture which, affected by the particular nature of the soil, could benefit from extensive irrigation. The map may then be said to provide less a description than an ideal figuration of the territory, which could be achieved if the scheme was accepted. It should be noted that this rendition of territory via its morphological features achieves two simultaneous goals: it suggests that it is possible to govern nature and it shows that agricultural productivity could be boosted by means of an irrigation scheme, which would also remove the need for land reclamation interventions around the lagoon.[51] Albeit partially and imperfectly, the lagoon is indeed featured in the map as an extension to the area, with a view to underlining the final objective of the proposed scheme. The drawing seems to argue that it is water management, and not just the presence of water or the nature of the soil, which makes an area suitable to the establishment of profitable human activities. Once again, therefore, we are in the presence of a strong iconizing intent: prospecting an irrigation scheme as the means and the way to make a positive contribution to the survival of the lagoon meant joining a public debate which involved the highest authorities of the State. The communicative result pursued here is the transmission of a theoretical kernel, hence an iconization, which finds its

[51] The management of mainland resources was a matter of debate in the Venetian Senate for the repercussions it could have on the viability of the lagoon. On this issue, see my: "State, Cartography and Territory in Renaissance Veneto and Lombardy," in: D. Woodward, ed., *The History of Cartography, Vol. 3: Cartography in the European Renaissance,* University of Chicago Press, Chicago, 2007, pp. 874–908.

ultimate expression in the landscape figuration that the map implements. Thus enlivened, the iconic junction draws its power from a system of differences that flows through and around social practices and ideological constructs. In the manner described, the map does not present itself as a descriptive tool reliable in its impartiality. On the contrary, the map emerges as a misleading symbolic system that harbors a shady, biased and arbitrary mechanism of representation behind a facade of artlessness and transparency.

It is precisely this conjectural aspect, present in all modern cartography, which leaves plenty of space to iconic junction and enables us to recover the process-like dimension of territory. The map does not return landscape, in its most banal sense, but the territorial dynamics resulting from the relationship between humankind and nature. Thus interpreted, the map reveals not only practical implementations but also intentions, items of knowledge, and projects of territorial intervention. It is irrelevant to find out that the tilled fields this map represents did not exist. What we should focus on is, rather, that such "distorted" message (as it were) can show us how technical workers used maps as a persuasive tool in public debates. It should be clear by now that to highlight and emphasize certain features in a map means to deploy a most effective type of rhetorical strategy. The map appears, then, as the product of a culture which in turn uses iconic junction to generate more culture, traceable at specific times in the transmission of knowledge. First, the map links up to the cognitive heritage of a society, increasing its territorial knowledge: cartographic communication entails compliance with the conventions that the mechanisms of social control have established. Secondly, maps present themselves as autonomous means of communication able to effectively envisage new territorial strategies. It is in this capacity that maps impose their own innovative interpretation of the world, even within the same mechanism of social control that produced them.

Comparison between these two maps by Sorte clearly shows that the use of landscape figuration in maps favors the adoption of a particular idea of the territory, which becomes all the more effective as iconic junction is allowed to spread and bolster persuasive communication.

Territorial Functioning and Iconic Disjunction

Over time, matters seem to become more complex. Instances of iconic junction take up a wide range of forms, at times quite unique, and may even be found in the form of separate documents. In this case their pairing allows us to reflect on what lies at the opposite end of iconic junction, namely disjunction between the two figuration systems. Although these use different points of observation, they still rely on the existence of junction. This is the case of two maps that reproduce a particular area neighboring the Venetian lagoon, a place of bustling economic activity and services. This is the area of Moranzani, present in two maps dating back respectively to the 17th and 18th centuries, both commissioned within the *Magistratura delle Acque* (Magistrature of Water), one based on a perspective view and the other on a vertical view[52] (Figs. 3.6 and 3.7).

The Brenta Canal, which flows into the lagoon near Fusina, at Mestre, is represented here in its complex layout, which was intended to meet the multiple needs of the city. The river Brenta features a system of locks, called Moranzan, which regulate the slope between the river upstream (the Brenta Magra) and downstream (the Brenta Salsa), to enable navigation between Venice and Padua, along the coastline that played a key role for Venetian settlement on the mainland. The lock mechanism adjusted the water level to allow boats to go upstream. A system built around the canal to divert waters bears witness to a wide range of uses. For instance, the Seriola, a fluvial diversion built at the upper end of the canal, ensures waters are kept clean by avoiding towns in favor of the countryside. Constant surveillance and special legislation prevented any illicit use of the canal, not least its use as landfill, since the canal ensured the supply of household water to Venice prior to the building of an aqueduct. The map also shows us that this diversion is autonomous from the ones of the river Brenta and of the Bondante, another canal featured on the map. The Seriola flows over the latter via a bridge canal, to reach a place where, through a system of pipes, drinking water was loaded onto special boat-tanks and transported to Venice. In winter, part of this water was diverted and conveyed into a *bucca per formare il ghiaccio* (a hole for making ice); in the summer ice could remain intact when conveyed into the *ghiacciaia* (icebox), which the perspective-based map represents as a hollow hump in the ground. Another artifact relied on drains to clean wool with water. It was called *il purgo* (the purge) and was located between the Seriola and the Brenta. That way, waters from the former could be used for processing wool and those from the latter for carrying it into the city. There is no doubt that the area overlooking the lagoon was essential to Venice, because it comprised numerous key facilities and marked a convenient cluster of suburban utilities. At the same time it is clear that the complexity of the waterway system, laid out on many levels via mobile locks that prevented the mixing of waters, required constant

[52] The first one is a perspective-based map, dated 1690, drawn by Matteo Alberti, "ingegnere alle fiumare" (river engineer) and archived under the tile "Mappa della Brenta Magra da Oriago e Fusina" (Map of Brenta Magra from Oriago and Fusina). Watercolor on canvas-reinforced paper, measuring 60.9 × 375.5 cm; Paduan perch scale 100 (ASVe, SEA, Brenta, 44). Here we will take into account the detail of the Moranzani area. The second document, from 1769, drawn by Stefano Foin, deputy surveyor engineer for the Venetian lagoon, is the area of the so-called Moranzan locks between the rivers Brenta Magra and Brenta Salsa. Watercolor on canvas-reinforced paper, measuring 54.2 × 77.5 cm; Paduan perch scale 100 (ASVe, SEA, Brenta, 82). For a fuller description see: F. Cavazzana Romanelli, E. Casti Moreschi, eds., *Laguna, lidi e fiumi…, op. cit.*, pp. 71–72.

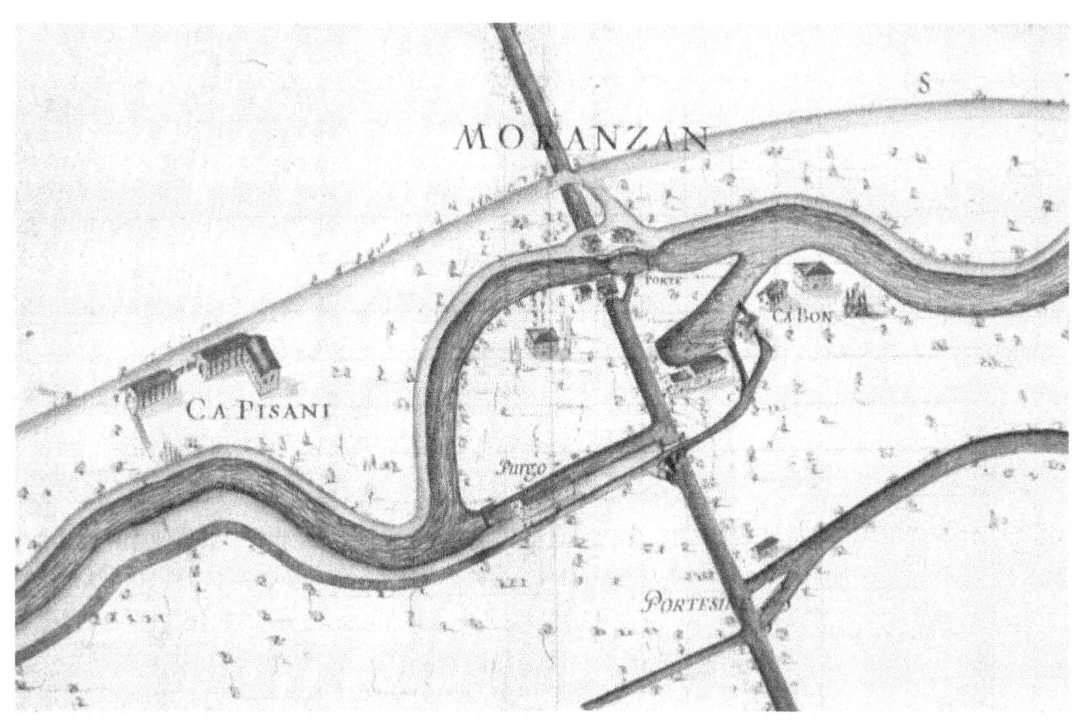

FIGURE 3.6

1690, Matteo Alberti, *Map of Brenta Magra from Oriago to Fusina 126.* (See a color reproduction at the address: booksite.elsevier.com/9780128035092.)

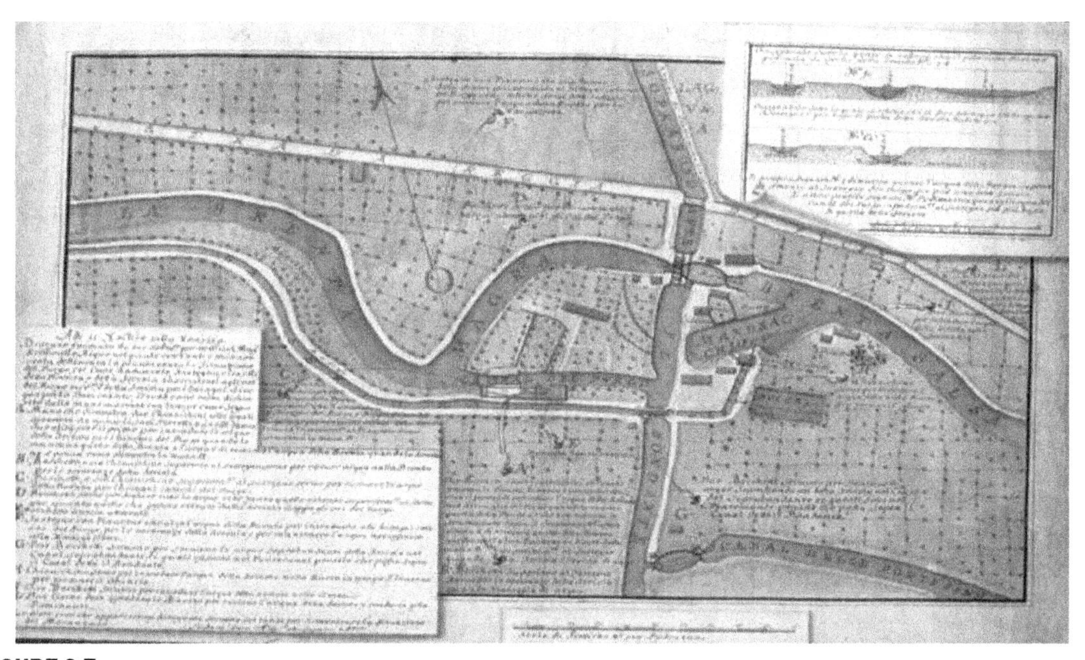

FIGURE 3.7

1769, Stefano Foin, *The Porte Area, otherwise named Moranzan between Brenta Magra and Brenta Salsa.* (See a color reproduction at the address: booksite.elsevier.com/9780128035092.)

supervision on the part of hydraulic engineers. The rapprochement of these two kinds of representation makes it possible to redress the lack of details in both maps and yields a comprehensive image of the area as a whole. It also shows the outcome of iconic disjunction in attaching relevance to the area alternately: both as a hydrographic system and a utilities hub.

The vertical map, intended to render the complexity of the hydrographic network (the canals are at different levels; there are tank bridges, filling pools, locks, etc.), returns quantitative technical data, supplementing them with profiles and measurements, hence abstracting territory from its visual connotations. It comes across as a drawing that represents waterworks and their dynamics of junction and disjunction in abstract terms, removed from the social value the area possessed. Even naming tends to underline water quality, using designators and long captions that refer to water flow (Brenta Magra) or salt composition (Brenta Salsa), which the *Moranzan* gates kept distinct.

By contrast, the perspective-based map aims to give an overview of various activities centered on this area (agriculture, industry, water supply and production and storage of ice), so it draws a visual outline that allows us to understand its functional interdependence. First, agricultural facilities (Ca' Pisani, Ca' Bon) are mentioned by designators, which recall the names of the owners and the layout of large estates and buildings; secondly, the volumetric profile of the *purgo* and of the *ghiacciaia* refer to the visualization of territory. Of course, what is not visible in perspective is not represented. The *bucca per formare il ghiaccio*, clearly outlined in the vertical map is not mentioned here, although its existence is beyond doubt. For even this map pays great attention to the *ghiacciaia*, despite an 80-year lapse between its publication and the publication of the first map.

While both remark on the centrality of this hydrographic hub, the two maps not only provide different information according to their aims but, alternatively, refer to either the territory or the landscape by using different points of observation. In the first case, vertical figuration emphasizes the process-based nature of territory, recalling the relationship which society establishes with natural elements – water, in this case – and qualifying territory as an open system which varies in time on the basis of contingent needs. Vice versa, landscape figuration refers to the visual form of territory and to the impact that reification has as an exclusionary practice ultimately grounded in matter.[53] In short, there is no doubt that the messages conveyed by the maps depend on the means used to convey them and must come to terms with such means in determining

[53] By reification here I mean a practice that can guarantee the self-sufficiency of a given society against environmental constraints, for the satisfaction of material needs. A. Turco, *Verso una teoria geografica della complessità*, Unicopli, Milan, 1988.

their iconic implications. When, as in this case, messages are transmitted in separate documents, then maps are in a position to convey a specific idea of what they represent by choosing a point of observation that upholds their aim. Both figurations do tend to a specific and partial view of territory. While the primary function of the map is to describe, the secondary function – clearly more important for our present purposes – is to conceptualize, to envisage the world from a particular point of view, giving us iconizing prescriptions about its mode of operation.

Cadastral Survey and Disjoint Conjunction

In the same period, we come across documents that give rise to very particular iconic junction: within the same document, we find a conjunction of figurations which, however, are disjoint one from the other. This is evident in a late map, from the beginning of the 19th century, which reproduces the town of Asolo for purposes of state property inventory related, in this case, to forestry resources[54] (Fig. 3.8).

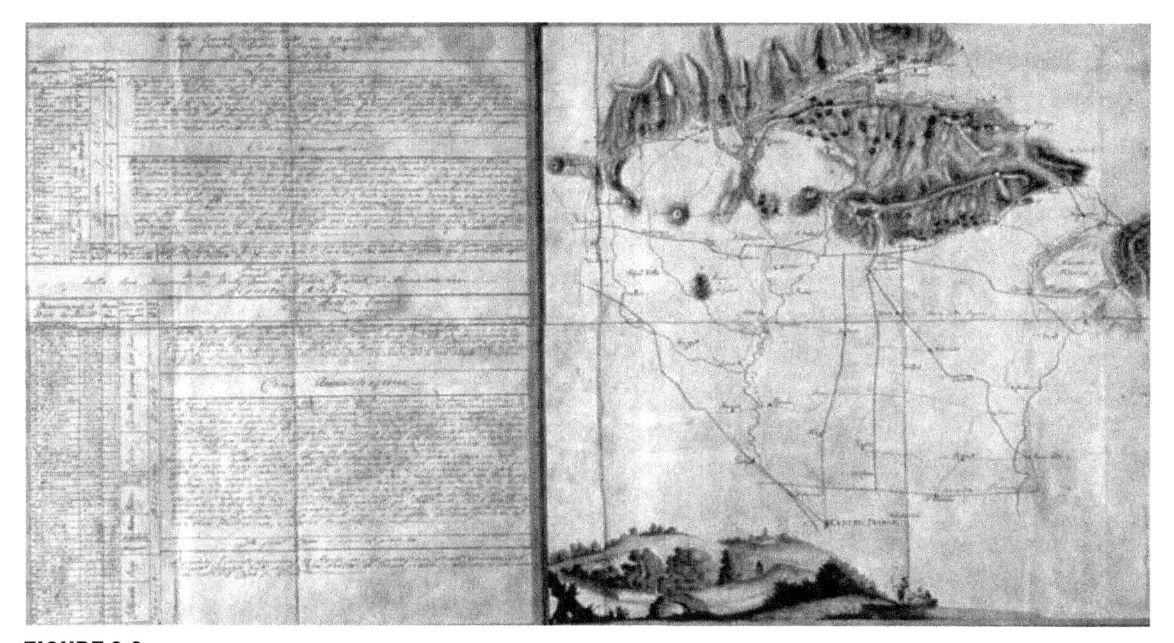

FIGURE 3.8
1805, *Asolo.* (See a color reproduction at the address: booksite.elsevier.com/9780128035092.)

---

[54] 1805, *Asolo*, Drawing on watercolored paper, measuring 50.0 × 78.8 cm; no scale given (ASVe, Provveditori sopra boschi, reg.171). See: E. Casti Moreschi, E. Zolli, eds., *Boschi della Serenissima. Storia di un rapporto uomo-ambiente*, Archivio di Stato, Venice, 1988, pp. 81–93.

The map is unique in graphical and decorative detail: to the right it depicts territory within a rich frame, in a cartographic figuration based on a vertical view. To the left, it features a chart that lists lots, localities, income, and the various forestry administrations in the area. Alongside, in the lower margin, there is a 19th-century style drawing: a landscape scattered with ruins and enlivened by rural scenes.[55] The combination of graphic outline, fine typeface and aesthetic appeal to landscape creates an iconic junction aimed to alert recipients to the importance of the territory represented. Asolo was a center of high symbolic value to Venice, and its relevance could not be underestimated, not even in an administrative record.

The territory of Asolo, which in the past had been a stronghold of the Roman defensive system, still expressed its antiquity and its authoritative status through its fortress. Within Venice's political and administrative framework, the Asolo region constituted a *podesteria* (seat of the podestà).[56] In the 16th century, when for a brief period (1489–1510) Asolo became the personal domain of Caterina Cornaro, Queen of Cyprus,[57] the area was in fact still ruled by the Venetian Republic, who exercised its power directly and simultaneously prevented Cornaro from reclaiming her possession.[58] And even in the later period – to which this map belongs – when Austria had taken over after the demise of the Serenissima, the appearance of the town was never neglected. On the contrary, what was guarded was continuity with the previous order, which was reflected in the will to retain the same values and keep all the facilities and offices in charge of territorial administration. Hence, within the inventory project that addressed all Venetian possessions, forestry assets were a primary index for assessing mainland resources in order to protect them. That is why such assets needed to be quantified and drawn with great accuracy. The plan shows a seamless strip of woods, covering the southern slopes of the hills, where the towns of Crespignaga, Coste and Cornuda are located.[59] Most of the woodland is, therefore, public property,

---

[55] It should be remembered that in the 19th century landscape had become an independent genre of painting, enriched over time with layers of meaning that were signaled by explicit references to history and to the cultural tradition.

[56] The mayor, a Venetian nobleman elected by the Great Council, lasted in office for sixteen years and embodied the civil and military authorities ruling in the name of the Venetian Republic since its expansion in the mainland. Full details about the context of this map may be found in: E. Casti Moreschi, E. Zolli, eds., *Boschi della Serenissima, op. cit.*, pp. 79–93.

[57] She was summoned to Venice, by injunction of the Council of Ten, with the promise that she would be given the castle of Asolo and an annuity of eight thousand ducats.

[58] Caterina lived in Asolo surrounded by a small court, which welcomed painters and writers including the famous humanist Pietro Bembo. That increased the prestige of the place, whose aesthetic value was guarded with great care, and sanctioned the forest as an important iconeme to express the harmony of landscape.

[59] The district of Asolo comprised 11 public forests and 44 private ones; the former covered an area of about 700 fields, the latter of about 350 fields. One Treviso field equals 5,205 square meters.

which demonstrates that even at that time, despite changes in legislation, forestry products were still largely considered a common good and, as such, reserved for the state. The chart provides detailed information on the type of woods, indicating, for both the public and the private ones, three forest "categories" based on the quality of the product. The map also points out that the timber obtained is intended for shipyard work, for social structures and for domestic uses.[60]

The criterion of public usefulness, on which the inventory was based, could not ignore the relevance the area had from the point of view of landscape. It would therefore be an error to interpret the landscape drawn along the bottom edge of the map as a merely aesthetic expedient of representation. On the contrary, the drawing serves the precise role of evoking the concept of landscape as a historical and social construct. As such, it expressed a commitment to maintain the layering of a long-settled tradition (as we have seen, the reference to antiquity is explicit). Because of its role, this drawing functions like an icon, able to shape the rest of the map and to substantiate the roles of the various figurations the map presents. Cartographic figuration aims to accurately reference and describe the area; the chart is meant to provide data and information useful to administrators; and the landscape serves to anchor the map in tradition and to stress its paramount iconic value. The conceptualizing propensity of representation is beyond doubt here. Nor can we ignore the fact that the combination of the three types of representation had the goal to communicate an idea of territory, whose visual form could refer to the historicity of the landscape. What we are dealing with here is an iconic junction issuing from landscape figuration, the communicative intensity of which affects and impairs any other interpretation. The iconizing role of landscape in maps is confirmed beyond any doubt.

SKETCHING IDEAS, CONVEYING CONCEPTS

The more intense is the iconizing aim of a map, the more relevant it becomes. This occurs because the systems deployed in a perspective-based view express the social values that territory has, so they can convey a message that aims to influence those who interpret it via communication systems that come close to first-hand experience. The result is a transactional system, which through various intermediations continuously circulates information and concepts that intersect on the double plane of cartographic communication: describing and

---

[60] The type of administration and the criteria that regulated public cutting were also clearly indicated. The 19th-century style chart also informs us that the product of private forests was meant almost exclusively for the respective owners, since these had been released from the obligation to make timber within their possessions available to the community. Thus, the Republic could make use only of the few oaks, marked as public, which still existed in those woods.

conceptualizing. In fact, my initial inquiry over the importance of distinguishing landscape from map gained momentum after my analysis of formal differences between the two made it possible to distinguish the functions of the former from those of the latter. It can now be reasonably argued that representations based on vertical figuration, ordered according to analogical referentiality or correspondence to reality and addressed to an interpreter on a double plane of communication may be counted as maps. Conversely, after excluding the prosaic level of visual/referential reproduction of reality, we can submit that the distinctive feature of landscape is a marked propensity for abstraction, whether artistic or scientific, capable of making a concept intelligible through iconization.

However, the most striking result of our survey is that we were able to detect hidden resonances and communicative outcomes interwoven in what we called iconic junction only after learning how to distinguish a cartographic representation from a "landscape." Iconic junction is indeed a semiotic figure of the utmost importance, not so much because it manages to convey the communicative outcomes of either figuration additively, but because it can autonomously process their combination to achieve a unique, unprecedented product. More precisely, iconic junction uses a *system of communicative interferences to promote the formation of syntagms, or in other words of relational clusters able to affect the iconization of the world.*

All this was made possible because we acknowledged the distinctive features of either figuration, neutralizing the image currently associated with the two terms, and recovering the specific nature of their communicative aims. What we found is that the identification of a map as such does not derive exclusively from its self-referential outcomes. It is true that we first tend to identify maps because experience leads us to consider them like other products, which were classed as such. The analysis we proposed, however, led us to the crucial recognition that maps present a double level of communication, able to meet informational and transactional needs towards areas that belong to other types of figurations. We will submit therefore that maps are well-suited to establish relationships with other types of descriptions, to forge mashups with other representations. Because of that, maps prove to be strategically crucial tools for developing novel forms of communication. This is what is currently happening for *cybercartography*, which relies on Information Technology (IT) advances to show its developments on multiple levels and multiple combinations. Cybercartography aims to recover the territorial sets of knowledge of local communities through direct involvement of inhabitants and Volunteered Geographic Information (VCI), through digital maps, based on hypertextual languages and arranged in atlases that can depict complex social, political and economic phenomena. Cybercartography considers material aspect, cognitive aspect and map as territorial parts of one whole social construct, able to show the depth of such knowledge sets and the various sensory forms with which they were expressed.[61] Clearly, the recovery of *chora* and, therefore, of the social significance of territory necessarily implies the recovery of landscape as a concept capable of expressing such social significance, but also depends on the ability to represent territory by recovering its symbolic meaning. In fact, the analysis we have just made would have been impossible without the prospect of recovering the value that territory had to the Venetian society in the Renaissance and of tracing sources that could explain the concept of landscape that the cartographer (in this case Cristoforo Sorte) had set out to achieve.

To conclude, the possibility of departing from topographic metrics and proposing the chorographic alternative may only present itself if we recover the social sense of territory, communicatively conveyed in the iconic meaning of landscape. What I intend to do in the second part of this book, however, is less to propose a generic pretopographic metrics than to envisage a chorographic metrics.* Taking stock of the social needs currently required of cartography and of the technologies available for its realization, chorographic metrics may ensure recovery of: 1) the sense of place expressed by landscape; 2) the plurality of parties who determine such a sense of place; 3) the pragmatic potentials offered by computer technology.

[61] D.R.F. Taylor, T. Lauriault, eds., *Developments in the Theory and Practice of Cybercartography. Applications and Indigenous Mapping,* Elsevier, Amsterdam, 2014 (1st ed. 2005). See also the contributions of some forerunners: M. Dodge, R. Kitchin, *Mapping cyberspace,* Routledge, London, 2001. But also: M.J. Kraak, A. Brown, *Web Cartography,* Taylor & Francis, London, 2001; K.L. Piper, *Cartographic Fictions: Maps, Race, and Identity,* Rutgers University Press, Fredericksburg (USA), 2002; J.W. Crampton, *The Political Mapping of Cyberspace,* University of Chicago Press, Chicago, 2004; M.P. Peterson, *Maps and the Internet,* Elsevier Science, St. Louis, 2006; A. Moore, I. Drecki, eds., *Geospatial Vision: New Dimensions in Cartography* (Lecture Notes in Geoinformation and Cartography), Springer, New York, 2008.

2
PART

CHAPTER 4

Technology in Action: Participatory Cartographic Systems

The world was so recent that many things lacked names, and in order to indicate them it was necessary to point.

G. Garcia Marquez, One Hundred Years of Solitude, 1968

This chapter is meant as a preliminary introduction to experimental landscape cartography. It sets an exploratory path towards a chorographic metrics able to convey the social meaning of territory. I analyze the profound changes that have occurred in contemporary cartographic production, especially those investing the role of the interpreter and those related to the unprecedented social needs that maps are called upon to address. First I will consider the outcomes that Information Technology (IT) has had on mapping in promoting a wide range of new production sites and new professional profiles for cartographers. Then I will survey the social landscape that shows cartography's involvement in the practices of democratization, both within governance policies or territorial planning and in terms of citizen access to data from public institutions. Finally, I will discuss the issue of participation in the collection of cartographic data and propose a model, pioneered in the field of environmental protection in Africa. I will present a methodology, called SIGAP (Geographic Information Systems for Protected Areas), which takes charge of the entire process of cartographic construction and produces participatory tools based on the use of geographic information systems (GIS). That will lead us to reflect on the semantic consequences of the availability of GIS tools online and, more specifically, to consider interactivity as a participatory potential for web-based cartography. This chapter will therefore provide a reflection on the importance of method for retrieving cartographic data; on the vital role of participation in the construction of maps; and on the potential WebGIS offer to promote such participation. These are essential conditions for rendering landscape in modes that pave the way to a chorography.*

CONTENTS

METAMORPHOSIS OF THE CARTOGRAPHIC WORLD

The first novelty that the world of contemporary cartography places before our eyes is the transformation of the figure of the interpreter: *cartographers* now comprise a wide range of individuals, both professional and nonprofessional, institutional and noninstitutional who participate in the construction of maps; *recipients* take on hybrid profiles and show an

Reflexive Cartography. ISSN 1363-0814, http://dx.doi.org/10.1016/B978-0-12-803509-2.00004-5

unprecedented propensity for cartographic interaction and intervention. Both new trends can be explained in view of the social change that has taken place in two specific realms: the political realm of democratization inherent in governance; and the institutional realm, marked by the adoption of geography-based Information Technology (IT) in relations with citizens. Let us consider both.

New Institutional and Technical Contexts

Democratization policies have come forcefully to the fore also following the gradual demise of the decisional role of the state. Having lost its exclusive role as a political agent, the state has in fact been partnered with other subjects, who come from a new form of citizenship no longer grounded in the idea that a citizen belongs to a national territory by birth. We are dealing with the so-called "mobility citizenship," centered upon values that reflect the functioning of an online world, where mobility and multiple relations, brought about by the collapse of boundaries, are the criteria whereby a given subject is identified as a citizen.[1] The redefinition of citizenship has expanded and diversified the number of political agents who want and can have a voice in the social forum. New places of territorial and environmental planning have therefore been established, where the presence of new "citizens" comes from their awareness of being able to weigh upon decisions: not only through the modes of indirect representation or the usual electoral systems, but also through forms of direct participation.[2]

The first outcome of this new approach to politics has been an unavoidable search for consensus, achieved by widening participation in decision-making processes that involve both large-scale globalization questions and the small-scale management of local territory. Within the increasing complexity of political scenarios, consultancy groups, which were originally promoted through grass-roots initiatives, caught the attention of international bodies. This eventually led to define the principle of governance, whereby decision-making on issues of public interest engages public institutions and private companies with

---

[1] Citizenship itself is now legally intended as an affiliation not based on national territory but on social participation (D. Zolo, "Cittadinanza: Storia di un concetto teorico," in *Filosofia e Politica*, 1, 2000, pp. 5–18. On the many senses covered by this concept and on its relation to participation, see among others M. La Bella, P. Santoro, eds., *Questioni e forme della cittadinanza*, F. Angeli, Milan 2011.

[2] Harvey noted how participation inheres in the postmodernist view of reality which marks a break with modernity when it states that we have no legitimate claims for establishing any one theoretical assumption as prevalent. Postmodernism therefore sanctions a wider decisional basis, built up on a sense of historical continuity and of collective memory. (D. Harvey, *The Condition of Postmodernity: An Enquiry into the Origins of Cultural Change*, Blackwell Publishers, Cambridge, MA, 1990, pp. 113 and ff.).

actively participating local communities.[3] Such principle, by now an inevitable aspect of democratic decision-making, currently stands as a testing ground for finding tools that may adequately promote dialogue in the management of public matters. What discourses on governance invariably highlight is the fact that standard modes of the decision-making have lost their legitimacy or effectiveness. It has become necessary to supply a set of tools aimed at enforcing the principles of transparency and sharing.

It is in this context that cartography stands out in its strategic role as a form of representation that aims at consensus. As noted by Jacques Lévy, cartography is a language virtually present in all stages of public action and particularly effective inside "hybrid forums."[4] In his words: "In these meetings, words often have the drawback of being overbearing markers of differences and inequalities within society. While not itself a harbinger of neutrality, or able in any way to exempt those who use it from pursuing competence and rigor, cartography has the advantage of enabling each to freely express themselves, favoring commensurability and the sharing of aims and data. Maps are nimble enough to evolve seamlessly along the communication chain between the formal and the informal. They can shift from a sketch to a diagram, from a project to a contract (and vice versa) while keeping actors in a state of careful alert. If used competently, maps can serve as the most effective and most egalitarian tools of expression for citizens. They enable one to place onto the same plane what is present, what is predictable and what is desirable. They allow citizens to have a voice not only in reaction to the documents submitted by elected representatives and experts, but also in the form of active participation at all stages of decision-making. Maps may undoubtedly become privileged vehicles of what was called 'actor-driven administration', a form of territorial administration that favors margins of freedom over static constraints, the stakes of society over 'turn-key' scripts, governance over sectorial policies, politicians over politics, in short, a practice that tends to establish a close link between public schemes and citizenship."[5]

---

[3] Over the last few years this principle has had remarkable success even on an international level, in the context of environmental policies. Appeals to governance are frequent in Environment and Development papers drafted in UN meetings with a view to coordinating protection initiatives and local involvement. On this issue and participatory systems, see: F. Burini, "Sistemi cartografici partecipativi e governance. Dalla carta partecipativa ai PPGIS," in: E. Casti, ed., *Cartografia e progettazione territoriale..., op. cit.*, pp. 178–192.

[4] In the words of sociologist Michel Callon. These forums aggregate a wide range of actors ("wise men," experts, professional politicians, citizens, etc.) with the aim to debate and solve issues of public interest (nuclear waste, AIDS, etc.). An in-depth analysis of the issue is found in: M. Callon, P. Lascoumes, *Agir dans un monde incertain: Essai sur la démocratie technique*, Le Seuil (collection "La couleur des idées"), Paris, 2001.

[5] J. Lévy, "La carta, uno spazio da costruire," in: E. Casti, ed., *Cartografia e progettazione territoriale..., op. cit.*, pp. 42–61, cit. p. 59.

It is also a fact that most administrative bodies did not hesitate to adopt Cartographic Information Technologies (CIT) as indispensable tools for addressing the complexity of contemporary societies. CITs are associated with a number of public and social advantages: improvement in social practices; optimization of decision-making; facilitation of communication processes; advancement in the actual exercise of local democracy. Geographic Information Technologies (GIT)* may be seen as part of CITs or rather as a special embodiment of CITs, because they rely on the same technical infrastructure and use the same communication networks, that is internet, intranet etc. More specifically, the use of GITs is twofold: on the one hand they serve to obtain, to assemble, and to manage geographical data; on the other they aim to analyze, formalize, and visualize such data in a graphic or cartographic form.[6] In a sense GITs express the relationship between an individual, a social group or an organization (of which they are the product) and territory, which can potentially modify such relationship.[7] It is a field where momentous choices are made, so the ability of local actors to integrate and manage these new information systems is crucial for the future of territories. Their scope of impact is vast and touches upon all the sectors involved in territory management (town planning, territory survey and diagnostics, network and infrastructure management, land management, transportation, etc.) as well as in environmental planning (protection and environmental conservation, local development, implementation of safeguards).[8] In time, GITs have come to dominate the practices of design, planning, and management of territory at all levels of public administration (international, national, regional, local, urban, and rural). They engage ever-wider and ever more diverse issues (development, economics, natural resource management, territorial marketing, crisis management, public health, town planning, etc.).[9]

---

[6] S. Roche, K. Sureau, C. Caron, "How to Improve the Social-utility Value of Geographic Information Technologies for the French Local Government? A Delphi Study," in: *Environment and Planning B*, vol. 30, n. 3, 2003, pp. 429–447.

[7] S. Roche, "Geographic Information and Public Participation: Research Proposal from a French Perspective," in: *URISA Journal*, vol. 15, APA 11, 2003, pp. 29–36.

[8] The U.S. Census Bureau, for instance, provides online access to a large amount of data that can be turned into maps and lets users decide the maps' framing, scale, theme and figures (www.census.gov).

[9] The widespread use of GITs rests on two social phenomena: the first has to do with the dematerialization of data, including geographic information (C. Écobichon, *L'information géographique: nouvelles techniques, nouvelles pratiques*, Hermes Science, Paris, 1994); the second is tied to the new practices of territorial actors. Never have so many graphical representations been produced in support of management and territorial planning (S. Roche, "Impiego sociale delle tecnologie d'informazione geografica e partecipazione territoriale," in: E. Casti, ed., *Cartografia e Progettazione Territoriale...*, *op. cit.*, pp. 164–177). Similarly, critical situations call for management tools and new skills in analyzing geographic data (S.B. Liu, L. Palen, "The New Cartographers: Crisis Map Mashups and the Emergence of Neogeographic Practice," in: *Cartography and Geographic Information Science*, Vol. 37.1, 2010, pp. 69–90).

GITs deserve recognition for ensuring wide and open access to a large quantity of territorial data, and for making these easily available even to inexperienced users. Over the last few years, however, scholars have pointed out two aspects that widen the gap between GITs and local participation: 1) GITs may either marginalize or consolidate a few specific social groups, since such tools are only accessible to a small part of the world, the ones that possess the means and the instruments to buy and update expensive software; 2) GITs may in fact induce a rift between representation and actual territory, because they produce cartographical representations that are granted objective status and are then used to lay down and to carry out territorial interventions. Ultimately, such systems may lead to viewing the local communities they process as "others." The other is essentially seen as "existing in a Euclidean space rather than in an actual territory, whereas the person who produces cartographic information is still centered on the actual world and feels entitled to make decisions by freely acting on territory." What critics have pointed to is the power GITs wield as symbolic systems that express a biased worldview: a supposedly objective interface between society and reality that is in fact designed along the prescriptions of those who are endowed with technical knowledge.[10] Starting in the mid-1990s, criticism of this kind was met with a renewed attention to creating cartographic systems fit to reclaim the role of local communities and to produce maps that take local interests into account.

We are thinking of participatory cartographic systems. Depending on the type of technology involved and on the level of participation, these comprise: 1) *participatory cartography*, produced by local communities upon request of an external agent;[11]

[10] For a discussion of the swift evolution GITs have been undergoing, from Community Integrated Systems and to Public and Participatory GITs to recent Web 2.0 cartographic technologies, see: W. Craig, T. Harris, D. Weiner, eds., *Community Participation and Geographic Information Systems*, Taylor and Francis, London, 2002; M. McCall, "Precision for whom? Mapping ambiguity and certainty, in (Participatory) GIS," in: *Participatory Learning and Action*, n. 54, IIED, London, 2006, pp. 114–119; T. Joliveau, "Le GéoWeb, un nouveau défi pour les bases de données géographiques," *Espaces Géographiques*, vol. 40, 2/2011, pp. 154–163.

[11] Participatory mapping involves the intervention of local representatives upon request of an external agent who steers the search for the theme to be mapped. Used in round tables on territorial planning, it relies on a variable scale to highlight the resources used by local communities for productive and symbolic activities (F. Burini, "Community Mapping for Intercultural Dialogue," in: *EspacesTemps.net*, 30.01.2012 http://espacestemps.net/document9252.html. We should also underline that, as we shall see in detail, advanced systems of participatory cartography online are gaining momentum. In these, communities not only participate but also largely control the process of mapping, thanks to easily accessible and economically viable frameworks which do not require professional skills to be used. These systems may radically alter the category of "other" as we intended it here. Yet, they now include a technological component that is widely available only within urban areas and is still largely absent in low-tech or non-tech contexts. Invaluable on this subject is the recent volume edited by D.R. Fraser Taylor and Tracey Lauriault, which shows the evolutionary changes new technologies have triggered in these type of maps (D.R.F. Taylor, T. Lauriault, eds., *Developments in the Theory and Practice of Cybercartography..., op. cit.*, 2014).

PARTICIPATORY CARTOGRAPHIC SYSTEMS	Participatory Cartography	CIGIS (Community Integrated GIS)	PPGIS (Public Participation GIS)
Technique:	manual design/ computerized graphics software	GIS systems	GIS or WebGIS systems
Issuer:	Local community/ external agent	External agent / local community	Local community/ internal or external institution
Recipient:	Local community/ external agent	External institution /local community	Local community/internal or external agent
Scale:	local/regional	regional/local	local/regional

FIGURE 4.1

Types and features of participatory cartographic systems.

2) *Community Integrated GIS* (CIGIS),[12] built and used by agents external to the community but on the basis of data collected in accordance with participatory methods; 3) *Public Participation GIS* (PPGIS), created and used directly by local communities in a dialogue with their administrators or an organization that presides over them[13] (Fig. 4.1).

I will return to these participatory systems later, because they are the field in which testing was conducted. Let us now instead reflect on the repercussions of such mapping innovations by considering the figure of the interpreter.

Reconfiguring the Cartographic Interpreter

The irruption of information technology into the field of mapping greatly affected interpreters and, in particular, the role of cartographers, who now

[12] These are GIS-produced maps which contain or combine information drawn from the local communities and fed into the system by an external actor. Addressed primarily to those who act upon territory (research institutions, national or international institutions, etc.), they are also used at negotiating tables (T. Harris, D. Weiner, "Implementing a community-integrated GIS: perspectives from South African fieldwork," in: W. Craig, T. Harris, D. Weiner, eds., *Community Participation and Geographic Information Systems, op. cit.*, pp. 246–258).

[13] PPGIS were initially conceived as tools meant for local communities, that is for "grassroots communities" capable of asserting themselves as active negotiators to administrative bodies and local institutions. Nowadays PPGIS are ready to address the wider scope of opportunities opened up by globalization. They are also based on information provided by local communities and are produced either by internal actors or by external actors who belong to the institutions involved in the solution of issues. See the website: http://www.ncgia.ucsb.edu/. See also the IAPAD website (Integrated Approaches to Participatory Development), almost entirely devoted to participatory mapping systems (*www.ppgis.net* - Open Forum on Participatory Geographic Information Systems and Technologies). A mailing list on these topics is included.

comprise a composite range of professionals from different backgrounds and affiliations (institutional or noninstitutional, professional or nonprofessional). On the one hand, we have cartographers who work for public mapping agencies, such as IGM in Italy, and who, despite their training in topography, increasingly take on the profile of IT engineers. This change followed the recent overhaul of such agencies which, technically, adapted mapping to the latest IT solutions and typologically diversified their cartographic offer in response to social contingencies. Social needs downplay the military purpose underlying topographic maps and make them available for socially relevant and humanitarian causes.[14] We also witnessed, however, a proliferation of private mapping agencies, which entered the market forcefully. To public administrations and private companies, these agencies offer a very sophisticated range of products, first intended for cartographers-programmers who possess IT and engineering qualifications but later enjoyed by professionals from various disciplines (forestry, botany, architecture, urban planning, geology, etc.).[15] To these we should add researchers-cartographers from university laboratories and research institutions who engage in applied research and produce maps on commission but also promote innovation and theoretical debate.[16] These latter figures are professionals who use their scientific competence to examine the setup of maps vis à vis the new challenges posed by society and IT. Finally, we have government agencies interested in the dissemination of geographic information systems, notably the ones in Canada and the United States, the drivers of innovation and cartographic expansion which most European countries have started to emulate.[17]

---

[14] An exemplary case is the change that occurred within the French IGN. See the website: www.ign.fr.

[15] Examples are, for Italy, the ASITA union, founded in 1998, which comprises the four national associations working with geographic information systems (SIFET, AIC, AIT, AM/FM GIS ITALY) and, in the case of France, the AFIGEO association, established in 1999. In the U.S., the Coalition of Geospatial Organizations (COGO) was founded in 2007 to promote cooperation between various bodies and associations engaged in the development of geographic information systems: ASCE, ASPRS, CAGIS, GLIS, GISCI, IAAO, MAPPS, NSPS, NSGIC, UCGIS, USGIF, URISA.

[16] It would be impossible to mention them all here, so I will limit myself to citing the ones that, in this highly prolific and changeable context, have been recently more proactive. *Geomatics and Cartographic Research Centre* (GCRC), Carleton University, Ottawa, Ontario, Canada; *Laboratoire Chôros* of the EPFL in Lausanne, Switzerland; CASA – *Center for Advances Spatial Analysis* at the University College London, England; *State GIS Technical Center* of Morgantown University in West Virginia, United States; *Laboratoire Image et ville* University of Strasbourg, Switzerland; CODATA-Germany *Committee on Data for Science and Technology* in Berlin, Germany; *Centre de recherche en Géomatique,* of Laval University in Quebec.

[17] *The National Center for Geographic Information Analysis* (NCGIA) was founded in the United States in 1988. It is a consortium of universities, aimed at promoting the use of such technologies in the study of territory. This promoted the creation of similar associations in Europe, such as EUROGI. On the spread of these technologies and the causes of the gap that exists between different parts of the world, see: T. Joliveau, "Tecnologie cartografiche per la partecipazione territoriale. Un approccio prospettico a partire dal caso francese," in: E. Casti, J. Lévy, eds., *Le sfide cartografiche...*, *op. cit.*, pp. 161–176.

In all these places of production, therefore, cartographers play a hybrid professional role that combines IT skills (interaction with mapping software) with skills related to the management of incoming data. Cartographers also possess, or should possess, a semiotic understanding of the communicative mechanisms of maps. In short, new cartographers are those who, being familiar with digital systems, are placed in a position to manage the entire mapping process. Nonetheless, it should be kept in mind that even before the process begins a number of binding decisions have already been made by the computer programmer, such as the mandatory selection of a topographical framework, or the puzzling choice of new graphical conventions that replace at least in part former codification rules.[18] Cartographers are thus compelled to comply with the opacity of the IT system which forces them to use a basemap and yet allows them to combine it with new options.[19] On the other hand, they are also entitled to configure the typology of geographic data and choose how to visualize it according to the options offered by the system.

Despite these obligations, cartographers can, with regard to theory at least, regain the autonomy which the institutionalization of topographic mapping had denied them for nearly two centuries. They can decide how to configure geographic data for input and which visualization mode to adopt. The geographer/surveyor who was under government dictates has been replaced by a cartographer/IT expert who is somehow able to independently affect the final product. It is unfortunate that in most cases cartographers should be found lacking both in geographical competence and in cartographic semiotics.[20]

---

[18] The programmer is either the person in charge of tailoring software to meet the needs of particular contexts or an agent involved in global marketing (ESRI, Pitney Bowes Business, Intergraph etc.). On this role and on the assembly and management of a GIS see among others: P.A. Longley, M. Goodchild, D.J. Maguire, D.W. Rhind, eds., *Geographic Information Systems and Science*, John Wiley and Sons, New York, 2011, pp. 425–450. With regard to the proliferation of conventions, it should be noted that, unlike the past, when cartographic agencies were part of a political and institutional plan which guaranteed terminological uniformity, private or commercial mapping ventures today are not interested in rigid codification, because their goal is to market the widest possible range of products able to meet specific and various needs. These objections echo the ones moved against the mandatory use of topographic metrics, on which GISs are largely based. Harley's ideas on both cartography and GISs have been used to critique their positivist assumptions (L. Harris, M. Harrower, *Critical cartographies special issue. ACME. An International E-Journal for Critical Geographies* 4(1), 2005. Last accessed 11 June 2015, http://www.acme-journal.org/volume4-1.html).

[19] This is the case of 3D cartography, the new front on which topographic prerogatives give way to visual-analogical ones, as we shall see in the following pages.

[20] The risks inherent in this new cartographic framework were exposed by many. Colette Cauvin, among others, has long advocated some form of control over the relevance and quality of cartographic products (C. Cauvin, "Des trasformations cartographiques," in: *Mappemonde*, 1998, vol. 49, pp. 12–15). On qualitative issues to do with Web 2.0 cartographic products and on the redefinition of the role of the cartographer in this context, see also W. Cartwright, "Delivering geospatial information with Web 2.0," in: M.P. Peterson, ed., *International Perspectives on Maps and the Internet*, Springer, Berlin-Heidelberg, 2008, pp. 11–30.

Today many read this as a sign that cartographic production has lost its force and question its future.[21] It should however be noted that, albeit hybrid in their role and markedly uneven in their level of competence, cartographers reflect the fragmentation of institutions, as we noted above. As such, they exclude the possibility of control "from above" via an indisputable cartographic model. On the contrary, cartographers' multiple profiles should be seen as a potential for building a multiform interpretation of territory. In this transitional phase, that may appear chaotic, but is bound to gain momentum in the competitive field of commerce, where products are adopted or discarded according to whether or not they can bridge the information gap in contingent situations. It is to be hoped that knowledge of the mechanisms whereby maps are built and communicate may in the near future become part of a generalized and widespread geographic awareness capable of rejecting *de facto* ineffective products.

Our inquiry would be incomplete if we failed to mention the particular role that participatory systems, and especially participatory mapping and CIGIS ascribe to the cartographers. In the first, the "cartographer" consists of a series of subjects from the local community who are asked by the proponent (a researcher or an external contractor) to build a map on a theme that is relevant to the community. Let me reiterate that this mapping does not aim to produce a description of territory as built by locals but, rather, to render one or more territorial interests of such community in a figurative manner. This needs to be stressed on account of the criticism anthropologists aroused about representations of the territory of the Other. Although moved by the best intentions, they argue, these maps produce cultural contamination, since the point of view of the intermediary irreparably undermines the authenticity of the representation of the place. In fact, as far as possible and when they exist, mediators play a passive role, limited to coordinating the drafting of the map document itself and promoting narration and the identification of icons to be included on the sheet. Conversely, the participatory intervention of the local population permeates the entire construction process: from data collection and, therefore, the assembly of information, to the graphical transposition obtained through participation of the entire community or of the social group involved in the issue. Only the last phase – namely digital translation or database setup and graphical layout – is entrusted to the proponent in the role of IT cartographer, but even at this stage, observance of the icons

[21] Franco Farinelli speaks of the crisis of cartographic reason: F. Farinelli, *Geografia. Un'introduzione ai modelli del mondo*, Einaudi, Turin, 2003, pp. 200–201; Id., *La crisi della ragione cartografica. Un'introduzione alla geografia della globalizzazione*, Einaudi, Turin, 2009.

proposed by the community and of their syntactic relationship is safeguarded.[22] PPGIS do not instead pose the issue of contamination, because the cartographer coincides with one or more members of the group and plays both the role of proponent and that of drawer of the map.

As we turn to the figure of the second cartographic interpreter, that is the recipient, we ought to mention the complexity of his demands, aimed at obtaining increasingly complex maps that are detailed, dynamic and targeted to specific action.[23] Paper cartography, but especially display cartography, has witnessed an expansion and diversification of cartographic users similar to the ones of images and multimedia. The rapid development of GIS and other software led to the transformation of the figure of the recipient. Through interactive and kinetic mapping, recipients demand to query maps (selection of themes, scale, data, discretization criteria, and ways of reading) according to their own personal taste. What is more, they demand to be part of mapping itself, building a database and assembling their own customized maps. This need underlies experiments on *Volunteered Geographic Information* (VGI). As mentioned previously, "cybercartography," a cartographic paradigm of the digital age based on the use of multimedia interfaces for presenting and organizing cartographic information, has seen the development of an open-source platform called *Nunaliit Cybercartographic Atlas Framework*, which increases opportunities for participatory mapping. Information is stored in "cybercartographic atlases," collections of online multimedia maps. These produce a knowledge set of the chosen territory, at various scales and using multiple languages, based on data produced and entered directly by local inhabitants.[24]

In short, the double figure of the interpreter cartographer/recipient looks increasingly symbiotic. It may be identified as a single subject who builds and interprets maps in a social, bottom-up informational context.[25]

---

[22] In participatory mapping, inquiries never refer to a single person but to the entire community or a coherent part of it in the exercise of a profession or function. As such it takes place in a public space, with the consent of the local authority. In sub-Saharan Africa, which I will address in the following pages, this mapping is done in the central area of the village, with the permission of the village authorities and following the ritualization of the palabre, or as Bidima explains, of the moment when decisions are made and upheld and the relationships between subject, interdiction and law, between culture and nature are expressed. In this context, information is produced by the dialogue between the participants, in which arguments are compared and examples given in order to secure sharing (J.-B. Bidima, *La Palabre. Une jurisdiction de la parole*, Michalon, Paris, 1997). Thus, the rift between interviewer and interviewee is softened in the communicative flow of narration.

[23] The most common ones provide address lookups or directions (for example, www.mappy.com; www.maps.google.com, www.viamichelin.com; www.mapquest.com etc.).

[24] See note 11 in this chapter.

[25] This attitude is inherent in participatory mapping systems because their use in consultation involves the exercise of empowerment. That does not exclude the urgent need to address a number of issues tied up with IT skills and involving the use or adoption of systems which include direct or indirect users (W. Craig, T. Harris, D. Weiner, eds., *Community Participation and Geographic…*, *op. cit.*; S. Roche, "Impiego sociale delle tecnologie d'informazione geografica," *op. cit.*) which I will discuss in the pages that follow.

Optimists see this as a sign that information control has finally been superseded. Pessimists instead lament the loss of meaning in information, caused by the fact that the reliability of such information cannot be validated. These are open issues, to do with the wider context of online information as such. They cannot be addressed, as some propose, by implementing information controls aimed at raising the level of technical skills possessed by interpreters, be they cartographers or recipients. It is rather the entire mapping process that needs to be reconfigured, refuting the authority of topographic metrics on which most IT systems still rely and reducing its claims to exactitude and exhaustiveness to embrace plural and limited cartographic information. Without a genuine break with the past, the paradigm shift fails and the claim to innovation is futile. Let us consider a working proposal in that direction which addresses mainly the type of territorial information to be included in the map.

THE GEOGRAPHIC INFORMATION SYSTEMS FOR PROTECTED AREAS STRATEGY IN W TRANSBOUNDARY BIOSPHERE RESERVE (WEST AFRICA)

Let us now look at one of many challenges cartography has to face today: the translation of concepts that underlie territorial policies into operational tools. The recurrence of such terms as "co-management," "participatory cooperation," "consultation" in environmental protection programs may certainly be said to mark the adoption of sustainable development principles. These were proposed in the course of the Rio Summit and expanded, in the direction of the co-participation, by the UNESCO MAB program. On the other hand, this expresses the need to find methods and tools immediately usable in practice, able to translate and substantiate theoretical aims into concrete results.[26]

Over a period of five years, a working group coordinated by me endeavored to meet this challenge by providing a model of zoning based on environmental issues, such as *landscape approach* and *community conservation*, which would take over the guidelines proposed in the international arena and translate them operationally. The analysis was carried out via a research methodology named

[26] It is no chance that the international organizations engaged in an environmental cooperation program should advocate tools that enable them to establish and manage actual planning, which now provides for the systematic involvement of local communities both in decision-making related to their development and in the management of resources. On this need, see: T. Joliveau, M. Amzert, "Les territoires de la participation: problème local, question universelle?," in: *Numéro spécial de Géocarrefour. Les territoires de la participation*, vol. 76, n. 3, 2001, pp. 171–174; G. Borrini-Feyerabend *et alii*, eds., *Indigenous and Local Communities and Protected Areas*, IUCN, Cardiff University, Cardiff, 2004; M. Batton-Hubert, T. Joliveau, S. Lardon, "Modélisation spatiale et décision territoriale participative. Conception et mise en œuvre dans des ateliers chercheurs acteurs," in: *Revue internationale de géomatique*, 2008, vol. 18/4, pp. 549–569.

SIGAP (*Geographic Information Systems for Protected Areas*).[27] It uses GIT (participatory mapping and CIGIS) to propose a participatory management plan for the periphery of protected areas and pursues two objectives: 1) to recover the territorial layout and the values of local populations as markers for the zoning that affects the area to be managed; 2) to build participatory mapping suited to convey the social significance of territory at a dual scale (local and regional) and, therefore, capable of influencing operational decisions for intervention.

The Strategy, tested under the *Programme Régional Parc W / ECOPAS (Ecosystèmes protégés en Afrique Sahélienne)*[28], recovers the set of specifically geographic competences in applied research and involves all phases of information processing: adoption of a theory which informs terrain methodology; interaction with the inhabitants for data reading; construction of interpretative models and their cartographic visualization, and targeted and mindful use of cartography. The final product is a GIS interactive multimedia system (Multimap), useful for field research, intervention strategies, capitalization and the diffusion of achieved results.[29]

Theoretical Framework and Scope of Territorial Application

On the assumption that conservation projects related to sustainable development in Africa generally advocate a reflection on the role of territory, the SIGAP Strategy uses field research to recover the territorial dynamics of local communities. This is a prerequisite for implementing a cooperation which posits the criterion of habitation as a premise for drawing co-management rules.

Territory is considered as the outcome of a process that joins natural space and society within a circular flow. While ostensibly obvious, such precondition paves the way to the adoption of a different perspective in addressing issues

---

[27] This is a strategy defined by the *Diathesis* Cartographic Lab at Bergamo University, which already had conspicuous operational outcomes. See: E. Casti, "A Reflexive Cartography to Tackle Poverty: A Model of Participatory Zoning," presented at the International Conference of Bangkok, IUCN, November 2004, http://www.iapad.org/publications/ppgis/Casti_IUCNa.pdf; Id., "Geografia e partecipazione: la Strategia SIGAP nella RBT W (Africa Occidentale)," in: *Bollettino della Società Geografica Italiana*, Serie XII, vol. XI, 2006, pp. 949–975.

[28] An environmental cooperation program supported by the European Union on FED/EU funds. See the website: www.parks.it/world/NE/parc.w.

[29] Field research was performed by the team of Bergamo University as well as by African researchers (University of Ouagadougou; Abdou Moumouni University of Niamey; University of Abomey-Calavi of Cotounou) and was supported financially by the PRPW/ECOPAS Scientific Committee (backed by the CIRAD center – Coopération Internationale en Recherche Agronomique pour le Développement of Montpellier) for three years (2001–2004), during which inhabitants of the park periphery were involved in the evaluation of their territory. This was followed by a phase of cartographic and digital processing which lasted two years (2004–2005) and was implemented by the team of the University of Bergamo for the production of a Multimap. Promoters for this last phase, in addition to the ECOPAS Consortium (bringing together all the countries involved in the protection program of the W Park) were IUCN (International Union for Conservation of Nature), Cooperazione Italiana and UNESCO.

of environmental protection. First, it definitely rules out the hypothesis that a conservation project may result in the preservation of a *status quo* of natural conditions, idealistically and aseptically untouched by human action. Then, according to specific patterns, it acknowledges the interrelationships between humans and their frame of existence. That means that to the human group that inhabits it, territory is a prerequisite for life not solely by virtue of its material resources, ensuring sustenance, but also on account of the symbolic, cultural and identitarian value it embodies. Ultimately, territory marks the place where social demands for the reproduction and the functioning of a given society come to the fore. Recovery of the mechanisms that regulate the circular flow just mentioned enables us to define and investigate the kind of relationship a community has with natural resources and is the first step in developing strategies of cooperation and environmental protection. This is especially true in specific areas, such as Africa, where populations who settle in the periphery of parks, and, therefore, far from urban centers, still base their functioning on political systems anchored in tradition and in a tight relationship with territory.[30] Reconstructing the fabric in which such powers intertwine is therefore indispensable in order to recover its political setup and the authority that it wields to regulate social relations around and within the protected area.

Therefore, it is the importance played by territory in participation that leads us to reflect on its representations, and first of all on cartographic representation, and on the likelihood that these may convey its social meaning. Research mapping was carried out by means of participation, which multiplied the actors involved using unconventional languages and graphics. And graphic rendering allowed researchers to attempt to discard the presumed objectivity of the map in order to neutralize iconization. More precisely, mapping was done with the participation of local associations, by deploying a figural language that could undermine and question geometric conventions and thus pave the way to new formulations, suited to convey the meaning of inhabited places. The goal, obviously, could not be achieved by imposing previously encoded methods of mapping. Rather, it was necessary to figurativize the narrative that long public meetings produced, to follow their thread without striving to comply with scale, metrics, conventions or other.

Participation produced a series of reflections on the role that cartography plays in shedding light on the multiple interests of the actors involved in territorial planning for developing countries, or, as they are now defined to discard the unwieldy meaning of development, for resource-rich countries,[31] within which

[30] Local inhabitants acknowledge and respect the right over land, hunting, breeding, in short over natural resources because the person who exercises it has family ties with the laman, the one who first sanctioned the person's lawful possession by divine covenant.

[31] These were juxtaposed to technology-driven countries.

maps play a central and undisputed role. The only way forward was to devise a model of participatory mapping addressing sub-Saharan Africa[32] which was grounded in cartographic semiotics and could make a solid contribution to a participatory approach.

Preparation of such mapping was preceded and accompanied by territorial analysis. The drawing up of participatory maps entailed an analysis of territory, carried out in the course of a long stay in the field and aimed at exploring the layout of territory and at understanding how the local society relates to its natural resources. Such understanding would assist in the process of mediating between the interests of the many actors involved. In particular, and with regard to specific aims – namely to propose a periphery zoning respectful of local territorial systems – we deployed a SIGAP Strategy comprising different phases, meant to provide interim knowledge tools to park managers.[33] Work was broken down into four phases:

- The first one aimed to establish a database of territorial systems for local populations based on field data visualized cartographically;
- The second entailed data modeling needed to plan zoning and a clear outline of the building phases of the project we were endeavoring to carry out;[34]
- The third indicated starting routes for working participatory proposals to follow;
- The fourth involved the design and construction of an interactive and multimedia system of data capitalization and cartographic processing, as an operational tool for project managers and, more broadly, as a cognitive tool aimed at everyone via the web (Multimap).[35]

Before considering the merits of these phases, we need to briefly look at the framework of the international principles that need to be met and to mention the territorial milieu in which the SIGAP Strategy was first tested.

---

[32] The countries involved in field research are: Guinea and Senegal (1995–1996) within the Programme Régional d'Aménagement des Bassins Versants du Haut Niger et de la Haute Gambie of the European Union; Burkina Faso, Niger and Benin (2000–2005) within the EU Programme Régional Parc W/ECOPAS; Burkina Faso (2005–2009) within the inter-university cooperation project *Outils pour la prise de décision dans la protection environnementale: le Complexe WAP* between the University of Bergamo and the 2iE-Institut International d'Ingénierie de l'Eau et de l'Environnement of Ouagadougou.

[33] These were substantiated in reports, training workshops for park managers, maps used by local populations, with a view to giving voice to their cooperation, and maps used in political debates for the validation and promotion of the plan.

[34] It is well known that the reliability of a model depends on a transparent illustration of all its steps: the methods with which data were collected; the criteria by which they were aggregated; indexes employed as processing elements; theoretical and analytical categories on which modeling can be based.

[35] This system, which includes all the cartography cited here, may be consulted at www.multimap-parcw.org.

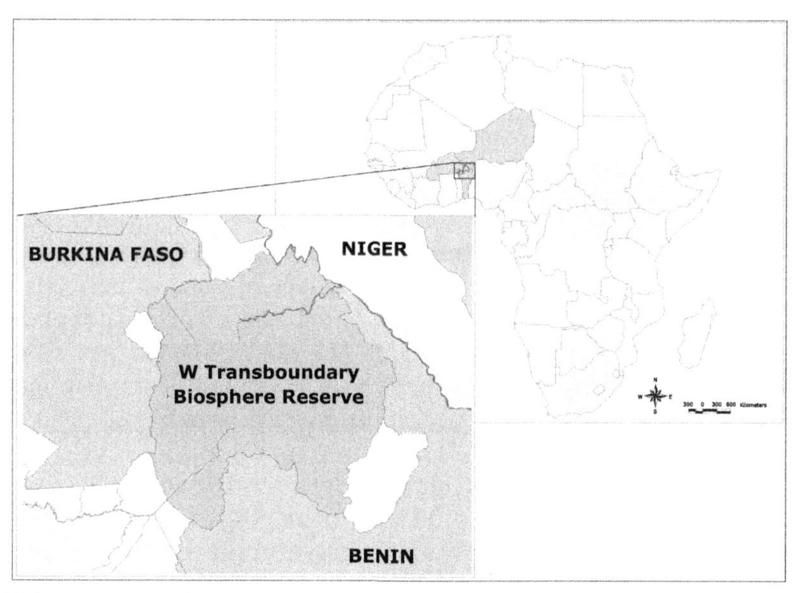

FIGURE 4.2
Localization of the W TBR (Transboundary Biosphere Reserve).

Having been officially recognized as a Transboundary Biosphere Reserve (TBR) of the MAB/UNESCO program since 11 November 2002, the W Regional Park[36] (Benin, Burkina Faso, Niger) adopted the SIGAP Strategy[37] to meet the needs for an effective management of the protected area and its periphery (Fig. 4.2).

The MAB (*Man and the Biosphere*) classification model was translated into a zoning based on respect for the social and territorial systems of local populations and, therefore, consistent with the objectives set by UNESCO.[38] In order to overcome the limitations that each zoning entails and, at the same time, place it in the context of participatory cooperation, the SIGAP Strategy

---

[36] The W TBR (Transboundary Biosphere Reserve) is a protected area of West Africa covering an estimated area of 31,000 km^2 characterized by great environmental, as well as social, diversity which required long and elaborate analysis for the implementation of a Protection program. See the recent: A. Ghisalberti, *Le migrazioni in Africa Occidentale tra ambiente e politica. La periferia del Parco Transfrontaliero W (Benin, Burkina Faso, Niger)*, L'Harmattan Italia, Turin, 2011.

[37] The body responsible for its management, the Programme Régional Parc W / ECOPAS, integrated the results obtained by this strategy in its *Plan d'Aménagement et de Gestion* (2006–2010). That showed that the plan is viable and makes it plausible for the strategy to become an intervention model for other UNESCO areas designated as Biosphere Reserves. See: A.S. Aladji-Boni, D. Bruzzone, M. Diallo, D. Dulieu, M. Falcone, O. Novelli, C. Paolini, "La Riserva transfrontaliera della Biosfera W: integrazione economica e conservazione per lo sviluppo locale. Una prospettiva per l'Africa Occidentale," in: I. Cresti, J.L. Tuadi, eds., *Il continente verde*, Bruno Mondadori, Milan, 2011, pp. 270–296.

[38] UNESCO, *Biosphere Reserves, special places for people and nature*, UNESCO, Paris, 2002; Id., *Five Transboundary Biosphere Reserves in Europe. Technical Notes*, MAB Programme, Paris, 2003.

operationally tied the triple division of the Biosphere Reserve (core zone, buffer zone and transition zone)[39] to the specific social features of the territory under examination.

The Modular Structure

I am going to follow the modular sequence of our research to present the results we achieved.

MODULE 1 - It coincided with an on-site survey of the area, for collecting *double-scale* data. A regional scale highlighted aspects of the population's ethnic setup, the number of inhabitants, the number of villages, seasonal and permanent migration, data that offset a virtually complete lack of knowledge of the territorial organization of local populations. This scale also provided for an initial assessment of human presence impact on the park. Knowing that a territory is inhabited by a very high number of people (610,000 in an area of about 16,000 km^2) belonging to different socio-linguistic groups (12 ethnic groups) and engaged in processes of integration and/or marginalization with respect to practices that respond to external needs (the cultivation of cotton, for instance), enabled us to draw a picture of the complex mechanisms that exert pressure on the protected area. At the local scale, the complementary idea was to explore in depth the social organization and functioning of the territory, in an attempt to recover specific features of the relationship social groups had with natural resources. We anticipated that their sets of knowledge and their protection system, tested over the centuries, would contribute substantially to the management strategies for the Park. This approach revealed the limits as well as the potential of the area involved in the protection program. For one, research highlighted the presence of traditional features that ensured balance between population and natural resources. On the other hand, it exposed phenomena introduced from the outside which can and must be addressed and made compatible with conservation.[40]

---

[39] According to the definition of the MAB Program, Biosphere Reserves are organized into three related areas that perform different functions and promote complementary activities of nature conservation and use of resources (UNESCO, *Biosphere Reserves, special places ...*, *op. cit.*, pp. 16–17). In a first implementation phase for Regional Park W, in the absence of field data, distance from the central core provided the criterion for delimiting zones. Subsequently, such distance-based definition was discarded and integrated with the sectors identified by the system of participatory zoning outlined here (A. Billand *et alii*, *PAG – Plan d'aménagement et de gestion de la Réserve Transfrontalière de la Biosphère W 2006-2010*, vol. II, Stratégie, Ouagadougou, PRPW/ECOPAS, 2004, pp. 34–37).

[40] The survey entailed extended stays in the villages, especially those chosen as sample villages. These were the object of an in-depth inquiry to collect information on features shared with other villages belonging to the same ethnic group, such as economic system, network of customs, or other. Field research reports may be consulted under the "villages cibles" section of the Multimap system: www.multimap-parcw.org.

Field observation and the data collected were made communicable by diverse means: initially by entering and sorting information in tabbed spreadsheets processed through Windows Excel software and later by translating data into maps via GIS applications in order to present results (even provisional ones). These maps were not merely the visual presentation of the results obtained during the survey. Rather, they were instruments the interpretation of which generated more information, in addition to that gained directly in field research. More specifically, the processing and cartographic visualization of data at a regional scale (CIGIS) produced a set of documents that showed the size of phenomena and their territorial dynamics, highlighting social mechanisms that are responsible for stability or change (Fig. 4.3). For the local scale,

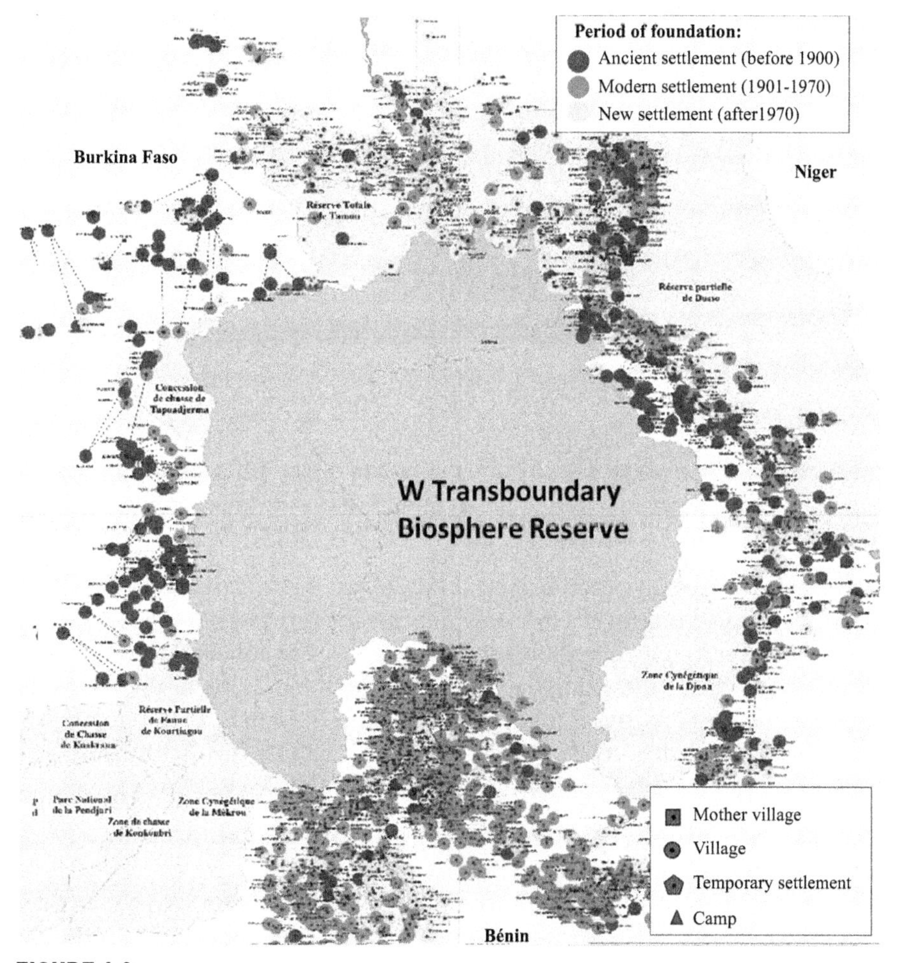

FIGURE 4.3

Regional-scale CIGIS example indicating the time of foundation and the type of settlements on the outskirts of W Regional Park. (See a color reproduction at the address: booksite.elsevier.com/9780128035092.)

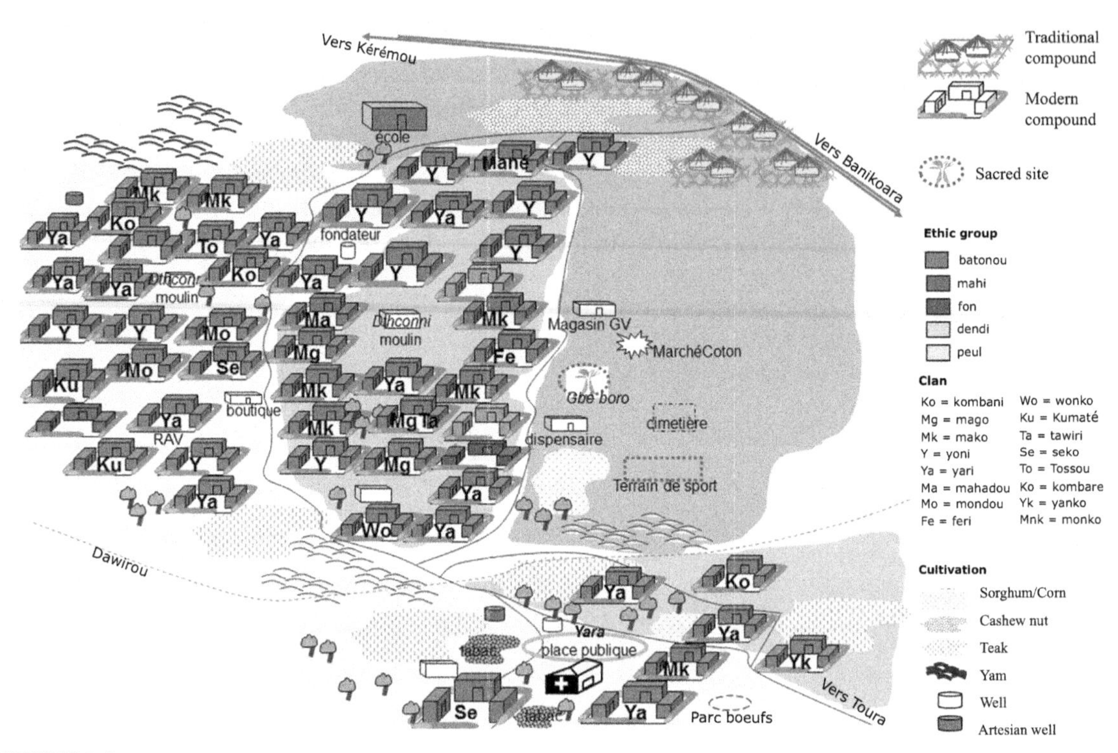

FIGURE 4.4

Example of digitized participatory mapping: the village of Gbeniki. (See a color reproduction at the address: booksite.elsevier.com/9780128035092.)

participatory maps were built. By translating the digital information produced by the inhabitants themselves, these revealed the social values entrusted to the territory (Fig. 4.4).

Methodologically, the procedure we followed used participatory cartography from the very first phase, as a tool able to generate information supplementary to that culled through observation and inquiry. On a regional scale, such a procedure provided for the sizing but also the cultural modeling of the territorial phenomena found in the region. On a local scale, it brought to the surface the values and knowledge items of local inhabitants, which highlight the importance of topical competence in the management of places.[41]

Once data had been collected, and the strong interrelationships between population, territory and protected area at a functional level were confirmed, it became possible to recover also culturally significant data, which in turn

[41] 1,574 settlements were surveyed. For the most part, these were unknown to administrators themselves. An in-depth analysis was instead carried out on 21 sample villages.

allowed for a more specific definition of zoning options for the periphery. We started from the idea that a given environment may present sensitive and intrinsically more or less fragile features, which define its propensity or unsuitability for change both in terms of land use and in terms of the identity stakes it raises.

This method tends to make zoning obsolete when intended as a division of territory into strictly functional areas, because it recognizes that, in its multiple valences, the environment cannot be reduced to the sum of its separate parts but must interact on multiple levels to ensure a balance is achieved between the resources and the individuals who use them (and not just to promote competition for resources).[42]

MODULE 2 - The next step involved the development and proposal of a sectoralization. The three zones outlined in the MAB classification scheme (core zone, buffer zone and transition zone) were taken as interdependent and connected areas that share the same objective: to ensure environmental protection of its natural and social components, in line with the directives of the UNESCO program which considers the core zone as the "engine of the alliance" between man and the environment. In turn, both the buffer zone and the transition zone are connected to the core area, which is both the legitimate territory where, over the centuries, customs have been regulated by tradition and also a functional place, essential for its physical-climatic as well as its symbolic features.[43] It was necessary to define a zoning that, by recovering the relationships that the periphery entertains with the central core, could point out their implications for the local populations. Our purpose was less to affirm the right of use for natural resources than to deploy arguments for securing agreements between the Cooperation Program and the local communities, such as would provide compensation for the non-use of resources and ensure the introduction of practices compatible with conservation. Particular zoning criteria were taken into account operationally (sizing and qualification of population, territorial organization and use of resources, elements of tradition and modernity) and three indexes were drawn from them: 1) Cohesion, which assesses the degree of homogeneity and coherence of territory and provided a foundation for determining modes whereby the conservation plan can harmonize with local issues[44]; 2) Pressure, which gauges the impact of human action on the

[42] M.R. Turner, R.H. Gardner, R.V. O'Neill, *Landscape Ecology in Theory and Practice*, Springer-Verlag, New York, 2001.

[43] Inside the park, for instance, there are several sacred sites the population appeals to in religious and social practices, which are now hampered by restricted access to the Park.

[44] The cohesion index was inferred from an assessment of resource utilization. We identified *eco-functional networks* by grouping villages which established and maintained mutual relations based on the presence of a shared natural resource used by different actors and on the type and prevalence of *productive activities* they promoted (hunting, fishing, harvesting, speculative agriculture, subsistence agriculture, farming, transhumance, trade).

conservation area and led us to quantify urgent needs and prioritize protective lines of action; and finally 3) Localization, which defines the proximity or distance of particular dynamics from Park boundaries and estimated the degree of risk to which the core zone is exposed.

Eventually we identified two different areas: *level 1 units*, defined around social-territorial cohesion and *sectors*, based on cohesion and on the pressure exerted on natural resources. Both were placed within the *localization framework* whose extension was made to coincide with that of the buffer zone.[45] The latter, which measures 20,000 square kilometers, comes therefore from the incorporation of 16 sectors and 83 Level 1 units, in which the first combine with the second to provide a transcalar view of socio-territorial structures. More specifically, *level-1 socio-territorial units* served to identify areas whose territorial interdependence produces homogeneity that reverberates in social identity.[46] *Sectors*, instead, grouped *level-1 units* into wider areas. Such areas were defined both by integrating the cohesion index found for level-1 units and by reference to pressure.[47]

This zoning yields a highly effective planning outlook and includes the possibility of mediation and consensus action, proportionate to the complex factors described above. It does so by recovering aspects such as the regulation of land use, the hierarchical networks woven into a village and between villages, the various planes of authority, the organization of economic activities, all factors that show the ways in which local societies relate to territory.

Mapping built for this purpose (Fig. 4.5) should be understood as a cartography for planning because, although it is based on shared information and, therefore, belongs to CIGIS, its informative purpose is territorial planning.

MODULE 3 - This phase of the research focused specifically on participation, that is the involvement of local actors in environmental protection, and adopted

---

[45] The *localization framework* develops, in a social sense, the distance that the MAB takes as a parameter to define the *buffer zone*.

[46] These were determined by the features of territorial cohesion. With regard to *population*, *ethnic distribution* was investigated to determine homogeneity or heterogeneity and to highlight the specificity of multicultural areas. As for *territorial organization* we looked at relationship networks (réseaux), both *hierarchical* and *functional*, between villages in order to reconstruct authority ranks and identify actors. Finally, to establish the *tradition* or *modernity* we set up indicators (type of housing, presence of associations, worship practices, agricultural systems, and infrastructure) useful for assessing impact on the protected area.

[47] The pressure exerted by sectors onto the Park was obtained by calculating: the demographic pressure of each sector in relation to the extension of the core area; the demographic pressure of each village expressed by settlement density compared to the stretch of park boundary overlooking the sector; and, finally, anthropic pressure defined as the ratio between economic activity and natural resources. These indicators enabled us to gauge territorial pressure, which is the overall pressure exerted by each unit onto the Park and its reserves. Territorial pressure provides a measure of a dual environmental risk: the one that concerns the specific sector area, and at a broader scale, also the one that affects the protected area as a whole.

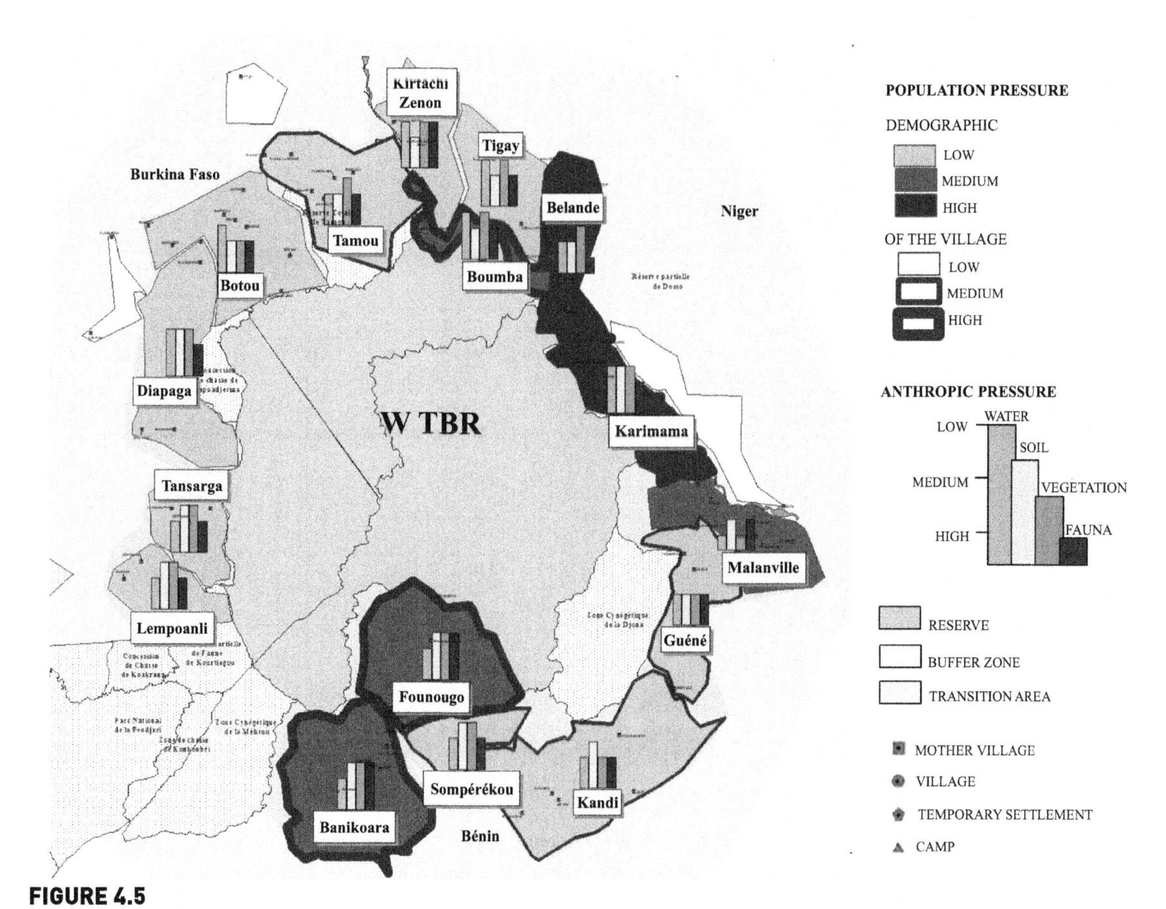

FIGURE 4.5

Zoning sectors on the outskirts of the W Regional Park. (See a color reproduction at the address: booksite.elsevier.com/9780128035092.)

community conservation as an approach expressly designed to achieve it.[48] While providing for different degrees of involvement, such an approach recognizes participation as the crucial factor enabling actors involved (local communities and organizations involved in the management of the park) not only to share

---

[48] The *community conservation* approach involves the participation of local populations during the phases of data collection and mapping. Today, attention to the ethical implications of the spread of information obtained by the local populations is very high, as attested by the mailing list: Open Forum on Participatory Geographic Information Systems and Technologies (ppgis@dgroups.org). In our case, however, I think I can claim that the rights of local people were not only stated, but vigorously defended. For example, adoption of the statute of tradition and, therefore, of territorial legitimacy as a platform on which to build zoning allowed us to acknowledge the cultural identity and the heritage of values that people bring into the circle with respect to natural resources. Besides, the local communities not only gave their knowing consent to the dissemination of information, but expressed a strong interest in making sure that it was circulated, claiming the strategic role that their values can play in conservation and, therefore, in making their rights heard at the tables of consultation and decision-making.

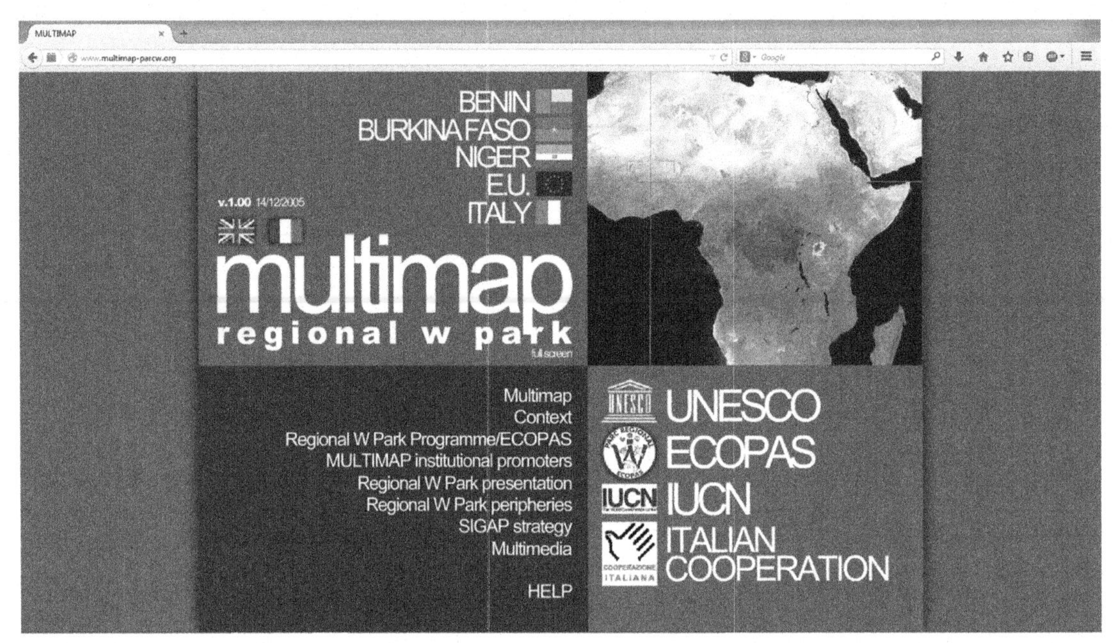

FIGURE 4.6
MULTIMAP Homepage. (See a color reproduction at the address: booksite.elsevier.com/9780128035092.)

basic choices, but to co-manage resources and take full responsibility for such management.

The SIGAP Strategy, in short, established a zoning of *environmental risk* that provides guidelines for ensuring sustainability. The proposed interventions aimed to: 1) contain territorial pressure; 2) resist the dynamics that promote environmental risk.[49] Even at this implementation stage mapping played a key role in assessing the type of risk and its intensity of occurrence. The analysis, which was merely outlined here, allowed us to develop working intervention guidelines addressed to the managers of W TBR.

MODULE 4 - The final phase of research consisted of the implementation of the MULTIMAP system (Fig. 4.6).

It is a system of online capitalization, based on the communicative implications of GIS, which combines the significance of territory with a participatory perspective. *De facto*, the added value of MULTIMAP is that it enables

[49] As concerns theory, we adopted the PSR OCSE intervention model (*Pressure/State/Response*) adjusted to the specifics of the African context. Indicators relating to economic activities were therefore supplemented with those tied to cultural aspects which are underlined in the systemic approach proposed by UNESCO.

users to make their own informational choices, putting them in a position to visualize the interdependence between territorial analysis and cartographic representation. What we are dealing with is an altogether particular system with respect to technique, to communication and to content, for this system boosts interaction, addresses a wide and varied range of users and takes up issues of social and political consequence for environmental protection.[50] From an IT perspective, it is an interactive and multimedia capitalization tool that presents a set of data and their digital processing via an apparatus of maps that yields a multiscale view. Multiple search filters allow superposition of cartographic data or, conversely, their selection and, at the same time, their communication via other forms: texts, photos, movies, music. Once posted on the World Wide Web (www.multimap-parcw.org), this information becomes easily accessible to any web surfer, even to those who are at times excluded from it due to the unreliability of network nodes in some parts of the world, including sub-Saharan Africa. Our efforts were driven by one underlying goal: to create a tool for participatory management on Africa and for Africa, able to handle technical issues through a light multimedia system (based on Adobe Flash Player) posted online and therefore usable in any part of the planet.

Overall, the most significant result the SIGAP Strategy achieved was the zonation of the W TBR periphery based on the complex territorial reticularity which innervates the region at various levels. This supplied Park managers with a model on which working strategies may be shaped with an eye on the holistic meaning of African territory. With respect to cartographic testing, the Strategy proved remarkably effective: first, because it made it possible to give an overall visual form to the diversity of the data collected and offered itself from the start as a cognitive and operational tool geared for communication; secondly, because its ongoing implementation brought to the fore self-referential data that served to reflect on the adequacy of the zoning plan that was being laid out; thirdly, because maps, namely participatory maps and CIGIS, provided the products from which to build a plan for environmental protection in terms of participatory cooperation and co-management.As we move on to consider the phase of posting the MULTIMAP online and setting up a screen for online consultation, we need to address a number of issues regarding PPGIS and online cartography.

---

[50] Operationally, MULTIMAP provides for: 1) dissemination and access to all the data regarding the periphery of the W TBR at regional and local levels; 2) extrapolation of data on individual national components involved in the Protection Program; 3) aggregation of data by sectors and units; 4) targeted consultation of data on each village (obtained by supplementation of maps, photos, texts, and graphics) placed in its territorial context; 5) creation of a knowledge base of pressures impacting the park; 6) diagnostics of periphery dynamics.

FROM THE SHEET TO THE SCREEN: PPGIS AND ONLINE CARTOGRAPHY[51]

Our analysis of participatory systems will now consider the third typology mentioned earlier, that is, PPGIS. We will reflect on the peculiarity of these with regard to the effects produced in the transition from paper mapping to online mapping. In fact, while participatory mapping and CIGIS may be created and used in either form – print or digital – PPGIS require online posting, either via a local network or on the Web. As such, PPGIS call for a mode of use that questions the core assumptions of paper cartography. Since I discussed them at large in the early chapters of this book, I will not enumerate them here. Rather, I would recall the limitations that on-screen cartography entails and later dwell on its most relevant communicative feature, namely the type of interactivity promoted by IT innovations that go under the name of Web 2.0.[52]

WebGIS: Limitations as well as Advantages

Adoption of WebGIS* as a mapping system that supports PPGIS compels us to investigate its technological framework since the early stages of implementation to see whether, and to what degree, the peculiarities of the medium affect the production of meaning in maps, in terms of codification, synthesis and mediation of geographic data.

First it is important to note that individual maps produced by WebGIS systems – which I shall refer to as "views" to comply with common IT parlance – originally shared the display issues that computer screens posed not only in the mapping, but, more generally, in any type of digital rendering. Limiting factors were present even before the networking phase or online transition, and they were caused by the adoption of the screen as a support tool.

---

[51] Much of this paragraph and the one that follows (4.4) was developed in collaboration with Francesca Cristina Cappennani, whose authorship I share. I take this opportunity to thank her publicly for her willingness to interact with me and capture the insight of my analysis, endowing it with communicable form.

[52] Web 2.0 generally indicates an evolution of online services based on all those online applications that ensure a high degree of site-user interaction (blogs, forums, chat, systems such as Wikipedia, YouTube, Facebook, Myspace, Gmail, WordPress, etc.). With specific regard to cartographic innovation, we witnessed the spread of Web Map Service (WMS), a technique, defined by the Open Geospatial Consortium (OGC), which can dynamically reproduce georeferenced layers of information with open-source access. Along these lines, the INSPIRE European Directive, which came into force in 2007, aims to promote the development and use of territorial data throughout the European Union. Web 2.0 technology differs markedly from the so-called Web 1.0, popular until the 1990s, which largely consisted of static websites, with limited interactive facilities to end users except for normal page browsing, or the use of e-mail and search engines. The use of cartography in Web 2.0 is illustrated by Google Earth, OpenStreetMap, Wikimapia, etc.

In order to think about the effects of these limitations on cartography, let us for the moment consider the *view* as an electronic map which, much like a paper map, appears fixed and invariable. One obvious fact is that a display monitor is more unwieldy than a sheet of paper: moving a map, drawing on a map or taking notes and measurements becomes more involved. Unlike its paper counterpart, a screen size determines the maximum viewing area of the map as a whole. What is more, first-generation display monitors entailed a drastic reduction of the map's resolution.[53] Unlike print maps, which vary from a minimum resolution of 300 dpi (dots per inch) to 1,200 dpi, on-screen maps were generally viewed at a resolution of 72 ppi[54] (pixels per inch, that is less than a third of a print map) and on display monitors of fairly limited size. In short: smaller figures and at lower resolutions.

To understand the technical issues this causes in cartography, let us assume we see a map as divided into a grid; the higher the resolution is, the denser the grid will be. As one can easily see, if we keep the size of the viewing frame unchanged, even the degree and sharpness of the represented detail will be directly proportional to the number and size of the dots/pixels used to define it. Hence, the resolution parameter limits the amount of information contained in a cell and on a wider scope, in a map. That is why, as a rule, the size of the smallest representable element in raster graphics[55] (the first one to be adopted also in digital mapping) must be significantly larger than the size of a single cell.[56] In essence, the larger the size of the map frame and the resolution factor, the more detailed

---

[53] Here we refer to the resolution of the medium used to visualize the map (either screen or sheet). Thus, the "resolution" indicates the number of cells (dots or pixels) needed to cover one measurement unit (inch) across the digitized surface of the sheet or screen. When instead we speak of resolution with reference to the layout of raster-based geographical data we mean something different. In GIS, the latter is defined according to the data acquisition method: it may be expressed in meters, in the case of satellite or plane surveys, or even in dpi (dots per inch), in the case of a scan from a preexisting paper map. In either case, resolution is established in relation to the direct source of data capture: for remote sensing, the direct referent is territory and the resolution indicates the size of the area corresponding to a single cell; in the case of raster images, the referent is the printed map source and resolution expresses how many points make up an inch on the map.

[54] Generally, in the case of data acquisition from existing cartography, the input resolution is set to 400 dpi, since this value "allows a good reproduction of all the map details and is a good compromise in terms of the size of the raster files generated" (G. Biallo, *Introduction to Geographic Information Systems*, MondoGIS, Rome, 2002, p. 78). This means that a lower resolution would lead to a loss of information.

[55] There are two main data structures in GIS: raster and vector. In the first, data are stored on the basis of a matrix. Such matrix calls for the segmentation of a surface into even cells, which are attached to an attribute in terms of either absence or presence. A vector-based structure, whose logic is closer to traditional drawing, classifies objects into points, lines or polygons, each of which is expressed by vectors, i.e., mathematical functions that return its geometry, and is tied to attributes referring to space, territory and topology. Unlike generic, non-GIS graphic formats, for example those of CAD, GIS data structures entail georeferencing. In other words, the elementary objects that compose them retain, among their qualifying features, the coordinates of their geographical reference system.

[56] Referring to statistical studies on this issue, Giovanni Biallo underlines the need to "use cell sizes of at least half the length of the smallest element taken into account" (G. Biallo, *op. cit.*, p. 50).

a raster map is. This "matrix-like" mode of graphic representation, which underlies the functioning of the display monitor itself (what is perceived visually on the monitor is always the result of a dense sequence of cells and pixels), is thus, somehow, the inevitable prerequisite of screen-based visualization.

To be sure, both the introduction of vector graphics, which define graphic information through mathematical functions, and the design of ever more powerful display monitors and graphics cards have virtually overcome all the visualization issues that a "cell-based" graphic layout entails. This, however, does not deny the fact that the limits of that layout had be accounted for in the early development of digital maps and that, as we will see shortly, such limits influenced the thinking and iconographic choices of the cartographers who first had to deal with the transition from paper sheet to digital media. That is all the more true if we consider that, although they do address most resolution issues (except for display monitor size), vector graphics formats are best suited to render simple graphics forms, made of geometric figures and flat colors. Complex images are instead hard to render vectorially, which tells us that vector formats will make it more difficult to include semantic codes such as *stumping* that are quite common in cartography. The raster format seems better suited to express those codes, despite the drawbacks we already discussed. In summary, the mere transition to a digital system forced us to come to terms with some technology-related issues and to question the advisability of adopting some of the sign conventions until then commonly used in cartography, thus opening the way to new solutions.

To this we should add that, at first at least, if one considers a map as a static object and the web page as its frame, the internet for its part failed to provide a solution to the issues outlined above. If anything, the web seems to have exacerbated them. In most cases, the embedding of a mapping in a website meant (and still largely means despite the advent of touchscreen tablets) that the map had to contend for space with the browser's menus and tabs, scroll bars and navigation bars, banner ads, logos; in short with whatever usability criteria[57] had dictated in order to facilitate site browsing and later became a permanent fixture of users' browsing. What followed was an even greater reduction of the map size with respect to the maximum size fixed by the display margins, at a resolution that, however, was still optimized for the display itself.

It should also be kept in mind that a web-based platform makes it impossible for cartographers to predict the recipient's hardware and software configuration with absolute certainty: the graphics card, display resolution,

[57] The term *usability* strictly means "ease of use." The term is now commonly used with regard to the web, namely thanks to the studies conducted by Jakob Nielsen, whose research aimed to isolate criteria for optimizing and simplifying web browsing for end users. His leading book is: J. Nielsen, *Designing Web Usability: The Practice of Simplicity*, New Riders Publishing, Indianapolis, 1999.

platform type, the browser used, represent variables that will determine the final appearance of the map on the user's display monitor. The only type of control cartographers can exert on the outcome of their work is to plan maps based on a rough estimate of the average surfer's technical skills or at least of the skills of their target users. With respect to resolution (but, as we shall see, not exclusively), this had considerable consequences, especially with files using the raster format, because if the recipient's display size clashes with the one expected, the visualization issues discussed so far actually increase. For instance, in the past it was quite common for the map frame to be too small. One was forced to scroll vertically or horizontally to visualize it, which was both tedious and ineffective. The map's communicative impact was marred, because part of it would be constantly beyond the frame, thus preventing users from gaining an overall view of the mapped territory, especially on small-scale maps. Again, we could think of the implications of having to "weight" each element of the representation. Smaller graphic features may become invisible if the map is reduced beyond an acceptable size, and that is not because a different reduction scale is applied according to consciously established criteria, as it happened with traditional maps, but due to the inadequate resolution of the user's display monitor, not optimized for the kind of visualization the map was intended for. These are issues addressed only in part by later advances in PPGIS technology, which allow users to resize web frames, adjust their layout and recalculate the map's new extension and an adequate scale factor. Even so, frame size and display resolution continue to hamper cartographers: the restrictions they impose can never be thoroughly circumvented.

Repercussions in Cartography: Changing Codification

What we have seen so far shows that the semiotic process of cartography was now subject to new conditions. They were restrictions at work even in the initial phase of mapping, when one's aim is to produce a web-optimized map: conditions which affected the cartographer's selection, simplification and codification of cartographic data. Cartographers were forced to adopt new criteria to build maps suited for the type of content transmission, readability and visual quality prescribed by digital documents. In addition to a simplification of detail, made necessary by the transition of conventional print maps onto computer screens, they had to cope with reduced map sizes, display resolution issues, web limitations and technical restrictions inherent in digital graphics formats. Design guidelines for maps, adopted to solve technological issues, eventually imposed themselves to outline a renewed, if not completely new, semantics of maps. They are essentially the product of discussions by cartographers who dressed the role of *Web map designers* and reflected on the implications that the IT medium and the Web have on cartography. These reflections pursue a century-old line of research, whose aim was and is to refine and perfect cartographic art and mapping techniques. Yet they have also led to the creation of a specific cartographic domain aptly called *Web cartography*.[58]

Within this specific field, grounded both in studies on web usability and on encoding rules drawn from cartography, researchers looked from the start for new modes to take full advantage of the possibilities the medium presented. Technological shortcomings were redressed with expedients offered by the IT medium itself. Action was taken along this path in two directions. First, variables at play in the encoding of traditional cartography were revisited and their suitability for web use appraised. Then new solutions were found, hardly used in print maps because considered either ineffective or uneconomical or impossible to achieve.

We are going to look at some examples of the new developments that invested the semantic codes of maps so that we can refer to them later, when in a more cogent line of argumentation, we will endeavor to understand if and how they intervene in the redefinition of icons and the communicative outcomes of maps. The particular technology cartographers have to cope with in order to produce digital maps optimized for computer screens entails first that they should strive to achieve a low level of informational densification.[59] Online maps, unlike their print counterparts at similar scales, call for restraint in the number of territorial elements or geographical features to be represented. To remove the combined shortcomings of resolution restrictions and reduced display size, and ensure good readability of all their features, digital maps should generally be designed to include a limited quantity of informational content. Such expedient provides for the use of larger, and therefore more easily identifiable, graphic symbols than the ones normally used for print maps. From a semiotic point of view this is no trifling matter, because it affects the selection and generalization of geographic data. It is a delicate intervention, which may well have destructive consequences if carried out arbitrarily, for the semiotics mechanisms that a map activates mean that the absence or omission of a sign on the map entail the negation of the designator that sign refers to.

---

[58] The foremost reference is to Menno-Jan Kraak and Allan Brown, who brought together contributions from several scholars to formulate standards of usability for online maps. The practical design guidelines mentioned in the pages that follow constantly refer to: M-J. Kraak, A. Brown, eds., *Web Cartography, Developments and Prospects*, Taylor & Francis, New York, 2001, especially pp. 87–106. An overview of the issues related to online cartography design and to the restraints imposed on the map when it is posted on the web may also be found in B. Jenny, H. Jenny, S. Räber, "Map design for the Internet," in: M.P. Peterson, ed., *International Perspectives on Maps and the Internet, op. cit.*, pp. 31–48.

[59] In the words of Jeroen van den Worm, who set out to find new design rules for online maps, "not overloading maps with too much content is a basic principle applicable to all web maps": J. van den Worm, "Web map design in practice," in: M-J. Kraak, A. Brown, eds., *Web Cartography, Developments and Prospects..., op. cit.*, chap. 7, pp. 87–106, cited p. 106.

As attempts were made to find a middle ground between technical requirements and mapping needs, technical stratagems were devised which eventually seeped into the very encoding of cartographic data. Even without touching upon the interactive functions typical of a web environment, we could consider for instance the use of *transparency*. This is a graphics effect which, in the specific case of online maps, can render a layering of levels the less marked of which are overlaid with a sort of transparent film (*fogginess*), making it possible to give more emphasis to prevalent icons on the levels that was not overlaid. Transparency meets the need to show fewer elements on digital maps than on printed maps but avoids the risk of losing information which, although marginal, may be useful to a thorough understanding of territory for the purposes of referencing. In this sense, transparency also comes across as semantic variable, because it derives from the use of color to establish a visual ranking between elements and thus promotes the creation of the connotative meaning of icons.[60]

Transparency is a strictly graphics tool, but among the most effective functions made available by the IT medium to compensate for the limited space maps are given on a webpage and thus avoid resorting to a drastic selection of data, there are some that show us the dynamic dimension of maps. Operations such as zooming and panning, very common especially but not exclusively in WebGIS maps, activate a sequence of views which do not fundamentally alter the concept of scale reduction or invalidate the principle of analogical accuracy on which maps are based. Yet, such views provide for a virtually endless widening of map boundaries within a fixed frame and they adjust to the level of detail sought by the user.[61]

The "interferences" of display monitors on the semantics of maps led to a redefinition of the usage rules for designators. Consider, for example, the typefaces and formats traditionally adopted in cartography, which the new digital environment tends to alter or discard. Italics, conventionally used to distinguish natural features from anthropic artifacts is generally avoided because it loses sharpness on a display monitor. Capitalization is used instead, because even when zoomed out, uppercase is more readable than lowercase. Printed maps, instead, would limit the use of uppercase to highlights. Finally, while

---

[60] Here I am referring to the communicative outcome of the interaction between cartographers and recipients, mediated by self-reference. This is bound up with the Euclidean setup of web maps, which aim to convey as a socially constructed and shared value what was in fact produced by topographic icons.

[61] I should recall that the basic structure of the map is the set of assumptions, rules, and graphic formats – including reduction scale – on which a denominative projection is built. It renders objects analogically as they are found in reality and, in topographic cartography, establishes that their metrical relations are defined accurately. In this case, although we are dealing with a sequence of maps at different scales, we need to consider whether that has repercussions on the communicative outcome. E. Casti, *Reality as representation...*, *op. cit.*, pp. 47–58.

in traditional maps names are positioned along the geographic features they represent, on the web it becomes necessary to limit their inclination as much as possible, even though the recent introduction of tablets with screen rotation options has made this requirement less urgent.

Besides these general issues, WebGIS-generated cartography presents other difficulties due to the automatic placement of designators on the map. Since the cartographer no longer intervenes manually to adjust or move captions to boost the map's legibility and improve understanding, problems arise when captions grow particularly dense. These may end up overlapping or, if the software allows, they may be erased or moved in accordance with initial settings largely entrusted to the machine and difficult to tweak after the initial setup. It is common practice to address this issue by programming the on-screen appearance of names at a preset scale interval so that some captions will only show beyond a certain zoom factor.[62] Another common expedient is the use of pump-up. The designator, and with it possibly other content, appears when the cursor hovers over the reference object. In either case, these two ways of handling designators already go beyond the notion of static maps and activate dynamic features which, as we shall soon see, have tangible repercussions on the mechanisms of cartographic communication.

To the first rule of Web cartography, which entails a restriction of data densification, a second principle is also added, whereby the cartographer involved in producing a web-optimized map must ensure that a number of features are recognized immediately. They are features which represent the primary data conveyed by the map but also activate functions made available by the IT medium: interactivity, animation, multimedia. So, for example, it is not uncommon to find a 3d effect applied to point symbols that serve as links to other information. That is done in order to give such links the appearance of buttons and, mimicking the experience of the average user, make them easily identifiable as "hot spots" (Fig. 4.7a).[63] That brings into play a visual variable, shadow, which did not have the same semantic import in printed maps, where hachures (or shading) were basically used to render altimetry contour lines of the terrain (Fig. 4.7b).[64] Now this graphics mode is adopted to convey, at least on a superficial level of interpretation, operational information to do with the

---

[62] Signal examples are Google's mapping services: Google Maps and Google Earth.

[63] It is a visualization of Rocky Mountain National Park valley in Colorado (USA), which allows the user to take advantage of multimedia. When the flag is clicked, an information pop-up appears. See: http://www.viamichelin.com/web/Maps/Map-United_.

[64] In this case, the figure shows an excerpt of a paper map of the area itself, where shading is used to render land elevation. See: *Rocky Mountain National Park, Trails* illustrated topographic map, 1: 50,000, National Geographic, 2008.

(a)

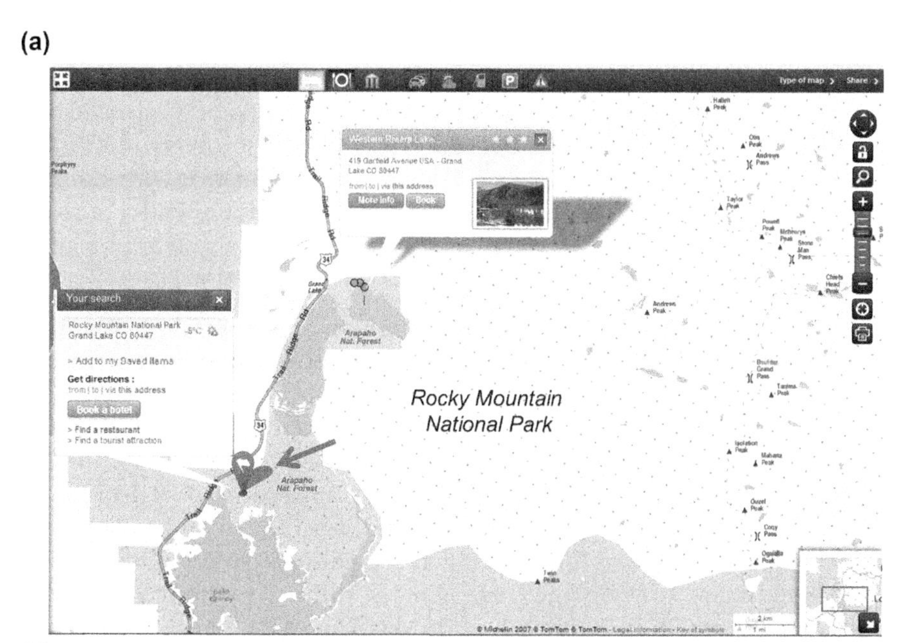

(b)

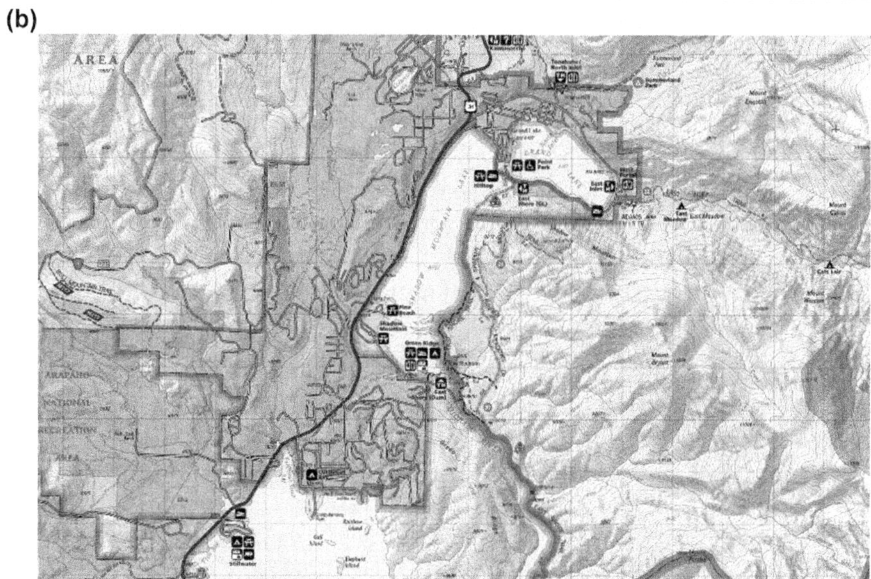

FIGURE 4.7

A multimedia map (a) and a topographic map (b) of the Rocky Mountain National Park in Colorado (USA). (See a color reproduction at the address: booksite.elsevier.com/9780128035092.)

hypermedia and interactive reading of the map, without reference to any distinctive, testable feature of the designator.

This graphics mode, as already seen for transparency, intervenes in the construction of the connotative meaning of map icons. When hypermedia functions are attached to features and visually marked, iconization is activated and the importance of that specific designator with respect to others is asserted. Transparency and shading are thus added to the icon variables of traditional cartography, such as the size of a point marker or the color intensity of areas, which by convention serve to signal a hierarchy, a ranking system or a given priority in numbers, economics, demographics, culture or others.

Technical requirements also affect the choice of figurative signs. Reduced display frame sizes for web maps and the use of vector graphics tend to favor geometric shapes or simple abstract figures over complex, figurative symbols. The former are easier to view even on screen even at comparatively small sizes, while the latter, although more aesthetically pleasing, would have to be much larger than their printed counterparts to be legible on screen. The same applies to line forms, areas, orientation and background color, which are ill suited for adoption as semantic variables in the drawing of a digital map. For one, their use is made difficult by poor screen visibility, which would make them hard to distinguish at times. For another, they make graphic processing more complex and increase loading time. The latter aspect, compounded by poor resolution and limited map size, needs special consideration particularly when dealing with WebGIS produced maps. For unlike other online mapping types, the *view* of a WebGIS exists only virtually, in the form of a data set, until user-specified settings trigger its production. The visualized map will be thoroughly processed by the system and rendered each time based on user specifications.[65] This requires intense computational power, over a generally large amount of data, to create and render the map graphically. Hence the need to simplify graphics, reducing processing time and, if no broadband issues intervene, yielding quicker search results to the end user.[66]

Ensuring swift processing response is even more critical in our case, for the time it takes for data to be visualized on screen has been shown to have

---

[65] As in all web-based operations, even web cartography is bound to the client/server web framework, based on a system of data distribution from a computer called *client*, which sends requests for data or services to another networked computer devoted to that purpose, called *server*. See: B. Plewe, *GIS Online*, OnWard Press, Santa Fe, 1997, pp. 63–95.

[66] Data flow between *client* and *server* (and back) determines traffic along web connections, metaphorically called information highways. Data transfer speed, expressed in bits per second, depends both on the physical line and on the volume and type of data being transferred. For up-to-date statistics on the subject see the NUA website.

repercussions on the communicative process. Even early studies on web usability conducted by Jakob Nielsen showed that beyond a 10-second wait time the user's attention level decreased and the propensity to leave rose. Today's fast connections have made this time lapse even lower.[67] Communication may then be broken even before the message reaches the addressee. Besides, if response times are sluggish the user loses the perception of interactivity which is instead a distinctive feature of the Web environment and represents one of its foremost potentials from a communicative point of view. All this suggests how important it is to use all the expedients that may variously contribute to improving response times in a web mapping service. And the most critical among these is arguably setting up simplified and streamlined graphics,[68] which require modest computational power and use the most abstract type of symbols available.[69]

Our overview of traditional cartographic codes and our inquiry of the changes that they have undergone on account of technology would not be complete without a reference to chromatism, and the type of issues involved in the use of color in a web environment. It should be pointed out that color adjustment was one of the most debated and thorny issues faced by early web designers.[70] Hence the so-called *Web-safe* palette of colors, comprising 216 colors that all browsers would have to support and render in the same way, seamlessly across all systems. These were the color values that graphic artists and, in our case, web cartographers would have to comply with to ensure a consistent color experience on the display monitors of end users. Quantitatively, this color spectrum did not impose heavy limitations on maps. After all a 216-color palette was more than a cartographer could use on a printed map and significantly more than the colors generally present or needed in a cartographic representation. In fact, limitations obtained even here, because Web-safe colors tended to have darker shades and provided for few pale colors, which are instead those normally used in cartography for backgrounds or for the rendering of less relevant themes. Adequate color contrast was thus difficult to achieve, and made it impossible to distinguish similar shades of color, an essential procedure especially for interpreting thematic maps.

---

[67] J. Nielsen, *Designing Web Usability…*, *op. cit.*, pp. 42–49.

[68] Another key expedient to improve response times in comparable software and hardware setups is to use a well-structured Geodatabase.

[69] That is all the more true for Web 2.0 applications where simplified graphics content typically facilitates access for a wider number of users. In these applications, the basic service provides a very "clean" map on which users insert their contributions. This is a binding need for those who attempt to innovate the language of cartography in order to render landscape, as we shall see in the next chapter.

[70] For an overview of the complex issues surrounding the use of color online, see: M.E. Holzschlag, *Color for Websites: Digital Media Design*, Rotovision, Hove, 2001; R. Pring, *www.color. Effective use of color for web page design*, Watson-Guptill, New York, 2000.

Even though the WebSafe palette was superseded by a 16-million-color RGB palette, supported by new display technology and advanced graphics cards, the variables involved in color rendering on screen (different graphics setups, users' brightness and contrast settings) make it virtually impossible even at present to fully control color rendition on the web. The user of color is encouraged by the fall in production costs for color displays on electronic devices, with images that look ever more vivid and captivating. Yet, the issue of color rendering online makes it crucial to reassess and rethink its scope of use, bound to context and to specific requirements, such as the need to mark a distinction or encode a hierarchy.

It is clear from what we discussed here that the encoding of territory information for web cartography is subject to interferences, which translate either in a reduction or in an expansion of the variables used in traditional maps. Such adjustments, made mostly to conform to the technological medium, are relevant for the communicative effects their combined presence has on mapping and on its interpretation. Nonetheless, they do not invalidate the semiotic dynamics we found in traditional maps. Whatever the gamut of available codes, propensity to opt for a specific one does not entail any actual disruption of the functioning of the cartographic communication system as such. For that will still rely on the concomitant presence of two distinct structures: analogical, which reflects, mostly on a two-dimensional plane, the arrangement of objects in a three-dimensional world; and digital, which, via a symbolic, albeit changeable apparatus renders discrete attributes of the designator it represents. Ultimately, the domains of semantics, syntax and pragmatics do not alter their patterns of intervention for establishing denominative cartographic projection, nor do they change their mutual relations. To that effect, and at to this level of reflection, the cartographic icon forms and creates meaning today on the web just as it did on paper in the past. The working tools and the pieces of the puzzle may vary, but the method remains unchanged. Cartographic semiosis, understood as "the process whereby something [on paper] functions as a sign for someone called upon to interpret it" remains substantially unchanged. If, therefore, the World Wide Web promises a "revolution" even in cartography, that does not consist of the interposition of the display screen or in the fact that maps are circulated online. That kind of interference would at most account for certain innovations that involve primarily the realm of graphics and of cartographic design. If there is a "revolution," then it lies hidden elsewhere, beneath more articulate layers. These we will analyze shortly, as we leave behind a technical notion of the web and of what it contains to embrace its complex reality at a wider scope, able to capture some of the dynamics whereby, inside it, meaning is produced.

ONLINE CARTOGRAPHY: INTERACTIVITY AND FIRST SEMIOTIC IMPLICATIONS

In order to understand the innovative thrust of interactivity we need to outline the broad features of online cartography.[71] Data digitization, their transfer over a network connection (namely the World Wide Web) and, finally, the possibility of rendering such data visually online, have led to the development of what is commonly called *online cartography*, which generally indicates the set of digital maps that circulate on the World Wide Web.

Speaking of online mapping means, in fact, referring to a multifarious universe, which comprises different types of representations, both in the ways and means of their production and in the final forms they provide to end users. For the sake of accuracy, we will thus attempt a classification of specific phenomena that fall within the broad realm of online cartography. To be sure, we are still dealing with relatively new and changeable phenomena, which have not yet been systematically defined in exact and consistent terms. In some cases, unambiguous definitions are hard to achieve and misunderstandings may occur.[72] Even in the past, this issue led researchers to set up classifications confined to their specific field of research.[73] In a context like this, still quite magmatic, what we propose is but one of the many possible orders, not at all final or conclusive and yet reasonably consistent with our approach.

One final clarification is called for. As we set out to differentiate one cartographic typology from another based on their level of interactivity, we must acknowledge that their boundaries are blurry, even though the classification we propose here establishes and upholds clear demarcations. In fact, using

---

[71] What our work necessarily leaves at the margins is a wide-ranging analysis able to capture the semiotic aspects of digital cartography and gauge their weight in determining perception of territory and modes of intervention.

[72] The lack of clarity still prevalent in this realm is attested to by the inconsistent use of terms such as *dynamic maps* and *interactive maps* to refer to the same type of objects or to different objects altogether. Even Dodge and Kitchin, who propose an articulate cartographic classification, in a few paragraphs of their *Mapping Cyberspace*, use the term "interactive" to refer differently to "dynamic" maps and "multimedia" maps ("hypermedia maps") (M. Dodge, R. Kitchin, *Mapping Cyberspace, op. cit.*, pp. 76–80).

[73] To see how these coincide or differ and to grasp the complexity of the issue see the classification proposed by M. Dodge, R. Kitchin, *Mapping Cyberspace, op. cit.*, p. 72; M.J. Kraak, A. Brown, *Web Cartography, op. cit.*, pp. 3–4; B. Plewe, *Gis OnLine, op. cit.*, p. 47. The classification presented here draws important suggestions from the work of these authors. Although we obviously recognize the relevance of these studies for our purposes, we cannot thoroughly subscribe to the typological distribution they advocate. The framework they propose is accurate but still presents significant shortcomings. We will attempt to explain why in the rest of this chapter.

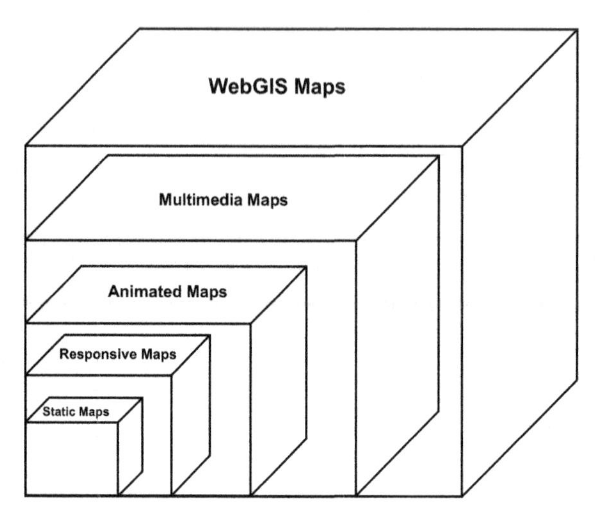

FIGURE 4.8
Systematization matrix for online cartography.

a metaphor to show the system of relationships they mutually entertain, we could think of Chinese boxes: one typology is, in some way, incorporated into the next and thus superseded by it (Fig. 4.8).

Thus, responsive maps may be regarded as enhanced static maps; an animated map, in itself an uninterrupted series of static maps with interactive features comparable to those of responsive maps, incorporates and extends the first two; a multimedia map may adopt some features and potentials of all the typologies that precede it, fine-tuned to its own goals. Similarly, and suitably to our analytical purposes, we find some degree of overlap even between multimedia maps and WebGIS maps, not least because a WebGIS system can have multimedia maps as its cartographic output.[74] Yet, as we compare multimedia maps to WebGIS we cannot fail to note that the geo-database system of mapping (which WebGIS borrow from GIS) combined with web-based services opens a marked rift between multimedia maps and WebGIS and consequently even between WebGIS and other online maps. That occurs both on the level of technology, and as we shall see, on the level of semiotics, due to the implications that such combination has with regard to the amount of data processing involved and the possibility of user intervention in the definition of the map itself. That confirms the absolute exceptionality of WebGIS, which emerge in the field of cartography in truly revolutionary terms.

[74] Some scholars welcome the ever-increasing alliance between GIS and multimedia cartography as opening up ever-wider possibilities, which finally release GIS from the preserve of technicians and pundits (W. Cartwright. M.P. Peterson, G. Gartner, eds., *Multimedia cartography*, Springer, Berlin-Heidelberg, 2007, 1st ed., 1999).

We need to stress once again that our systematization must be viewed within our scope of analysis, grounded in the assumption of the semiotic and communicative aspects of maps. What matters to us is to establish the meaning that the cartographic object and its individual components have within the relational process between an issuer/transmitter and a receiver. We believe such systematization should rely on analysis, both of the ways in which maps are generated, a phase largely controlled by the issuer/transmitter, and of its modes of presentation, perception and use on the part of the user/recipient, a feature of the utmost importance for electronic maps in general. The dynamics differs from that involved in traditional mapping because once the map had been drawn, it was subject only to interpretation but not transformation on the part of end users. Cartographers produced physical objects, with features and shapes that remained unchanged after their creation. The arrival of the digital world changed the rules and the new context in which these maps operate cannot be ignored. Together with multimedia, the web environment anchors its own existence and expansion in a type of communication that is no longer one-way, as in the one-to-many scheme of the arts, television or cinema. Communication is now multi-directional, based on the many-to-many scheme which favors communicative models where the action of the transmitter and that of the recipient no longer allow sharp distinctions.[75] Roles mingle and with them, new mechanisms emerge, which profoundly affect even the latest developments in cartography.[76]

Starting with these assumptions, we find that the user-centered perspective adopted especially in the past by some scholars is limiting, because it envisages an order for web cartography based solely on how the map appears to

---

[75] In fact, information control on the part of the cartographer is ensured today by the presence of classification scripts: metadata. Metadata are an integral part of the semantic web, a term coined by Tim-Berners Lee to refer to the possibility of arranging web data in a structured form to make user search and queries more effective (T. Berners-Lee, M. Fischetti, *Weaving the Web*, Harper, San Francisco, 1999).

[76] In a Geoweb 2.0 environment, the terms WikiGIS*, Geo-wiki, and wiki-mapping indicate solutions based on the principle of geo-collaboration. This way of building geographical content favors a dynamic and collective production of knowledge. It links users transversally, making it possible to work democratically with the aim to share information effectively and constructively. A WikiGIS may be defined as a GIS built online which conveys collective knowledge over a territory or a given spatial phenomenon. Unlike many other cartographic services, based on the volunteered creation and enrichment of geo-localized content without an actual intervention of users on the database, WikiGIS technology ensures recorded traceability at all stages of contribution. In a Wiki perspective, maps become instruments for reading the dynamics at work in the construction of spatialized knowledge and those implied in the traceability of spatial representations. On WikiGIS see: L.D. Ciobanu, S. Roche, T. Badar, C. Caron, "Du Wiki au WikiSIG," in: *Geomatica*, v. 61, n. 4, 2007, pp. 455–469; S. Guptill, "GIScience the NSDI and GeoWikis," in: *Cartography and Geographic Information Science*, v. 34, n. 2, 2007, pp. 165–166; S. Roche, "Leggere e scrivere la carta sull'onda del Geoweb 2.0," in: E. Casti, J. Lévy, eds., *Le sfide cartografiche...*, *op. cit.*, pp. 47–64.

the end user in terms of its static or dynamic viewing, with no regard for the times and the modes of maps' production.[77] That runs the risk of sidelining solutions for online cartography which show their potential for innovation at that very phase.

We would favor a systematization that tends to isolate types via an extended application of the notion of dynamism, not simply seen as the counterpart to visual stasis, but played over a dual level, engaging both the time of map visualization and the moment of map production. In this light, dynamism combines with interactivity and redeems the latter term from general misuse. In the digital world anything, or nearly anything, may be defined as "interactive," but it is our aim here to recover the word in its new and more meaningful valence. To be properly acknowledged, such valence requires calls for an identification of multiple levels of interactivity, which would make it possible to differentiate one electronic type of cartographic object from another.[78] As we reject a generic and meaningless definition of interactivity, we rely on what was suggested by Lev Manovich, who talked about "closed interactivity" versus "open interactivity."[79] The former refers to that type of interaction in which the user plays an active role only in determining the order in which a digital object is explored, that is the access path to features which were previously produced and placed in a fixed system of relationship. "Open interactivity," by contrast, refers to a far-reaching interaction which enables the end user to issue commands not only to set up an order, but also to alter and change both the structure and the elements of a given digital object. In the context of geography, that yields a digital tool for planning and managing territory.

Along the argumentation lines deployed so far, we reach a definition of a structure built on three fundamentals, derived even in this case from our choice to trace the semiosis activated by the cartographic medium. First we have the axis of *semantics* (or *the construction of signs*); then we have the axis of *syntax* (or *relationship between signs*); finally we have the axis of *pragmatics* (or *usability of signs*). Each axis runs along a range of increasing values for specific features. Thus, the axis of semantics follows a course that stretches from *fixity* to *mutability*. The axis of syntax stretches from *immobility* to *dynamism*, via intermediate phases such as *transformation*, followed by *animation*, which mark the shift from an essentially metamorphic to a properly cinematic quality. Lastly we

---

[77] The reference is still to M.J. Kraak, A. Brown, *Web Cartography…*, *op. cit.*

[78] In fact, M. Kraak and A. Brow also include map interactivity as a parameter for second-level classification, but they fall short of pointing out what to us is essential, that is, a different degree of interactivity.

[79] L. Manovich, *The New Media Language*, MIT Press, Cambridge, MA, 2001, p. 40.

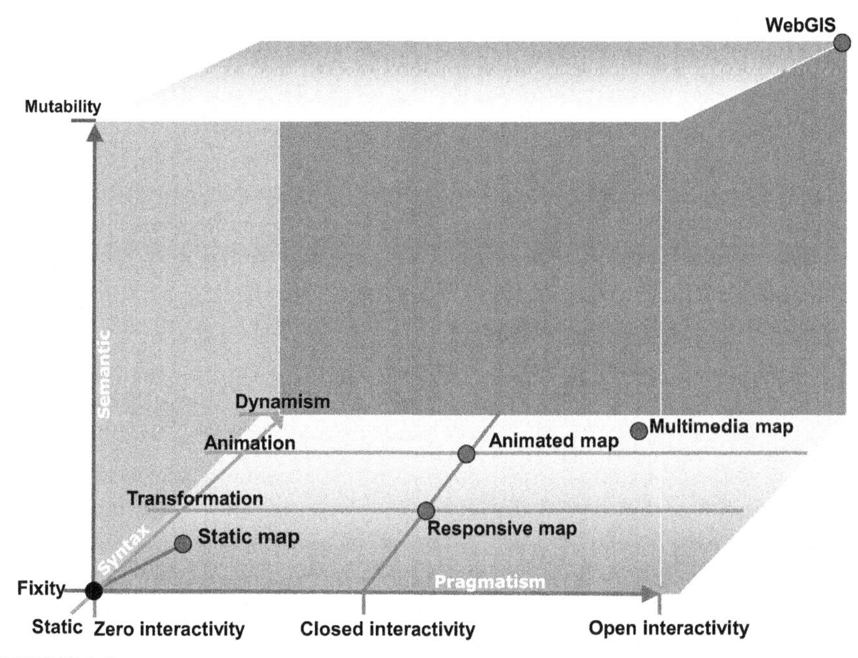

FIGURE 4.9
Incorporation relationships between online cartographic types. (See a color reproduction at the address: booksite.elsevier.com/9780128035092.)

have the axis of pragmatics which – to borrow Lev Manovich's terms – goes from *zero interaction,* i.e., the absence of interaction, to *closed interaction* and finally to *open interaction*.

Five types of online maps may be derived from this scheme: *static, responsive, animated, multimedia* and lastly *WebGIS* (Fig. 4.9).

Static maps – When we talk of static maps we refer to common digital images which simply rehash conventional maps. These were not originally produced for web publication, but rather derived from scans of geographic maps preexisting on paper, or drawn or reworked with graphics programs for purposes other than their online use. The end result is still a cartographic product which by format and size is suited for web publication. Static maps were fully produced before being placed online or visualized by a web browser as standalone objects or as features of a website. Thus, to the end user, they appear fixed, and lacking in dynamism. In addition, they provide a zero level of interactivity. This type of map is commonly intended for passive viewing and differs from a conventional map only by virtue of its medium, the sheet of paper having been replaced by a computer display. Along with conventional maps, static maps lie thus at the origin of the three axes of typological definition of maps: transposition onto the web led to no substantial innovation in semiotic terms. Once scattered anywhere on the web, also because of their low cost, ease of implementation and web transfer, such representations were a popular form on archival or commercial sites and are still present on the web pages of agencies or public institutions that make their databases or their cartographic assets publicly available to end users. One instance is the collection available on the *Perry-Castaneda Library Map Collection* website, a subsection of the website of the *University of Texas Library Online*, which offers a wide range of this type of digital map, collecting maps from around the world both at small and at large scales.[80]

Responsive maps – In a sense identical to static maps, except for the implementation of a technique which enhances their interactive complexity with users and their propensity for alteration.[81] Even in this case we have common digital images in web-optimized graphics formats, different from the previous category essentially because they are "clickable," or reactive to the click of a mouse. At the lowest interactivity level, responsive maps can be used as mere tools for navigating the global hypertext, since the whole image is usually clickable and links to other web pages, to an enlarged version of the image itself, or to another digital object (audio, video, etc.). A more advanced level of interactivity, still achievable within the most common W3C[82] web standards, divides maps according to coordinates[83]: portions of the image (in general of any image, including, therefore, the one that reproduced a map) called "responsiveness areas" or "hotspots" are each linked to other web pages[84] or objects of different types, as in the case described previously. The link is often to a variable scale enlargement of the selected image. The results achievable by applying this knock-on method to a sequence of maps led some to claim that these maps were the simplest GIS technology online. A semiotic analysis of the cartographic tool, however, as we shall

[80] See http://www.lib.utexas.edu/maps/index.html (University of Texas at Austin Library). It should be noted that this website is a telling example of a special type of WebGIS service, or *Static Map Display* in the definition of Brandon Plewe (B. Plewe, *op. cit.*, pp.73–75), which is largely considered the simplest and most cost-effective service. Note, also, that the same site may still feature services that offer different types of digital maps.

[81] The term *responsive*, borrowed from common usage, is used in a web environment to indicate a link to another document or web page, activated by the mouse cursor.

[82] The World Wide Web Consortium, that is, the body devoted to defining and recognizing web standards.

[83] Specifically, in the field of web graphics and web design, *image map* refers to images marked by this interactive feature.

[84] Interesting examples of responsive maps with links to web pages that provide information on the areas selected may be found on the Grosseto Online website, the civic network of the province, visible at http://www.gol.grosseto.it/puam/comgr/stor/natura/mappa02.htm. The website uses responsive maps with informative or promotional goals, providing short historical, environmental or geographical notes on the medieval sites of the Grosseto province or environmental data for each single town.

see, would see this as an undue comparison, because it neglects the fundamental structural difference between the two typologies. In fact, responsive maps entail a fairly limited degree of interactivity, definable as the lowest possible level of interactivity within a closed interactivity context. Nor can responsive maps activate the semantic and syntactic dynamics which are instead typical of Web-GIS. This type of cartographic product allows, therefore, only minimal navigation, and is essentially based on imitating zoom capabilities. An example in this regard is the site of the Civic Network of the City of Pisa, showing a general map of the city divided into quadrants: the selection of each brings up a higher scale image that makes the road network of the selected area clearly visible.[85]

Animated maps – These maps introduce an element until then unknown in mapping: animation. They are characterized essentially by their mobility, which mimics vision. These maps "move" because they result from a chain of static maps, usually in GIF format. Each map is slightly different and their fast, seamless sequence produces a film-like effect of dynamism. Because of that, animated maps are well-suited to show diachronic spatial phenomena, journeys, movements or layering of multiple phenomena within the same region. Their areas of application are, predictably, many. Animated maps are used to represent diachronic changes in demographics,[86] forecasts of meteorological variations or, for example, in the field of history, the movements of armies in the reconstruction of battles.[87] Even the virtual world of the web is not foreign to this type of map. Think, for example, of the maps developed as part of the PingER[88] project, which use localized animated histograms to show the monthly trend of web performance (package loss, delivery time, etc.), based on the monitoring of numerous sites spread across all continents. The main limitation of animated maps is that the user has no control over the object and the recipient's actions are limited to viewing, possibly following links to the global hypertext or accessing other data with the click of a mouse. This type of electronic map gives no control over frame speed, so animation may be alternatively too fast or too slow. That does not facilitate understanding, and seems often dictated less by a need to meet the cognitive and communicative goals of mapping than by an indulgence in special effects on the part of an amateurish cartographer.

---

[85] Responsive map of the municipality network of the city of Pisa: http://www.comune.pisa.it/doc/pisa1.htm.

[86] For instance, see the map that shows the demographic trend of humankind at http://desip.igc.org/mapanim.html.

[87] One example of this is the Cedar Mountain Battle map at http://bhere.com/plugugly/1862/6209cma.htm.

[88] *Ping End-to-end Reporting* is the title given to the *Internet End-to-end Performance Measurement* (IEPM) project, aimed at monitoring web traffic between member sites. The maps, which were processed from data from major web research and study centers (CERN, NUA, etc.), may be reached at http://www-iepm.slac.stanford.edu/pinger/perfmap/.

Multimedia maps – Even more than animated maps, *multimedia maps*[89] exert a very powerful aesthetic captivation. Here, however, aesthetics combines with ever more sophisticated interactive features. These maps offer at times functions that trigger graphical changes, or button-controlled animated forms.[90] At other times they are presented in static form but with supplementary interactive features. In a sense, multimedia maps combine all the cartographic types we have considered so far, yet they also supersede them quite markedly. For one, these cartographic products provide for different mouse behaviors (not only clicking but also hovering) which correspond to different actions. These actions range, at least potentially, from the visual transformation of the map itself, to pop-ups containing information, photographs, virtual reality environments, or sounds. That, however, can already be obtained technically (although rarely and with less convincing results in practice) with responsive maps animated via links.[91] However, in the development history of electronic cartography, multimedia maps signal genuine innovation, or the achieving of a form of interactivity which involves the internal components of the cartographic structure. In other words, the interactive functions of multimedia maps do not merely consist in linking a map online to other pages, objects or websites but in activating dynamic behaviors between the very constituents of the map itself. Hence legends which vary according to the themes displayed or, on the contrary, icons highlighted on the map based on a selection from the legend entries;[92] data and information that appears when you hover on a point of interest,[93] and even the possibility of choosing themes to

---

[89] These are objects built as specific software for creating digital movies, multimedia animations and objects for the web. Generally based on vector graphics, they usually need a plug-in for viewing. The most widespread formats include AVI, QuickTime, SWF. File format should not, however, be taken as a distinctive feature for each map type, both because some of the effects and functions that characterize these maps can also be achieved with development tools or script language (JavaScript, Dynamic HTML, etc.), and because these formats may also be the end result of proper WebGIS systems.
[90] An example of this type of map comes from the book site of the volume *Web Cartography…, op. cit.*, which at the webpage http://kartoweb.itc.nl/webcartography/webmaps/dynamic/di-example2.htm offers a multimedia map that traces the growth of the built area in the town of Enschede over time. In the form of a movie with button-controlled animation, this map allows the recipient to slide the paper forward or backward, or to freeze the map to a year of special interest.
[91] A successful example of multimedia cartography online are the cards of the town of St. Augustine found at http://visitstaug.com. It is a series of Google maps, depicting various resources: local interest points (attractions), recreational resources (recreation), receptivity (lodging), catering (dining), commercial activities (shopping). Every map can be zoomed and contains informational pop-ups that can be recalled by clicking on a point of interest.
[92] A dynamic topographic map of the United States, available online at the NationalAtlas.gov website on the page http://nationalmap.gov/small_scale/100topos/index.html, allows users to select one locality on the map and to visualize an interactive topographic map which provides informational details on any given topographic feature via a drop-down menu (select a feature).
[93] The NationalAtlas.gov website of the National Atlas of the United States of America, on the page: http://volcanoes.usgs.gov/ provides an example of this type of map, which shows the localization of potentially active volcanoes in the USA. By clicking the icon representing each volcano a pop-up appears containing a photograph and an indication of the alert level.

be mapped.[94] Nevertheless, no matter how advanced their level of interactivity, such maps lie still within the confines of a closed interaction, unable to probe the dynamic possibilities offered by open interactivity. As complex as multimedia maps may appear, everything that happens, or is made visible by interacting with them, was predetermined and built in its entirety[95] prior to their inclusion in the webpage that contains them. They reside in the server's memory as self-defined objects just as conventional maps, once compiled by the cartographer, are preserved in archives. As a result and, from this perspective, exactly as in the case of conventional maps, any changes in the information or in a single datum to be represented in a multimedia map imply an intervention *offline*, if not a complete makeover of the map itself, before the maps is made visible to the end user once again. Multimedia maps involve therefore laborious maintenance and updating. That also implies that the recipient's freedom of action remains marginal, not only because the user is unable to make changes to the object, but also because, even when the user is able to explore the map's contents via multiple links, such links are still fixed and can be numbered, since they result from a series of combinations between a defined list of options and possible functions. In some cases, the user is even left with one sole alternative: it is the case, for instance of cartographic "movies" which show the history of a phenomenon and allow the user to act only on the *backward* and *forward* buttons.[96]

WebGIS maps – These maps derive their name from the technology they are based on, namely WebGIS. As can be seen easily these systems result from the integration of two advanced IT technologies: the Global Network, or rather, the *World Wide Web*, and Geographic Information Systems, conventionally named GIS. The attributes of interactivity and dynamism are better suited to this type of map than to others. Only for this type of map can we talk, at least in potential terms, of open interactivity. And only WebGIS maps seem to fully meet the criterion of dynamism in the broader sense in which I have used it here. Like multimedia maps, these maps do not maintain one single visual form, but that is not all. WebGIS maps are generated "dynamically" or "on the fly" in response to specific user-defined parameters, corresponding to operations performed directly on the map with the

---

[94] A very simple instance of this interactive function may be found at http://mappinghistory.uoregon.edu/english/EU/EU15-04.html. It is a map titled *Travel of selected intellectuals in more detail* representing the movement of some Greek philosophers and intellectuals, including Hesiod, Thales, and Anaximander. A context menu that opens when you click on the map allows you to choose views of the various paths of the philosophers, showing their place of origin and the regions they traveled to.

[95] One could rightly argue that any digital object is "prepackaged," because everything that happens on a screen or within a computer is based on strict rules of algorithmic programming. This granted, however, we would still want to establish how much leeway is left to users and their possible degree of intervention.

[96] One instance is the Enschede map (cited above) but even the Multimap we presented earlier belongs to this typology.

mouse (clicking, dragging, etc.) or in another form (for example, by typing in geographic coordinates of a point of interest or choosing from a number of options). This is made possible by the fact that such maps are designed and intended from scratch as the visual result – computed by a dedicated piece of software we may broadly call a *map generator* – of automated processes triggered by querying one or more digital archives of geo-referenced data, or data tagged via a geographic reference system to specific coordinates on the earth surface. This geo-database is in fact part of a GIS system, or of one of its "abstracts."[97] What matters here is no longer simply to establish a "reading" path via user intervention but, rather, to variously contribute to the building of one or more maps, in sequence, on the basis of the informational needs of each recipient. The margin for user action, albeit constrained by operational functions made available by the site provider hosting the WebGIS, is expanded, from a simple query to possible changes in the database from which WebGIS maps are generated. The most intriguing aspect of this cartographical type of maps is the fact that they are dynamically generated via a Web browser interface that is linked to a database. This has far-reaching implications. First of all, it means that possible changes will not be merely cosmetic, but may well be recorded as variations – albeit temporary ones – of data within the database. This in turn leads to a potential real-time refreshing of information displayed on the map. Moreover, a dynamic mapping service allows recipients to produce a virtually endless number of single maps, because the possible combinations of queries made by them tend to infinity, even if they were supplied with very basic functionalities for a GIS, such as zooming and image re-centering. Even this rapid overview shows that, by virtue of their potentials, WebGIS maps are in the forefront of web 3.0 cartographic systems online. I will come back to them, therefore, when I advocate a new cartographic metrics able to restore *chora* by using their innovative features.

It is clear, at this point, that WebGIS systems are truly revolutionary and mark a communicative "paradigm shift" that makes comparison with printed maps obsolete: they sanction the possibility of recovering the multiple meanings of territory because the interpreters who intervene in their definition are multiple. Their new and, in some ways, revolutionary feature lies in the fact that they allow multiple parties to intervene in the drafting of documents and, with that – by virtue of the power that the map has to direct praxis – to guide the choices and policy decisions and at the same time redefine the meaning of the territory to be transmitted. Through interactivity that meaning is structured on the basis of the attitudes and needs of the subject, because unlike conventional cartography, WebGIS assimilates the perspectives of its user, namely the *web user*.

---

[97] The managers of a GIS database may decide to make only part of the entire archive accessible from the web or to set up reserved areas available exclusively to a small user group, usually granted access via passwords.

What is particularly significant is the fact that, even before questioning metric criteria and the symbolic system, WebGIS redefines the meaning of territory that can be conveyed by maps through interactivity, indeed marking a break with the prerogatives of topography. The aura of authority that surrounded topography and the absolute unquestionableness of its information are undermined by the new figure of the cartographer, no longer single and institutionalized, but plural and independent. The user, free to query the system and to combine or even process data at will, produces maps which can indeed open up new critical perspectives. The set of producible maps has the potential to establish an arena where different interpretations are compared and single representations combine multiple points of view, avoiding unilateral readings. In short, WebGIS opens up unseen horizons of communication and envisages the recovery of a new dimension of territory, a *chorographic* dimension, promoted by the intervention of web users. The hackneyed meaning attributed to territory by the rigid system of topography is replaced by the need to recover its cultural aspects and the identity policies of its inhabitants.

WebGIS is, therefore, a system fully ascribable to the practices of democratization, such as planning participation, transparency of co-management and shared decision-making, although many WebGIS systems currently in use fail to enforce in practice what they proclaim in theory. That, however, is not due to system constraints, but rather to the difficulty of enforcing practices of governance, although the swift and pervasive spread of such policies would suggest that the way to participation and concerted action has been paved.[98] Web GIS maps are currently enjoying wide circulation on the net, not only on privately owned sites designed for the territorial promotion but also on institutional homepages: ministries, research organizations, public agencies and public associations. To all, this new way of managing information appears as a very powerful technical and political tool.

---

[98] The principle of *governance* in the realm of territorial and environmental planning in international institutions, governmental or otherwise, dates back to the early 1990s (United Nations, Agenda 21, 1992), but its translation into operational tools and good practices of participatory governance of territory is more recent and has given rise to interesting initiatives only over the last few years. In Europe the principle of *governance* was sanctioned in 2001 (Communication from the Commission of the European Communities, *European Governance. A white paper*, COM (2001) 428 final - Official Journal C 287, dated 12.10.2001); for documentation on its application to urban areas of member states see: *A European handbook for participation*, Programme URBACT, 2006: http://ec.europa.eu/ewsi/UDRW/images/items/docl_9693_87937950.pdf. With specific regard to Italy, see the participative policies promoted by the Rete del Nuovo Municipio, an association created to implement governance projects and "bottom-up" planning in a number of Italian towns. G. Allegretti, M.E. Frascaroli, *Percorsi condivisi. Contributi per un atlante di politiche partecipative in Italia*, with CD-ROM, LUOGHI, Alinea, 2006. In the English-speaking world, the promotion of participatory urban initiatives has a more solid tradition. See in this connection the reference manual which lists over 50 techniques and methods: N. Wates, *The Community Planning Handbook*, Earthscan Publication, London, 1999.

To conclude, the significance of the transition from printed sheet to screen lies in the fact that WebGIS relativized the indisputable views of a territory "seen from the sky," reclaiming the point of the view of subjects who inhabit the place and open it up for planning and design. That is not enough, however, to mark the end of topography. For the time being the potential of this innovation remains unspoken, because no adequate graphics mode has yet been found to render its new spatiality. In WebGIS, such rendering is the element on which the role of participation is staked. Ongoing trials are promising because they are relying on advances in geographical theory that restore precisely that dimension of territory, but the crux of the problem is the spatial and iconic assumptions that WebGIS embraces.[99] The unthinking assumption of Euclidean spatiality and the use of a symbolization that, although changed, is still modeled after the one used for printed maps, expose the limit and the gap between interactive potentials and the possibilities to convey the values of the place and the sense of belonging of those who inhabit it. WebGIS has great potential and is well-posed to show the subjective dimension of society. For now, however, it is forced within the confines of unreflecting cartographic rendition. This is the gap that must be filled: to translate into action a topological dimension able to render the sense of place and the value of landscape.

As in a game of mirrors, the importance of the subject and of its cultural values refers us to landscape and to the possibility of rendering it cartographically.

This chapter overlooked the third dimension and its modes of representation to focus on the urgency of recovering the social sense of territory by investing in data to be included in maps and on participation practices offered by new cartographic systems. I will now turn to finding possible metric criteria able to innervate a chorography, with the goal of rendering landscape.

[99] One significant forum for these experiments was the conference held in Bergamo, 23–24 April 2009, under the title *Le sfide cartografiche: movimento, partecipazione, rischio* (www.unibg.it/ diathesis). In this transdisciplinary context, the leading theories in the field of geography (among which are J. Lévy, ed., *L'invention du Monde: une géographie de la mondialisation,* Presses de Sciences PO, Paris, 2008; F. Farinelli, *L'invenzione della Terra,* Sellerio, Palermo, 2007) were associated with the ones produced in IT management and with tests being conducted with a view to adapting maps to emerging social needs.

CHAPTER 5

Chorographic Horizon: Landscape Cartography

We must urgently move from a monocular vision of the world to a more subjective vision.

David Hockney, *That's the Way I See It*, 1993

Recalling the profound innovation cartography has undergone, this chapter advocates the creation of a new model called chorography, aimed at recovering the social and cultural sense of territory. Clearing the field of meanings that were attached to the word in the recent past, I will specify what chorography should be taken to mean, by tracing the assumptions on which its metrics is based and focusing on aspects that make its realization possible. Specifically, I will consider the semiotic nodes on which my proposal rests, namely: 1) topological space and landscape-based logic, viewed as the foundations for achieving a chorographic metrics; 2) the figural ways whereby we can recover the sense of landscape, by envisaging an icon built around the concept of archetype. This proposal will be exemplified on a specific cultural context, the Gourmantché context in Burkina Faso, showing data collection, the transfer of landscape values and their cartographic rendering. The goal is to substantiate the meaning of chorography and turn it into a concept which, by discarding topographic perspective, may introduce other points of view and other ways of figurativizing territory (Fig. 5.1).

CONTENTS

SEMIOSIS AND CHOROGRAPHIC METRICS

By *chorography* I mean *a cartographic representation that recovers the cultural and social sense of territory within the relation that the individual establishes with a place, expressed by the reality of landscape.* I am going to focus on two aspects for its implementation, namely: 1) the rendering of landscape; 2) the recovery of the

Reflexive Cartography. ISSN 1363-0814, http://dx.doi.org/10.1016/B978-0-12-803509-2.00005-7

INTERPRETER	**Cartographer**: plural (either professional or not; either established or not) **Recipient**: active (able to interact and intervene)
METRIC CRITERIA	**Topological space:** (expansion/contraction, reticularity) **Landscape-based logic:** (placement within a social space where phenomena to be investigated are symbolic, material, and functional)
DATA SURVEY	**Space measurement:** remote sensing **Phenomena:** recording of their social value
GRAPHIC RENDERING	**Observation point:** multiple (zenith-based, perspective-based, 3D-enabled) **Metric scale:** interaction of multiple scales **Icons:** plural (figurative, abstract), multimedia inserts (photos, graphics, other)

FIGURE 5.1
Chorography and the recovery of the social and cultural sense of territory.

subject as a social actor who represents and communicates the cultural function of the place where he or she lives.[1]

Even in this case, as I did for the term *topography*, I am going to try to clear the field of misunderstandings and distance myself from the meaning that over time was attached to the word *chorography*: a typology of maps "which reproduce part of the earth surface – called region for such purposes – and are different from geographic maps, which instead include the surface as a whole." This sense of the word, upheld in neo-positivist classifications, identifies the scale ratio as the only criterion for recognizing the different cartographic types and their attendant informational densification.[2] Thus it was determined that

[1] In the 16th century, Giovanni Antonio Mangini took up Ptolemy's classical distinction, stating that chorography is the science aimed at the recovery "of the quality of places, local areas, and landscapes analyzed and described in their historical, anthropological and emotional features" (G.A. Mangini, "Della Geografia di Claudio Tolomeo Alessandrino," in: *Geografia, cioè descritione universale della terra*, Libro primo, cap. 1, Venice, 1598; to trace the connotations that the term took on in Renaissance cartography see: M. Milanesi, "Leandro Alberti, Historicus Itinerarius," in: M. Donattini, ed., *L'Italia dell'Inquisitore. Storia e geografia dell'Italia del Cinquecento nella Descrittone di Leandro Alberti. Atti del Convegno Internazionale di Studi (Bologna, 27–29 maggio 2004)*, Bononia University Press, Bologna, 2007, pp. 249–272.

[2] It is a hackneyed and unreflecting assumption of meaning, implemented in the 19th century, over the Greek opposition between chorography and geography: the former attributed to the comprehensive description of places and regions, the latter to cartography as a selective and ordered representation of Earth. M.C. Robic, "Epistémologie de la géographie," in: A. Bailly, R. Ferras, D. Pumain, eds., *Encyclopédie de Géographie*, Economica, Paris, 1992, ch. 3, pp. 37–55.

a chorography was based on a scale ratio ranging from 250,000 to 500,000 and consequently showed a "region" with a varying degree of detail.[3]

The perspective adopted here, instead, uses *chorography* to define a map that focuses on socially produced qualitative aspects and envisages landscape by recovering both its visual appearance and its content, in ways and forms that differ profoundly from those of topographic maps. In this sense, the meaning of *chorography* is opposed to that of *topography*, since it aims to eradicate the metric conditions of the latter. In the first place, its notion of space stands up against the one theorized by Descartes. Two-dimensional (2d) Euclidean space is rejected and the recovery of the third dimension envisioned: the criteria of uniqueness of the point of observation, metric accuracy and linear distance are discarded in favor of the criteria of subjectivity, relativity, and topological distance. For the social dimension cannot possibly be detected in an abstract space seen as the container of phenomena. Rather, it must be sought in a space built according to social evaluations to do with metrics. In fact, the spatiality proposed in the field of geography is rooted in mathematical studies which clarified that distance cannot be defined in itself, but depends on the survey system adopted and, therefore, on the type of metrics used. The result is that there is no single distance but, rather, as many distances as are the metrics proposed to measure it.[4]

Similarly, regarding the value to be attributed to a given phenomenon, it is evident that chorography cannot abide by Cartesian logic. A Cartesian perspective lays exclusive emphasis on the physical quality of geographical phenomena, while chorography aims to recover those features that convey the phenomena's social complexity. Among those possible, I advocate here a landscape-based logic,* *understood as the recovery of the cultural substance of territory. While such recovery accounts for the material qualities of phenomena* – it could not be otherwise – it *also addresses those qualities that are intangible, symbolic and functional, in a hierarchy that is established via an assessment made by local inhabitants.* In short, it aims to represent landscape as an element that allows the recovery and preservation of the social features of territory. Only by abandoning the myth of a descriptive map whose authority is based on its presumed objectivity can one be in a position to engage a cartographic universe where a given message is inseparable from an explicit project. As Giuseppe Dematteis notes, since "the relationship between reality and cartographic representation may be subject to different interpretations, it is fundamentally ambiguous, in the sense that it is multiple. The same reality such map represents in this way may be

---

[3] This scale-based classification was certainly applied in Italy and largely accepted by various countries with slight variations. In any case, it was universally assumed that the chorography identified a region of variable reach. The term continued to be used over time to identify maps representing sub-national or provincial areas. With the beginning of the 18th century, however, the term fell largely into disuse, and was for the most part replaced by "topography" or "cartography." (http://en.wikipedia.org/wiki/Chorography).

[4] On this point, see the next chapter.

legitimately represented otherwise, just as those who observe a map or painting or those who read a geographical description may draw dissimilar ideas from them."[5] That entails giving up any presumption of objectivity or impartiality and acknowledging that our representational viewpoint is itself the product of a given interpretation and as such is based on conjecture.

Chorography relies on transparency and aims to produce maps closely linked to "the conditions of enunciation." In other words, chorography states its goals and enables recipients to make critical sense of what they find on maps. By doing that, chorographic maps avoid iconizing prescriptions. A tendency to iconization is, after all, ever present, especially when maps are taken uncritically as instruments for making sense of objective reality. If, however, the interpreters of maps are made aware of the fact that maps are invariably biased and must be interpreted, the iconizing tendency of maps subsides.

While taking into account all of the models, which social and technical advances in cartography have produced, and before addressing the operational aspects of the realization of a chorography, I will consider the theoretical framework from which we can start to establish it: hence the semantics of constructive assumptions found in the metrics, the type of data and their graphic rendering.

Chorographic Metrics: Topological Space and a Landscape-Based Logic

Chorographic metrics envisages a topological space* and a landscape-based logic. Firstly, two-dimensional Euclidean space and its attendant criteria of a unique observation point, metrical accuracy and linear distance are abandoned. Chorographic metrics embrace tridimensionality, thereby opening up to subjectivity, relativity, and topological distance.* Secondly, icons are developed to render less a cluster of properties selected by virtue of their material quality than a set of landscape values that are socially produced. Let us consider these two aspects.

Topological Space

We set out with the assumption that a chorographic metrics, based on topology, first calls for a consideration of the concept of space that underlies it. That entails first clearing the ground of *spatialism*, or the implicit line of reasoning that conceives space as a self-enclosed object, to be handled according to

---

[5] G. Dematteis, "Elogio dell'ambiguità cartografica," in: E. Casti, J. Lévy, eds., *Le sfide cartografiche..., op. cit.*, pp. 13–16. It should be remembered that this author was the first in Italy to draw attention to the importance of representation in the construction of the concept of spatiality and to reflect, in particular, on the phenomena of globalization (G. Dematteis, *Le metafore della terra*, Feltrinelli, Milan, 1996; but also: *Progetto implicito*, Franco Angeli, Milan, 2002).

general laws that disregard its social substance. It is also a line of reasoning that sees the material quality of phenomena as ultimately responsible for the organization of social practices.[6] We need therefore to specify, albeit briefly, what we mean by topological space in light of the most recent geographical studies.

As they reconstruct the changing senses attributed to space in the field of geography and retrace their philosophical implications, Jacques Lévy and Michel Lussault see topology as the spatial feature that can finally reject the constraints of Euclidean geometry. Topology advocates a new idea of space based on the relationship between individuals, to be contrasted to the notion of empty space as an abstract container of phenomena.[7] The two scholars are convinced that the recovery of the social dimension of space can be achieved by taking into account the relation of the actors that shape it and give it substance. Following a diagonal trajectory, which elevates the value of space in a social perspective, they replace an "absolute/positional" notion typical of Euclidean space, with a "relative/relational" idea, ascribed to topological space. Their underlying idea is that a worldview in which space is built *a priori* on the basis of the material expression of the visible rules out the possibility of recovering a great number of features. Such features are virtually inseparable from each territorial setup and from relations not predetermined by localization. Finally, they come to consider movement as the primary factor of a relative and relational spatiality.[8] Movement is a superior concept because of its aggregating potentials, which come from considering space as a dimension of a complex universe which functions as an open system and is able to integrate the most challenging features of other confined approaches coming from the social sciences. Their choice favors Leibnizian space which, in their view, is the first step towards the problematization of a space determined by human beings. And this space becomes a tool for keeping track of social facts and rejecting the idea of a closed territory, self-sufficient in its extension. Topological space actively opposes this self-referential notion by introducing "substance,"*

---

[6] Michel Lussault, who has long explored this issue (M. Lussault, *L'homme spatial. La construction sociale de l'espace humain*, Seuil, Paris, 2007; Id., *De la lutte des classes à la lutte des places*, Grasset, Paris, 2009), sees *spatiality* as a quality made up of a combination and correlation that comprises the material, the immaterial and the ideal, actors, objects and environment, and space. And within space, he uses the concept of *situation* which, freed from its metaphorical meaning, promises to be the inescapable dimension of any social phenomenon.

[7] Discussion over the following pages is anchored in the research of these two authors, whose works I will refer to for details in the course of my argument. An overview of their research, however, may be found in: M. Lussault, J. Lévy, eds., *Dictionnaire de la Géographie et de l'espace des sociétés, op. cit.* I refer to this volume for definitions of the concepts I discussed in these pages.

[8] The authors reached this conviction via different routes, which took into account: movement and its spatial representations for Lévy (*L'invention du Monde..., op. cit.*); the city and the urban implications that its flows determine in the case of Lussault (M. Lussault, T. Paquot, C. Younès, eds., *Habiter, le propre de l'humain: villes, territoires et philosophie*, La Découverte, Paris, 2007).

that is, the necessary characterization of each geographical space obtained by recovering its nonspatial dimensions.[9] Thus, space is not recognized as a completely autonomous and self-contained system. Rather, it is seen as osmotic to the social dimension, which must be explored in its dynamic spatial setups according to criteria established by *topology* (which examines the ways distance between phenomena is conceptualized); *scale* (as a discontinuity threshold in the measuring of distance and the evaluation of phenomena); and *substance* (the value a given phenomenon is endowed with).[10]

So, the adjective *topological* applied to space specifies a metrics that measures the distance between phenomena rejecting the "rigidity" of territory as a Euclidean space (a continuous space, marked by strict boundaries and tending toward self-containment) in favor of the "flexibility" of a *network*, a space built upon a system of relations, whose features are discontinuity and incompleteness. Lévy and Lussault use mathematical theory (which recognizes in topology the notions of *continuity* and of *limit*) to their own ends, introducing the notion of "vicinity" (that is a specification of the distance used when the spaces under consideration cannot be taken as contiguous) and acknowledging *extension/contraction* and *reticularity* as distinctive features of topological space.[11] Extension and contraction are closely linked to the layout of a topological space that is based on complex relationships and the existence of multiple paths other than the linear paths typical of topographic metrics.[12] The measurement of this space needs a new metrics, such as temporal metrics, which *de facto* distorts space (unlike the case of Euclidean space). Space is thus contracted or extended according to the importance of relational links that the social layout creates or of the attractive relevance of certain sites, which pervade territory and shape it either functionally or symbolically. Relying on network links means, ultimately, emphasizing the connectivity of space and recovering the heterogeneity of its social components. It means refusing to accept *contiguity* as a factor relevant in itself to the topological perspective of space. Topology is indeed brought to the fore by reticular distance, since connectivity provides

[9] Jacques Lévy repeatedly stressed the disregard of spatiality on the part of the social sciences which, he claims, "must take the spatiality of society seriously and consider it as a transversal [quality] potentially useful to approach the 'societality' of the world, and turn it into a solid concept rather than a plastic metaphor" (J. Lévy, "Un événement géographique," in: J. Lévy, ed., *L'invention du Monde...*, *op. cit.*, pp. 11–16).

[10] Subsequently Lussault added configuration which, he notes, is the "formal expression of the relational economy between spatialized objects" (M. Lussault, *L'homme spatial...*, *op. cit.*, p. 88). I will not be addressing this issue here because it is marginal to my purpose.

[11] As a matter of fact, they have less recourse to mathematic theory as such than to pre-topology, more suited to the social sciences because it rests on a less stringent set of axioms than does mathematics.

[12] Inside a space, whatever it is, distances are structured, even though there is no movement, but only virtual motion, since the localizations themselves are considered relative, in relation to each other, and thus fall within the same problem area of metrics.

a theoretical grounding based on the link/no-link binary dichotomy versus the proximity/distance dichotomy. Similarly, measurement uniformity may be itself removed by adopting multimetric systems that subdue it, distort it and introduce the use of multiple scales.

However, it should be noted that, in the line of argument pursued by these authors, the term *scale* is not made to correspond to the size reduction ratio between territory and its representation – typical of cartographic scale* – but is taken in the sense of the different dimensioning between geographical phenomena, in which both metrics and substance are taken into account. Therefore, the word *scale* means the production of a discontinuity threshold both in the measurement of distance and in the appraisal of phenomena. In the first case, a change in scale is identified with a change in analytical perspective over a space: a site or *area* may be identified as *large* or *small* depending on how they are measured. If the social dimension is defined in relation to topological distance, one must define the value of this distance by adjusting the use of scales, which will vary in accordance with the value ascribed to the visible phenomenon, rather than on the value of its topographical dimension.[13] For example, if we compare Sweden to Japan based on their topographic surface they are both small countries, but if we do it based on the number of inhabitants and the degree of urban development they differ profoundly. Everything depends, in fact, on the question being addressed. And yet, ascribing value is not a simple matter of adopting a different set of criteria (number of inhabitants, urban development, etc.). Rather, it is related to the different interactive potential between the actors that these situations envisage. An uninhabited space, even if it stretches over an important area (Greenland, for example), is considered small when compared to a densely populated area, although the surface of the latter may be limited (the island of Java, for instance).[14] In the assessment of phenomena, the specificity of social spaces leads us to distinguish between levels of size and levels of complexity, to state that the larger space encompasses smaller spaces but does not contain them, that is it is unable to express some of their specific features. Thus, the *global* is not the sum of the *locals*, because although they both express a social dimension they need different (in many cases quite opposite) forms of management. It is clear that such considerations may open new perspectives that make it difficult to identify scalar thresholds or scale

[13] That entails abandoning three theoretical precepts about space: *fractalism*, which makes scaled data redundant; *geographism*, used in classic geography and geopolitics and still based on the often implicit assumption that there is only one accurate way to measure space (namely, the Euclidean one); and *spatialism*, which deems it possible to deal with space on the basis of general laws, irrespective of its social makeup.

[14] J. Lévy, "Echelle," in: J. Lévy, M. Lussault, eds., *Dictionnaire de la Géographie…*, *op. cit.*, pp. 284–288.

changes, since the latter does not appear as a self-referential data, but as a tool related to a socially defined situation.

As a matter of fact, together with metrics and scale, *substance* is clearly the third component of spatial configuration. One cannot define a space except with reference to something that is both territoriality and spatialization.[15] Jacques Lévy defines substance as "the non-spatial component of spatialized objects."[16] Through their spatialization, social substances become visible and their existence within a society is crystallized. Therefore, by virtue of its abstract connotation, substance emphasizes the crucial spatialization of social phenomena. The notion of substance aims to reject the tenet whereby social phenomena may be taken as existing outside space. Substance endeavors, instead, to account for their spatialization. Unless it can conclusively be shown that the contribution of space to the description of reality is irrelevant, all phenomena must be seen to enclose a cluster of spatial qualities not only in their form but also in their content. Substance is not, therefore, an adjunct to reality or a secondary feature. It is, rather, a perspective which allows us simultaneously to detect territoriality and to incorporate into a spatialized universe.

To embrace this new idea of space in cartography means to enjoy a number of advantages, related both to action and to a change of perspective in content. Since it is open to other types of metrics, this new space paves the way to a reconfiguration of the basic structure of maps – that is, the set of assumptions, rules, graphics codes which address the relative distance between phenomena. Furthermore, this concept of space, rejecting the rigidity of the Euclidean model, assumes relationality as a prerequisite for recovering the individual and, therefore, opens the way to a landscape-based logic.

Landscape-Based Logic

As stated above, the adoption of the concept of landscape responds to an acknowledgement and a consideration of its two territorial functions: establishing a territory's visual form and expressing its cultural and social value. More precisely, we take a semiotic approach that defines landscape as "a geographic

---

[15] In addition to the physical and natural relations between objects within a social setting, the concept of *territoriality* refers to all the actions, techniques and tools that are mobilized in order to act territorially. That is connected to what, in the field of language theory, was defined as cognitive spatialization, or the procedure whereby objects are invested with spatial properties in relation to an observer who, by creating a network of territorial references, is in turn, "spatialized" with respect to them (A.J. Greimas, J. Courtés, "Sémiotique," *Dictionnaire raisonné de la théorie de langage*, Hachette, Paris, 1993, pp. 358–359).

[16] J. Lévy, *L'espace légitime*, Presse de la Fondation Nationale des Sciences Politiques, Paris, 1994, p. 68.

formation that the gaze organizes into a whole, emphasizing the crucial role of the observers who interpret the site they inhabit."[17] Such whole, therefore, is a construct, a genuine intellectual achievement. Through it, individuals interpret what they see and communicate it in a narrative that shows landscape as endowed with anthropological value, historical legacy and planning possibilities. We should also note that the action of the observer is cooperative. It does not convey merely what the eye records, but reconfigures perceptive data according to his or her values, performing an intellectual process that Augustin Berque defines as *médiance,** that is, the way in which individuals speak of themselves, of their past and future expectations in a social dimension.[18] It follows that the landscape is in some way a form of identity but is also, at the same time and above all, an iconization of territory thanks to which the value of territoriality acquires visual cohesion. Therefore, landscape takes on the properties of a mimetic operator able to "render" reality in communicative terms and to prescribe the rules whereby reality is to be grasped. This is the level on which the social significance of landscape unfolds. And having been released from a merely aesthetic-perceptive function, landscape is adopted as a social and identitarian device. We could claim with Eugenio Turri that its communicative power derives from the fact that it dramatizes a play, in which human communities, protagonists on stage, tell of themselves and of their relationship with the world. We should also stress that iconization is the outcome of an interpretation of landscape signs, ordered and made semantically consistent, in accordance with what Turri calls *iconemes,** emergencies that can endow everything else around with meaning and communicative coherence.[19]

---

[17] Adalberto Vallega claims that landscape is not a self-contained phenomenon but, rather, an intellectual construct injected into the communicative scenario. Semiotic analysis allows the assessment of such scenario by detecting the two levels of transmission: denotative and connotative. The former produces information on material processes, natural and social, which relate to visual landscape. The latter pertains to the processes linked to symbols and values, which belong to intangible culture (A. Vallega, *Indicatori sul paesaggio*, Franco Angeli, Milan, 2008).

[18] To shed light on the development of this intellectual construct, Berque turns to the concept of *médiance*, that is "a complex form of representation, at the same time objective and subjective, phenomenological and physical, ecological and symbolic" able to address both the scientific and the artistic issues ascribed to landscape (A. Berque, *Médiance des milieux en paysages*, Editions Belin, Paris, 2000, p. 32). This concept comprises also the proposal to acknowledge landscape as capable of restoring the bridge that modern science severed, that is, the relationship between science, art and morals (p. 86). A systematization of this concept may be found in the recent: Id., *La pensée paysagère*, Archibooks, Paris, 2008.

[19] The iconeme is defined as "the basic unit of perception, a sign within an organic set of signs, a synecdoche, as part which expresses the whole, or which expresses the whole with a primary hierarchical function, both as an element that best embodies the genius loci of a territory and as a semantically charged visual reference to the cultural relationship that a society establishes with its territory" (E. Turri, *Semiologia del paesaggio italiano*, Longanesi, Milan, 1990). This concept is taken up and developed in: Id., *Il paesaggio come teatro. Dal territorio vissuto al territorio rappresentato*, Marsilio, Venice, 1998, especially pp. 170–175.

These may then be the semiotic tenets we would start from to envisage the construction of cartographic icons able to order, rank and consistently communicate landscape. Before addressing their graphic rendering, we need to dwell on those features that substantiate landscape data to be included in a chorography.

We need to reflect on the type of data to be surveyed because we need to draw maps that are no longer restricted to conveying information based on the visual aspect of landscape and recover instead information related to social values. These are the only ones able to envisage *médiance,* as we endeavor to achieve what Berque explicitly invites us to do, namely to "persistently place the map in relation to territory."[20] Cartography was long based on the assumption that, in order to understand the space used by individuals, it was necessary and sufficient to focus on a set of material features related to topographic metrics. Chorographic metrics differs: it leads us to consider the interest of individuals who express themselves through landscape and qualify it as a socially shared public space. Practically speaking, we would need to be able to record the *territoriality* individuals assign to landscape, since territory reflects the values, the beliefs, and the knowledge that, as a local community, individuals have entrusted to it over time. To employ the notion of "landscape" in cartography means to highlight the cultural substance of territory, thereby dispelling the mistaken illusion that topographic and social spatiality may be one and the same.

To enforce this perspective we need first to give up our claims to exhaustiveness, but also, and most importantly, our notion of cartographic information as uniquely accurate. And of course we must make our objective explicit: we do not set out to map landscape as an absolute value in itself, but rather intend to record it as a set of shared values, coming from a specific cultural context, and aimed at supporting specific interests and a specific viewpoint. Our underlying assumption is that the idea of landscape is derived from the idea of *nature* and from the relationship a given society has with nature. Augustin Berque has convincingly shown that the various meanings of landscape devised by different societies – and the multiple values that go with them – come in fact from a primeval cultural choice made by societies in their relation to nature. He pointed out how this choice gave rise to diametrically opposed approaches: the one whereby man and nature are symbiotically taken as part of a cosmic sphere (*cosmogonization*) or, alternatively, the one in which such relationship was severed. The latter severance, which occurred in a number of cultures including

---

[20] As he explores the territorial practices which rules out the possibility of recovering *médiance,* the author insists on the map as an essential instrument to be related to territory: A. Berque, *Médiance des milieux, op. cit.,* p. 148.

the one of the West, is named *decosmogonization*, and consists of having separated more or less neatly the *land* from the "cosmos."[21] The presence or the absence of such distemper would then have given rise to different social values ascribed to nature and to a different way of making sense of landscape.[22]

Berque claims that Western modernity is characterized by the loss of a unitary and axiological order in the human/universal relationship. And such loss in turn inevitably mars our relationship with places, now commonly seen as mere terrain on which to build human activities. For Berque, all that depends on the fact that the word *cosmos* changed meaning over time and became synonymous with *universe*, irreparably moving away from what the Greeks ascribed to *kosmos* and the Latin to *mundus*. Today the word *cosmos* rules out the human world and especially the inner world of subjectivity. Today a cosmologist is an astrophysicist whose object of study has nothing to do with human action. Conversely, in the past both *kosmos* and *mundus* referred to an overall order where humans occupied an essential place. Environmental participatory policies of the kind promoted today presumably come to redress the imbalance Berque describes: landscape is now seen as an arena where subjects meet to measure and to record knowledge they have acquired over time in their contact with nature, a knowledge that will eventually become the values expressed by landscape: hence the need to consider landscape not only with regard to the distinctive cultural features of the society that produced it, but also in its dynamic configuration. This is produced by the continuous changes and fluctuations of spatial/temporal relations between nature and society and its analysis can reveal its primeval link to *kosmos*, in most cases now lost.

We should be aiming at a kind of cartographic rendering of landscape that recalls the foundational values of a given society. At the same time, however, we should trace the values that have accrued, multiplied, or were otherwise lost over time with regard to the stakes a given community put up in order to ensure its survival and its social reproduction. In sum, a landscape map will allude less to social *identity* and the generic values of the past than to the identity discourse*: the process whereby the set of values that emanate from

---

[21] A. Berque, "Trouver place humaine dans le cosmos," in: *EchoGéo*, n. 5, 2008, published 9 April 2008, http://echogeo.revues.org/3093.

[22] A close examination of the studies that have investigated such relationship shows that the various philosophies that underlie the human history of nature (American *wilderness* and *deep ecology*, European *conservatism*, *mythical thought* in Africa, and *fudosei* in Japan) attest to the wide-ranging ramifications of sense inherent in this concept. For Japan see: A. Berque, with M. Sauzet, *Le sens de l'espace au Japon, vivre, penser, batir*, Ed. Arguments, Paris, 2004; for Europe, of all: C. Raffestin, *Dalla nostalgia del territorio al desiderio di paesaggio. Elementi per una teoria del paesaggio*, Alinea editrice, Florence, 2005; for the United States: M. Schmidt di Friedberg, *L'arca di Noé. Conservazionismo tra natura e cultura*, Giappichelli, Turin, 2004; for Africa: A. Turco, *Configurazioni della territoriatà*, Franco Angeli, Milan, 2010.

territory are endorsed with a view to sanctioning the social identity that is communicated and iconized through landscape.[23]

To be sure, such cartographic perspective requires primarily a research method able to reclaim the substance produced by territorial action. We need to proceed by collecting terrain data which, interpreted in a participatory way, may serve to envisage and to promote such values within the local community. We should not forget that if landscape represents and conveys the values of territory, the latter is the product of the society who acted upon it and it is the condition whereby it exists.

From a practical standpoint, we simply cannot ignore the need to recover the territorial dynamics of local communities, for those make it possible for us to gather data on landscape from an analysis of habitation. The SIGAP Strategy, reconfigured to this specific end, aims to recover: 1) the metaphysical heritage, or the set of value assumptions (myth, tradition, beliefs, etc.) which underlies the very setup of society in its relation to nature; 2) the perspective of the individual and of the cultural features that inform his or her action. Hence, attention is directed to the type of data to be inserted in maps and to the techniques of participation. These multiply the actors involved and recover the presence of individuals as an essential prerogative for representing landscape. Concurrently, by virtue of its direct investment in IT, such strategy entails the disruption of traditional cartographic assumptions: the objectivity of maps is questioned and maps are opened up to a whole range of creative directions made possible by a relativistic informational approach. An example of such experiments aimed at advocating landscape as a "planning unit" may be found in West Africa, a reality pervaded by a mythological conception of nature.

GOBNANGOU: A CLIFF THAT "ENWRAPS"[24]

To an absent-minded traveler, coming from Diapaga, the sudden appearance of the Gobnangou cliff is visually striking. Its outline on the horizon, breaking the slow pace of the plateau's agrarian landscape, draws the eye

[23] It is not our purpose here to delve into the controversial issue of identity, especially in the field of anthropology, where the notion has long betrayed artificial and generalist assumptions. We are of course invited to reject a monolithic reading of culture and of the past (B. Anderson, *Imagined communities*, Verso, London, 1983; E.J. Hobsbawm, T. Ranger, eds., *The invention of tradition*, Cambridge Press, Cambridge, 1983; F. Remotti, *Contro l'identità*, Laterza, Bari, 2001). Our goal here is to recover the role landscape plays in communicating a sense of belonging, making visible the signs of the territorialization processes built up over time as references to evolving cultural values.

[24] A more thorough treatment may be found in: E. Casti, "Il paesaggio come unità di pianificazione del Parco d'Arly: la falesia di Gobnangou (Burkina Faso)," in: E. Casti, ed., *Alla ricerca del paesaggio nelle rappresentazioni dell'altrove*, L'Harmattan Italia, Turin 2009, pp. 21–68.

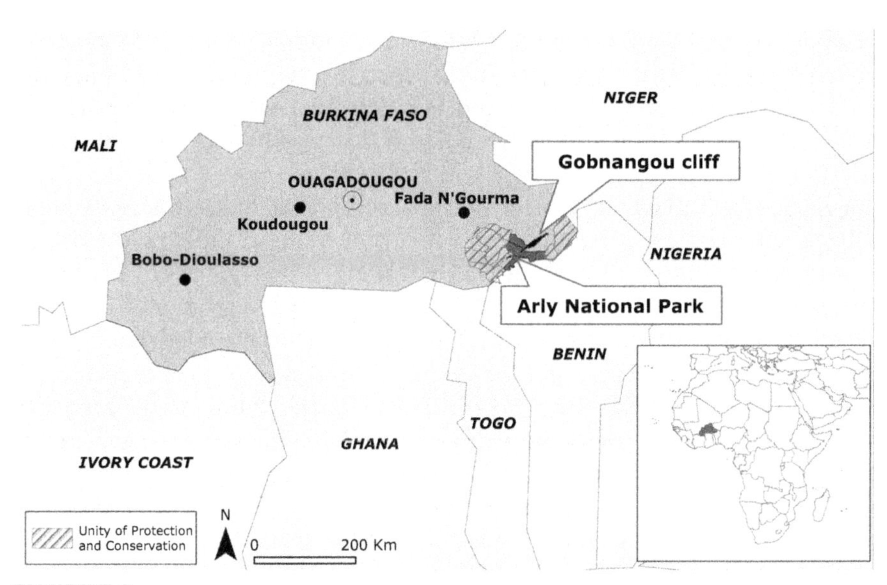

FIGURE 5.2
Localization of the Cliff.

towards the wall that stretches for about 70 km, from Tambaga, a large village located at its feet to the east, up to the boundaries of the Arly National Park, to the west[25] (Fig. 5.2).

The Gobnangou is the element capable of ordering and communicating the landscape experienced by an observer, whose understanding will depend, however, on the investment that he or she is willing to make. If one is not content with visuals or confined to aesthetics but wishes to access its social significance, one must retrieve the values that were assigned to it by the local communities.

In Gourmantché language[26] Gobnangou means "approach, I will protect you" and the string of villages that speckle the surrounding plains bear witness to the attraction that the cliff has exerted over time. Everything contributes, therefore, to indicate its role as a catalyst both toward the outside observer and to the local

[25] On its northern side, the cliff appears as a steep rocky wall which, with its 230-meter drop, makes communication difficult between the villages on either side. Conversely, on the southern front the Gobnangou slopes appear less steep, and the path leading to the top is less difficult. Lengthwise, after a short break of a few kilometers inside the Arly National Park, the cliff reappears in the west, in the area of Madjoari, with lower elevations, except for a few isolated reliefs, like the one called the *Tantatuari*, near the village of Tambariga, which culminates at around 270 meters.

[26] This is *gourmantchéma*, a language spoken by the ethnic group politically structured around the ancient kingdom of Gourma (or *Gulmu/Gulma* in the most common phonetic transcriptions) located in Eastern Burkina Faso.

inhabitant. If to a stranger the cliff may be said to vectorialize space, showing the direction to take, to local inhabitants it is much more: it is the one element on which their physical and social reproduction depend. Ease of access, ensured by the smoothness of the plateau where the cliff lies, may have been an attractive factor for the choice of sites in which to settle villages. Such choice was determined by an awareness of the most favorable conditions for foothill settlement: enjoying the ease of communication of the plain, without suffering the risk of flooding or landslides typical of steep areas; having springs at one's disposal that provide water supplies throughout the year; using natural vegetation, a powerful ally for survival; having an elevated area of retreat and defense nearby in times of war and conflict. Yet all this, albeit strategic for survival, is negligible compared to the powerful symbolic role that the cliff plays for its inhabitants: Gobnangou is the generator of the founding myths of the villages; Gobnangou guards the sacred values that inspire tradition; Gobnangou sanctions norms for a balanced relationship with nature; Gobnangou hypostasizes the exercise of traditional power. In short, Gobnangou is a designator that condenses and conveys various social meanings: by what has been called a "topomorphosis process," cultural values, which over time were attributed to the cliff, now reflect upon society, establish its organization and direct its functioning.[27]

The cliff, therefore, may be taken as the ideal device capable of recovering the social significance of landscape and of voicing the identitarian demands of those Gourmantché people who elected it as the foundation of their lives. This perspective is not aimed exclusively at verifying the cognitive power of landscape, but endeavors to advocate it as a "planning unit" able to ensure both the environmental protection of the Arly National Park and the local development of the region around it[28] (Fig. 5.2).

Myth and Landscape

The relevance of myth in social regulation, which has come to the attention of geographers over the last few years, leads us to appraise the territorial process of the Gourmantché people with regard to the sacredness of the cliff. Research has extensively shown that in the construction of territory, as well as in the attribution of meaning to the environment, these African people see spirituality as both the moral and the logical foundation of life and of social reproduction.[29] The principles used to sanction the natural order derive from the

---

[27] On topomorphosis, already addressed in Chapter 2 as a sophisticated instance of the symbolization process, see: A. Turco, *Terra eburnea. Il mito, il luogo, la storia in Africa*, Unicopli, Milan, 1999, especially pp. 128 ff.
[28] The Arly National Park, which covers an area of 1,530 km^2, near the Cliff, covers a vast area comprising the Arly Protection and Conservation Unit and the system of protected areas located in the region of Eastern Burkina Faso.
[29] On the recovery of myth in geography see among all the special issue *Cahiers de Géographie du Québec*: V. Berdoulay, A. Turco, eds., *Mythe et géographie: des relations à repenser*, v. 45, n. 126, 2001.

values society has retained in its own metaphysical "safe." These are conveyed by myth, a narration that translates such principles into norms whereby communal life and its actions are ensured, above all in the construction of territory.

It is here impossible for us to consider adequately all the manifestations of the Gourmantché territorial process. We can, however, dwell on the persistence and the centrality of myth as a narrative model among those people. On a descriptive level, myth informs the principles and on a normative level it prescribes the rules whereby the territorial process must be fulfilled. With regard to the former, myth intimates the presence of a supernatural entity that made the settlement possible. And about the latter, myth ensures the legitimacy of territorial action in accordance with that supernatural link. It should be recalled that, in myth-based societies, the relation established with the world is the very expression of the relationship society has with the deity. One is not allowed to do as one wishes with the land or on the land. There must be, instead, an ongoing process of transformation compatible with the representational system used by myth. Consequently, territorial action takes on a sort of double meaning: on the one hand it is a sign of divine goodwill granting its fulfillment; on the other it is an invitation to act responsibly and in harmony with nature. For myth is based on the fact that a given geographic layout carries an intrinsic property: it is a frame for human action which fully exercises its autonomy by observing and interpreting divine will. As such, myth ensures the transition from a mythical to a historical universe. And the territorialization process marks the shift from a grateful acknowledgment of divine munificence to an ethical view of human responsibility towards nature. Ultimately, being the lawful inhabitant of a place does not depend merely on the original pact sealed with the gods but from human action performed in accordance with divine will.[30]

Clearly, "African animism" goes well beyond the manifestation of a "naïve" acknowledgement of nature as endowed with a soul.[31] Rather, it appears as an elaborate tool for grasping reality and, therefore, a tool of rationalization and knowledge offered to the entire social group. Hence individuals are not left on their own to cope with the risk of failures, but helped by the whole society, which takes on such risk and avoids failure by constantly relying on tradition, a true social grammar for translating myth into history. A very elaborate set of knowledge ensures the functioning and reproduction of the community and minimizes the risk of error. It does so by adopting a social structure anchored to the genealogy of lineage and, therefore, to the ancestors/founders of the village

[30] H. Blumenberg, *Work on Myth*, MIT Press, Cambridge, MA, 1988.

[31] As long argued by some, including: R. Lenoble, *Esquisse d'une histoire de l'idée de nature*, Albin Michel, Paris, 1969. On the progress of research over this issue see among others: P. Descola, *Par-delà nature et culture*, Gallimard, Paris, 2005.

and, through them, to the elderly specialists of the word, the guardians of sites of worship and the technicians of the sacred (prophets, visionaries, foretellers, etc.). The seniority principle is thus the cornerstone whereby myth is passed down in order to maintain society within the cosmic order, to form social relationships and, ultimately, to hypostatize the power relations that ensure its functioning.

Along the same principle, even the cliff is seen as a sign of myth and its natural meaning is transferred from a denotative level to a cultural, connotative level through the meaning that is sedimented in its name. Those who activate the topomorphosis process mentioned earlier, which turns natural space into a source of social legitimization, promote such process of sedimentation: established norms, inspired words, decisions taken at a given site take on the characteristics of the sacred through a mythical narrative. Ultimately, topomorphosis comes across as a powerful ally to myth seen as a construct of the tradition that topomorphosis upholds. As such, topomorphosis perpetuates the setup and the political hierarchy of a given society.

To recover the social meaning of the cliff landscape is thus to consider the ways in which its mythical value, the foundation of Gourmantché territoriality, is represented. In effect, we repeatedly stated that the relationship between territory and landscape obtains at the level of communication. Territory is the outcome of a process of spatial transformation, brought forth by a social agent and rooted in multiple actions that are not always made manifest visibly. Landscape, on the other hand, is the empirical manifestation that an observer conveys through representation.[32] Landscape is closely related to the presence of an observer, on which element we need to dwell (Fig. 5.3).

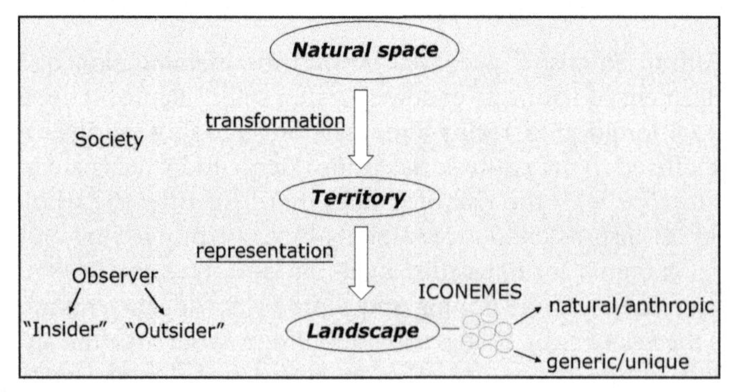

FIGURE 5.3
Landscape as the representation of territory.

[32] To avoid going into an elaborate explanation of the distinction between territory and landscape, I shall adopt the one best suited to my analysis, found in: C. Raffestin, *Dalla nostalgia del territorio al desiderio di paesaggio…, op. cit.*

The elaboration and communication of the landscape concept may be in the form of an articulate and conscious activity or, conversely and more generally, in the form of practical knowledge, a sort of understanding shared by those who partake in a given cultural system. Be this as it may, such understanding is the source of an awareness of landscape. As a cultural asset, territory ultimately becomes a value in its landscape form via the intervention of the observer, who consciously interprets it. We stated it before: the action of the observer is cooperative. It does not convey merely what the eye records, but reconfigures perceptive data according to his or her values, performing an intellectual process. This is indeed an identity narrative, in which the set of values, meanings, knowledge and interpretations refer to three semantic levels (present, past, future) which show landscape as endowed with anthropological value, historical heritage and propensities for future planning. Landscape then conveys the values of territory by casting them as a cultural whole, as heritage able to preserve and communicate the identity of the community that inhabits the place. It should therefore be clear that landscape, as the result of a communicative act, comes from a specific set of skills and values, both physical and metaphysical, which are typical of a given society (Fig. 5.4).

Landscape and *Gourmantché Culture*

In this sense, Gobnangou is a powerful identitarian factor which may be read on two levels: 1) that of external cohesion, since the cliff is taken as the

FIGURE 5.4
A village at the foot of the Gobnangou Cliffs. (See a color reproduction at the address: booksite.elsevier.com/9780128035092.)

FIGURE 5.5
The Pagou sacred site. (See a color reproduction at the address: booksite.elsevier.com/9780128035092.)

distinctive feature of the ethnic group one belongs to across the *Gulmu*; 2) that of internal distinction, since the cliff recalls the foundation myth and identifies each single village which is granted partial use of the cliff's side for its symbolic practices. To be more exact, each village holds an exclusive relationship to the cliff, which at the same time helps to consolidate the wider network of relations across the whole Gobnangou area. The interpretation of belonging, unequivocal and binding, takes shape among the inhabitants across a wide gamut of local nuances. So the social role of belonging is elevated to a transcalar dimension. To each village, the cliff provides the grounding that supports development by tuning it to the pressures of modernity. At the same time, each village is anchored to the origins and its belonging to the *réseau* of tradition is confirmed.[33] At the regional level, Gobnangou holds a sacral meaning somewhat related and yet distinct from the one recognized to Pagou, a small hill nearby invested with a powerful mystical aura (Fig. 5.5). Pagou, whose name (*honor and you shall obtain what you seek*) evokes the expectations ascribed to it by the Gourmantché throughout West Africa, plays a different role from the cliff: it is the elective site of the ultimate sense of sacredness, a source of inspiration and solution to the difficulties that individuals must face in their existence. Conversely, the sacredness of Gobnangou refers to the founding myths and to the regulation of the relationship with the Earth and, therefore, it is the environmental model one must refer to in the exercise of one's autonomy towards the World. We should recall that identity discourse responds to a dual model: environmental or social. By referring to such a model, we can understand in which contexts and by what procedures the sense of belonging evolves and consolidates, and especially from what

[33] By *réseau* of tradition I mean a cluster of villages that are politically structured and placed under the authority of chief of the head-village, from which they originated.

values it derives.[34] In the case of the cliff, it is clear that it emanates from the natural values expressed by territoriality: the set of cultural aspects which, by shaping behavior, establish the social backdrop of the sacral relationship with nature and with it the right to inhabit the place.

We may recall that Gobnangou marks the place where the spirits of nature reside. The cliff's vegetation and its steep walls allow for several iconemes in the form of *obulo*, or sacred sites: *Utanfalu* (the mysterious cave), *Pundougou* (the fall used in initiation rites), *Tanfoldjaga* (the place for the endowing of mystical powers), *Aguanda* (the rock hidden by vegetation), *Kuoli* (the stone struck during sacrificial rites). The cliff is the site of myth and ritual *par excellence*. Few are the privileged ones who are entitled access to it. Although the cliff is plainly visible to everyone, only a few elect may access its secret meaning: the village authorities, those who wield religious power (*perkiamo* and/or *parkiamo*) or political power (*bado* and/or *nikpélo*), who are entitled to maintain an ongoing relationship and to draw advice for human action.[35] These are the places that were revealed to the *nikpélo* by the *divinity of nature* through an incarnation in animal form or the appearance of signs which, deciphered by the geomancer (*tambipwaba*), indicate the ideal place for the foundation of the new settlement. In this way, the cliff attests and affirms the social status of the villages that lie at its feet, perpetuating the tradition and loyalty to the values on which Gourmantché society is based.

The symbolic role of Gobnangou is traceable even in territorial actions that followed the founding of the villages: the presence of cultivated fields crossed by the network of paths that ensure access to the various settlements is the outcome of a territorial action that sets the populated area apart from the one left in its original state. In fact, while routes of communication twist along the foot of the cliff, its slopes are characterized by signs of nature: only a few paths venture through vegetation, giving access to the few villages located on the top or in the south through narrow gorges such as *Utanfolbu*. The founding legend holds that the latter was discovered by a hunter, who is invested with divine powers in the exercise of his activity (Fig. 5.6).

[34] What is recovered here is the *process* whereby the discourse of identity takes shape. Territory, which is a *cultural asset*, communicates the conscious acceptance of its *value* through landscape, which in turn sanctions the *model* to which social identity refers. When it is linked to the values pertaining to human action (history, material artifacts, the transformation of space), this model is defined as *social*; when, by contrast, it focuses primarily on the nature and conservation of its resources, it is defined as *environmental* (A. Turco, "Environnement et discours identitaire dans l'Apennin abruzzais contemporain," in: *Montagnes Méditerranéennes*, n. 1, 1995, pp. 53–60).

[35] The *perkiamo* is the first reclaimer and, therefore, he who made the land suitable for settlement, while the *parkiamo* indicates the ministrant of sacrifices in sacred places. The *bado* is the political head both of the kingdom and of the village. In turn, the *nikpélo* is the authority of the village in his capacity as patriarch. When the latter exercises religious authority in a head village he takes the name of *parkiamo*.

FIGURE 5.6
The Utanfolbu Gorge. (See a color reproduction at the address: booksite.elsevier.com/9780128035092.)

Gobnangou also communicates a wide range of performative information derived from safeguarded knowledge, which people carefully hand down from generation to generation. So the cliff tells us of the relationship established with the villages over time: as well as determining their ideal conditions of settlement, it played the role of a retreat and hideout during internecine warfare. The protection guaranteed by the metaphysical presences of nature was enhanced through artifacts such as *Sapiakpéli* (ancient defense stronghold) or *Soguilafoli* (archaic barns in the rock with the function of hiding people and/or things) (Fig. 5.7). The cliff tells of future villages, because Gobnangou also means "reserve of land." To villagers, it offers fertile areas on its top, only partially tilled, and ready to stem ever-higher demographic pressure.

In short, the Gourmantché culture managed to produce a seamless transition from myth to history through the prism of territoriality, which reflects a real culture of activity. It is a political culture, deeply integrated with spirituality, which sets the geographic conditions for physical existence and the legitimization of identity.

FIGURE 5.7
The ancient Soguilafoli grain silos. (See a color reproduction at the address: booksite.elsevier.com/9780128035092.)

SOCIAL IDENTITIES AND ENVIRONMENTAL PERSPECTIVES: THE ARLY PCU

As noted above, Gourmantché identity is anchored in a model, the environmental model, which sees to the maintenance of a balanced relationship between the anthropization of space and natural resources. This model is not a prerogative of this ethnic group or of African society in general – although it must be stressed that, by emphasizing the value of nature, myth protects that society from wasteful action. Rather, it is one of the issues at stake in globalization, although in that case it arises from a different set of principles and concerns. The West adopted the environmental model with a view to countering pollution and conserving biodiversity, threatened by an economic development understood as growth, which prevailed over the past decades. The sharing of the same identity model between these different societies envisages their possible cooperation in the achievement of environmental conservation.

The territory we analyze here confirms that. We noted that the Gobnangou region hosts the Arly Protection and Conservation Unit, which today constitutes one of the primary areas on which the environmental policy of Burkina Faso focuses in the interest of preserving and enhancing its natural resources. This policy, aimed at protecting the environment of that country, in line with the global conservation strategy, sets three main goals: 1) the maintenance of essential ecological processes and fundamental systems which sustain life; 2) the protection of genetic diversity; 3) the sustainable use of species and ecosystems.

It is in this context that the theoretical potential of landscape can lend coherence to both the environmental protection project undertaken and to the local development of populations who inhabit the peripheries. As a conceptual construct, landscape can bring together conservation strategies and, at the same time, highlight the contrast between different logics at work: the logic of protecting agent and, therefore, of the Park with its reasons, its general and specific protection aims, its conservative approaches to multiple interpretations (technical, legal and institutional, etc.) aimed largely at maintaining landscape in its aesthetic-perceptive value[36]; the logic of the local communities, who act in that area according to the rules of production and use, sedimented in the course of a long cultural and functional tradition, which is expressed through landscape in its social and identitarian valence. What needs to be stressed is that, although competitive, such logics are not necessarily antagonistic. There is margin to avoid conflicts which may arise in a perceptive dissociation of the environment in question. For one, in their view of things local dwellers tend to reduce the Park to a system of prohibitions, so much so that this space of tradition runs the risk of being perceived by agents as a place of degradation, where partly dissipative practices are upheld. However, both the latter and the inhabitants strive to achieve the same goal, which is to preserve the ecosystem and, with it, landscape. Let us pause, then, to assess the practical implications of landscape based on the recognition reserved to that concept on the international arena.

The Systemic Approach: Landscape as a Planning Unit

The importance of the landscape has been attested by the major international organizations involved in environmental protection and land. UNESCO, IUCN, and the Council of Europe place landscape at the center of their interests in environmental protection projects related to local development. The sharing of this perspective by international bodies comes from the acceptance of scientific results, which submitted landscape as an all-encompassing concept within what is called *landscape ecology*. It is a discipline that adopts a systemic logic, hence a solipsistic view of the environment, of its process-like quality and of the constant interaction between biological and social processes. Landscape is seen as a concept of great applicative potential, which shifts the focus from the outcome of a single intervention to its systemic consequences, thereby envisaging the evolution of the world system as one single phenomenon. Because of its potential, landscape enters planning, which more and more insistently takes on ecological processes as the platform of anthropic activities.[37]

---

[36] Private individuals involved in protection may be associated with the State, because their presence depends on a revocable contract, the conditions of which comply with the principles of environmental rules currently enforced.

[37] Processuality gives rise to the idea of landscape as a planning unit able to analyze the interactions between society and the environment and, therefore, to take charge of both territorial planning and environmental protection (R. Gambino, *Conservare, innovare: paesaggio, ambiente, territorio*, UTET, Turin, 1997).

More specifically, *landscape planning*, which is the application aspect of the ecological approach (*landscape approach*), focuses on two aspects: 1) interdisciplinary management of natural areas; 2) landscape as the field of planning action for producing conservation and development. Planning is thus conceived as the set of interventions that does not neglect economics but is also anchored in the associative culture of nature and development. Similarly, environmental issues are addressed in the name of sustainability and of the load limit, taken not so much as a given data, but as dependent on the specific representations that define it. That reinforces the conviction that nature is not a definitive given, or a set of stable and enduring features. On the contrary, nature is addressed according to the process of socialization to which it is subjected and, therefore, to the representation which humans make of it. Operationally, *landscape planning* engages with the biological and socio-cultural systems of a given area in order to optimize the use of territory without affecting the renewability of its natural resources. A transcalar analysis is carried out first (of the region, the local situation) to recover each organizational level in its specific properties. At a regional level, the logics of the various actors involved and the policies that entail are explored. At a local level, analysis is directed to obtaining the representation that the communities have of nature as well as the activities and programs that they intend to follow for their own development.

When *landscape planning* is aimed to a protected area such as a park, the areas that make it up (the central zone and peripheral area) are taken as parts of a single landscape entity, which must be addressed by mobilizing multidisciplinary know-how able to grasp each sectorial feature. The integration of the various components, natural and social, at the base of the environmental approach, encourages a new kind of analysis. The problems of a region and the links between the landscape and the political-economic structure, both local and national, are analyzed as highly integrated entities. At the same time, the many actors involved are compared, with their different logics and landscape representations but also in their convergent goals. This leads to assuming landscape as a unifying paradigm throughout all phases of planning (from diagnosis to management). It is from these premises that landscape becomes the Unit of Planning (UP), with a view to integrating the features of the protected area and the social demands of the communities which populate its outskirts into an inseparable aggregate. In an effort to devise new forms of environmental protection – which, for the ordinary goals of conservation, may be tied to those of the socio-economic development of local communities – geographers have reached a new conviction. It is the belief that landscape is also the platform of social consultation, a public space, where the legitimacy of intervention is ensured by sharing landscape values. Such values guarantee environmental conservation while at the same time envisaging possibilities of development for the local communities. So, the *landscape approach* is supplemented by *community conservation*, an approach that, we submitted earlier, acknowledges participation as the crucial factor enabling the actors involved (local communities and bodies engaged in the park management) to compare their representations of nature in order to co-manage resources and share basis decisions over environmental protection. It is on this operational level that landscape shows its political significance and its potential as an instrument aimed at participation and consultation.

Participatory Methodology and Operational Tools

Translation of the landscape into UP involves the creation of tools to collect, represent and effectively communicate its social value, or the role it plays in transmitting the cultural values of territory. Therefore, implementation of this concept requires field research aimed at recovering the setup of territory. That is to be followed by: an eloquent interpretation of the data collected; a representation of the same data, through specific tools used as a platform on which to start consultation; and finally the drafting of guidelines for making operational decisions consistent with the system of shared values. In this perspective, through an inter-university cooperation between 2iE (*Institut International de Ingénierie de l'Eau et de l'Environnement*) of Ouagadougou and the University of Bergamo, a group of researchers conducted research on the outskirts of the Arly National Park, over a period of three years, using the SIGAP (Geographic Information Systems for Protected Areas) Strategy method. The research aimed to investigate landscape as UP and addressed the territorial aspects outlined in Chapter 4, while also pursuing the goals of: 1) recovering the landscape values of the local population; 2) communicating these values through a participatory and landscape mapping.

Field Research and Participatory Diagnostics

The first phase of research coincided with the on-site survey of the area, aimed at collecting data on the socio-territorial organization of populations who inhabit the foot of the Gobnangou cliff and, later, of those from the Madjoari Department enclave.[38] The lack of previous analysis required the collection of basic data and their arrangement into a framework able to record both local

[38] Data collection was carried out by students under the supervision of Olivier Lompo, a PhD in Geography of Development from the Università degli studi di Napoli "L'Orientale," with a thesis entitled *Légitimité territoriale et décentralisation en pays Gourmantché: dynamiques et enjeux socio-territoriaux à la périphérie de l'UPC-Arly/Burkina Faso* later published in book form: O. Lompo, *Burkina Faso. Pour une nouvelle planification territoriale et environnementale*, L'Harmattan, Paris, 2011. I take this opportunity to thank him publicly for his generous help in processing the data collected.

specificities and the regional context. A regional scale highlighted aspects of the population's ethnic setup, the number of inhabitants, the number of villages, seasonal and permanent migration, data which served to substantiate territorial organization and provided for an initial assessment of human pressure on the park. This was followed by a quantification of ongoing phenomena, which cause change and innovation. Such phenomena were placed within the general framework of the complex relationship between tradition and modernity, which needs to be managed. At the local level, the aim was to conduct an in-depth examination of the social setup and the mechanisms of territorial functioning, by linking them to the value of the landscape and the relationship that the communities have with the natural resources.

As for the qualitative side of research, we used participatory models based on evidence provided by members of the rural community and on data collection. This was carried out in accordance with a fixed scheme which allowed us to gather relevant information: 1) knowledge of the physical and morphological features of the area; 2) perceived localization of the village with respect to others (especially those inhabited by different ethnic groups) and estimated distance from the protected area; 3) cultural roots of the various villages along with the history of their foundation and the historical record we have of them; 4) social setup, i.e., the status and the role of the many social members (sex, age groups, genealogy, and guilds); 5) examination of the relations that exist between genealogies, estate policies, and the criteria legitimizing authority; 6) policies of territorial appropriation and utilization via symbolic or physical schemes that regulate collective life and ensure social reproduction; 7) potential conflicts with Park Management and possible causes; 8) mutual interests, and psychological or behavioral attitudes on which peaceful partnership between the agents involved may be based. This was done with a view to recovering the values or the issues which may affect, enhance or threaten the identity profile of Gourmantché society. Having identified received sets of values that ensured harmony with nature, external agents became convinced of the need to involve local people – who for centuries had ensured the preservation of the Arly PCU and of its peripheries – in the conservation plan itself. The persistence of a traditional system was shown to have great potential, insofar as it can successfully steer the changes introduced by external projects.

At a later stage, the rendering and cartographic visualization of territorial data yielded information quite unlike that culled through observation and inquiry. On a regional scale, they promoted the sizing and the cultural modeling of the territorial phenomena found in the region. On a local scale, they have promoted the emergence of values and local knowledge sets that highlight the importance of topical competence in the management of places (Fig. 5.8).

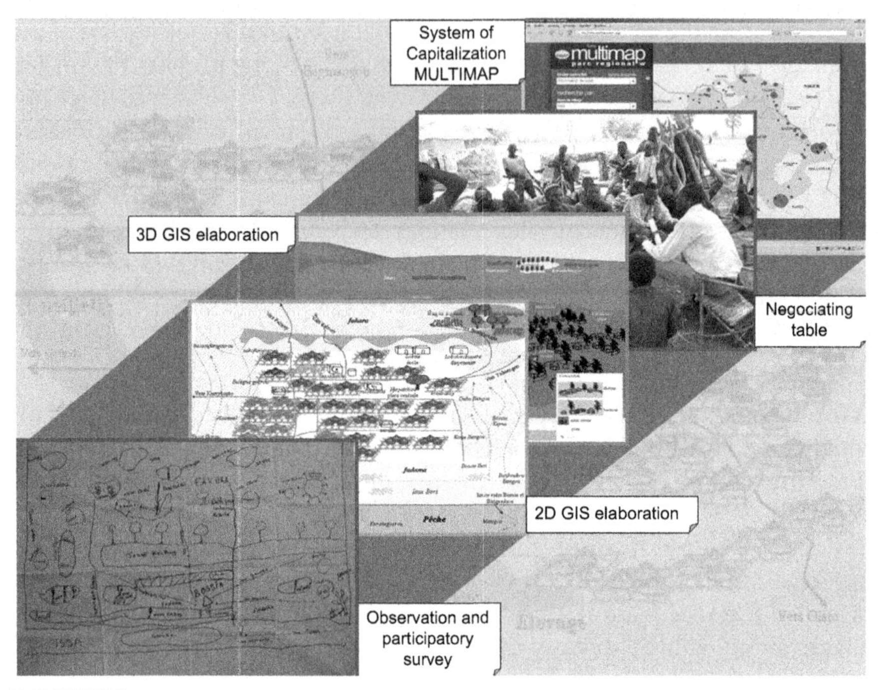

FIGURE 5.8

Building a participatory map. (See a color reproduction at the address: booksite.elsevier.com/9780128035092.)

PARTICIPATORY LANDSCAPE CARTOGRAPHY

Before illustrating how we set about constructing landscape cartography operationally, we ought to dwell briefly on the issues involved in the construction of representation of the Elsewhere by intermediaries who belong to an exogenous cultural context, albeit with the participation of local people. Postcolonial studies cautioned us against the presumption of being able to interpret the culture of the other without smuggling in our own mental categories, logical procedures, or social conventions.[39] As current *cultural studies* have shown, intercultural relation is a dynamic construct that is reminiscent of the process involved in the construction of the cultures themselves.[40] These studies address issues that open an important

---

[39] The issue of understanding diversity without absorbing it into our world comes from the awareness that any culture seeking to understand another will somehow change it, although it simultaneously changes its own as well. Semiotics has clearly shown that cultures are constructs, but even their study is a construct, as are the maps drawn from that same analysis and ultimately even their interpretation.

[40] Poststructuralism deeply undermined confidence in the reliability of a sterilized analysis of cultures. Scholars usually ascribed to poststructuralism include at least Michel Foucault, Edward Said, Pierre Bourdieu, and Arjun Appadurai.

chapter on the risks of misrepresentation or reduction of meaning intercultural mediation entails. They also insist on the need to achieve that awareness, which allows for a more balanced perspective and highlights a number of methodological choices to be made in an attempt to contain risks. In the experimental situation I am presenting, participatory mapping saw the intermediary in the exclusive role of facilitator, easing the graphic transposition of the meanings that participants ascribed to the landscape. Local actors involved in the operation could use his help to autonomously establish the graphic form best suited to express them.[41]

We made it clear that the method we employed in our research was arranged in various phases, in the course of which participatory cartography underwent an IT transformation. Later, the researchers attempted to transfer the social value of landscape deposited in the maps with complete respect for their original graphics forms. By focusing on the multiple logics and on the plural languages which can convey a sense of place, they discarded conventions to experiment with new solutions in the construction of the map. Attention was thus paid to the specifics of data to include and to their graphic transposition. Thanks to the participatory techniques outlined above, we obtained a large amount of data which was tapped for information about landscape.[42] We used unconventional languages and graphics formats to open expressive possibilities that could recall the visual substance of the place, but also its unutterable quality, which was conveyed by abstract icons (sphere, prism etc.) referring to concepts rather than to the features. In short, by focusing on IT rendering we tried to open up the map to the subjectivity of plural discourses. In sum, cartographic mapping was conducted by involving local stakeholders through their dialogic styles, respecting their choice to use figurative language, which was reflected in IT graphics. All this was done not under the naive illusion of reproducing the culture of the other, but in the belief that by reducing claims to exhaustiveness, undermining current cartographic conventions and pointing out the fact that they can be disputed, we would create the conditions to produce a representation that could respect the *médiance* of local communities.

As for the efficacy of our intervention, we leave judgment to others. Now, however, it should be noted that operationally this mapping proved suitable to play the role of symbolic operator able to contribute to the decision-making process, to intervene in the achievement of *governance* at the negotiating tables and to envisage *empowerment* of the various actors involved in planning.

---

[41] Achieving a climate of collaboration and synergy was the outcome of an intense work to infuse reliability in researchers.

[42] The maps were made using information obtained by respecting the collective narratives typical of the African context of the villages. In that context, the information produced by the debate between participants, through a comparison of various opinions, was addressed to the precise and limited target that we had set out to achieve, namely coping with a specific issue or one of their relevant interests.

Results of Research on the Outskirts of the Arly PCU

Research on the outskirts of the Arly PCU produced: a diagnosis of social and territorial features; the identification and interpretation of its problem areas and of its potentials; the drafting of maps as tools for the management of the park and the consultation with local people. Analysis of the functioning mode of those communities and of their sets of traditional knowledge in relation to nature showed that the balance, so far ensured by the statute of tradition, especially in land issues, is now being threatened by external logics such as the speculative economics of cotton farming or the political logic of decentralization. Potential for tourism development and environmental protection of the area's landscape resources also emerged.[43] These results were visualized via a new type of cartography, whose features may be summarized as follows: 1) the data used refer to both the referential and the social aspects; 2) the graphics rendering uses a mashup of 2d and 3d; 3) GIS systems combine with other interactive and multimedia systems. Let us consider a few of these cartographic products in more specific terms.

The map of the historical evolution of the villages (Fig. 5.9) shows the vast timeframe in which their foundation needs to be placed. A few particularly momentous periods stand out, corresponding to specific events of a social or natural order. The first refers to the precolonial period; the second is located at the time of the classification of protected areas, followed by a number of forced displacements of the villages that were in them; the third, which dates back to the drought of the 1970s and reaches the present day, is characterized by migratory movements that in some peripheral areas led to land saturation.

Even the decentralization process that Burkina Faso undertook in the 1990s affected, albeit only indirectly, the distribution of settlements, and not always in positive terms. This political and administrative reform was supposed to further democratization and the participation of local populations, but in some cases it caused an antagonistic overlap of the administration structure and the traditional statute of the villages. De facto, decentralization does not always acknowledge the ties of dependency between the two and tends to overlook the plane of traditional authority, thus curtailing what could have been an opportunity (rather than an obstacle) to the practical translation of such tradition. For example, administrative status was assigned less on the role that the village played within the *réseau* of tradition than in accordance with alien criteria, such as population count. That disrupted old territorial hierarchies, which were thus replaced by new ones. In some cases, the nikpélo village, whose chief is the patriarch or eldest and as such enjoyed but relative autonomy from the head-village in the traditional statute, was raised to the level of the latter, and attached

[43] Research results are thoroughly recorded in: E. Casti, S. Yonkeu, eds., *Le Parc National d'Arly et la falaise du Gobnangou (Burkina Faso)*, L'Harmattan, Paris, 2009.

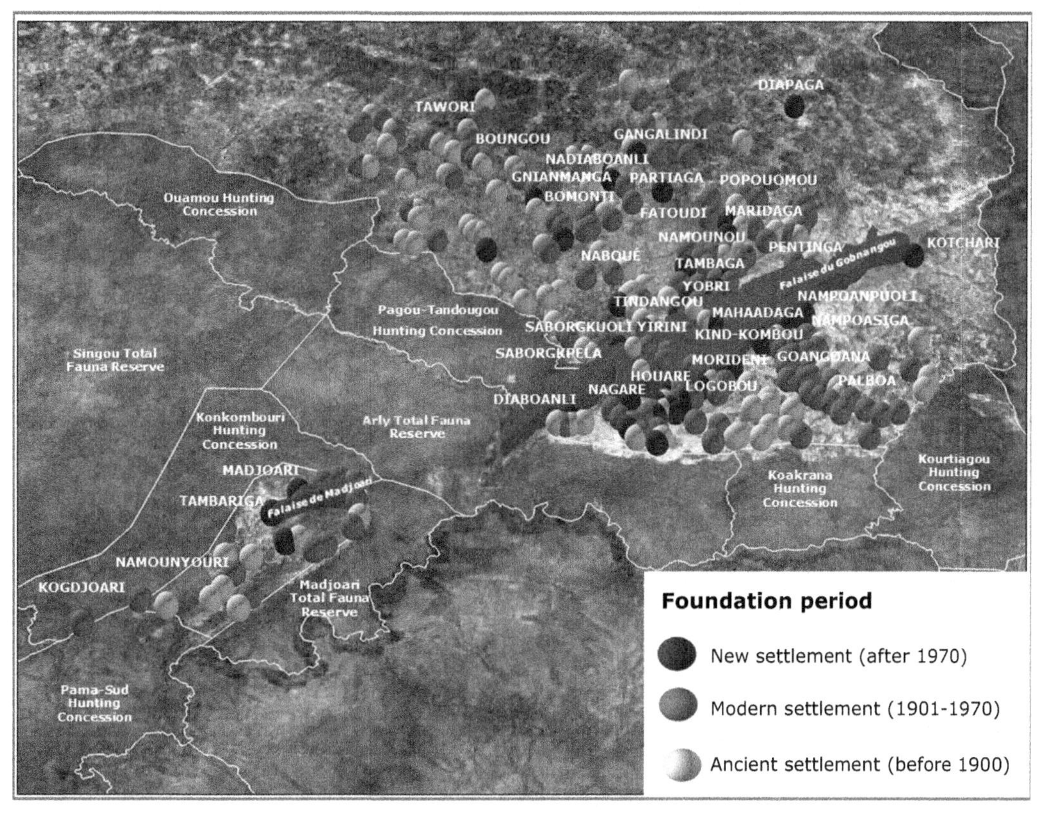

FIGURE 5.9

Historical evolution of villages on the outskirts of the Arly PCU. (See a color reproduction at the address: booksite.elsevier.com/9780128035092.)

functions and roles to people who used to have no authority either recognized or legitimized by tradition. In the Yirini réseau (Fig. 5.10), for example, the administration statute ruled that the head-village (Yirini) and those that depended on it (Thioula, Gnouambouli and Bolbigou) adopt the same statute, thus undermining the political authority of the bado, the head-village chief.

It is worth noting that regional cartography shows that decentralization also affected the balance in the distribution of settlements. In fact, the number of administrative villages varies greatly between the two sides of the Gobnangou cliff, at the expense of the Southern side, which shows a marked propensity to settlement centralization.

Finally, two remarks are in order about the graphic rendering of these phenomena and the fact that it alludes to the visual aspects of landscape. First, we would recall the rendering of the third dimension to recover the landscape volume and its morphological features, similar to the satellite photos of Google Earth. Secondly, we would underline the use of multiple points of view

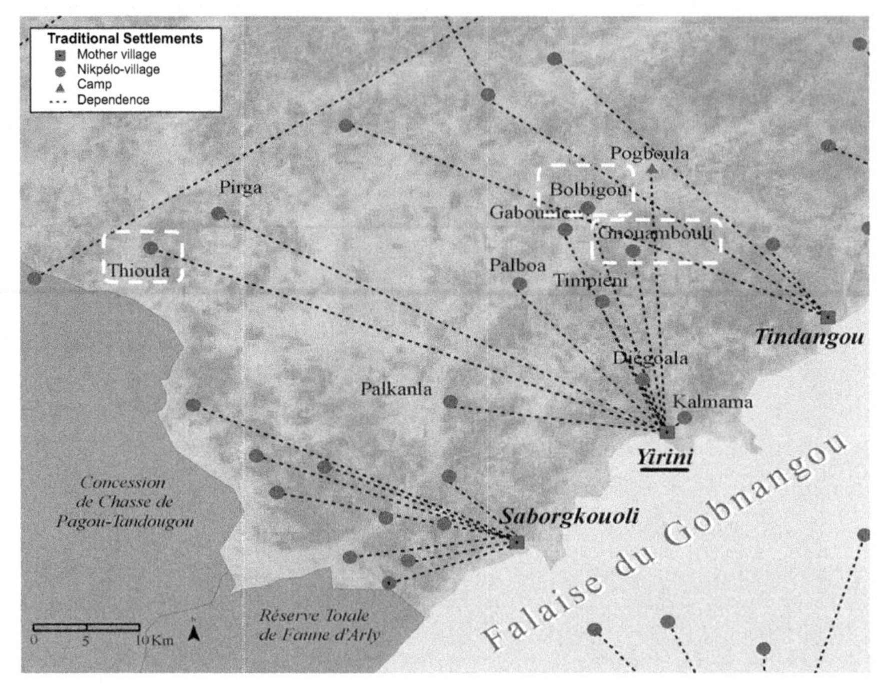

FIGURE 5.10
The Yirini *réseau.*

and oblique angles to bring out the centrality of the cliff and its attraction role for the villages. This is an attempt, still under development, to move away from the classic topographical view and envisage the value of the landscape, aware of the difficulty of representing what Franco Farinelli called its "arguzia" (ingenuity).[44]

If analysis shifts to the local scale, our task is even harder because cartographic information is directed to recovering the cultural aspects of the territory. The first map that bears witness to such intention is the one of the Yirini village (Fig. 5.11) which shows the prevalence of traditional values in the layout of built-up areas together with their hierarchical distribution. The types of concession and their distribution in areas marked off by natural features such as riverbeds or vegetation underlines the presence of different genealogies, which the map signals with different shades of gray for each concession. The distribution and the localization of production activities, aimed mainly at agriculture, show the basic sets of knowledge and land-related skills. The skills to do with fields devoted to subsistence farming (millet, sorghum and maize) respond

[44] By that he meant the ambiguity of a concept that straddles reality and representation (F. Farinelli, "L'arguzia del paesaggio," in: *Casabella*, n. 575–576, 1991, pp. 10–12).

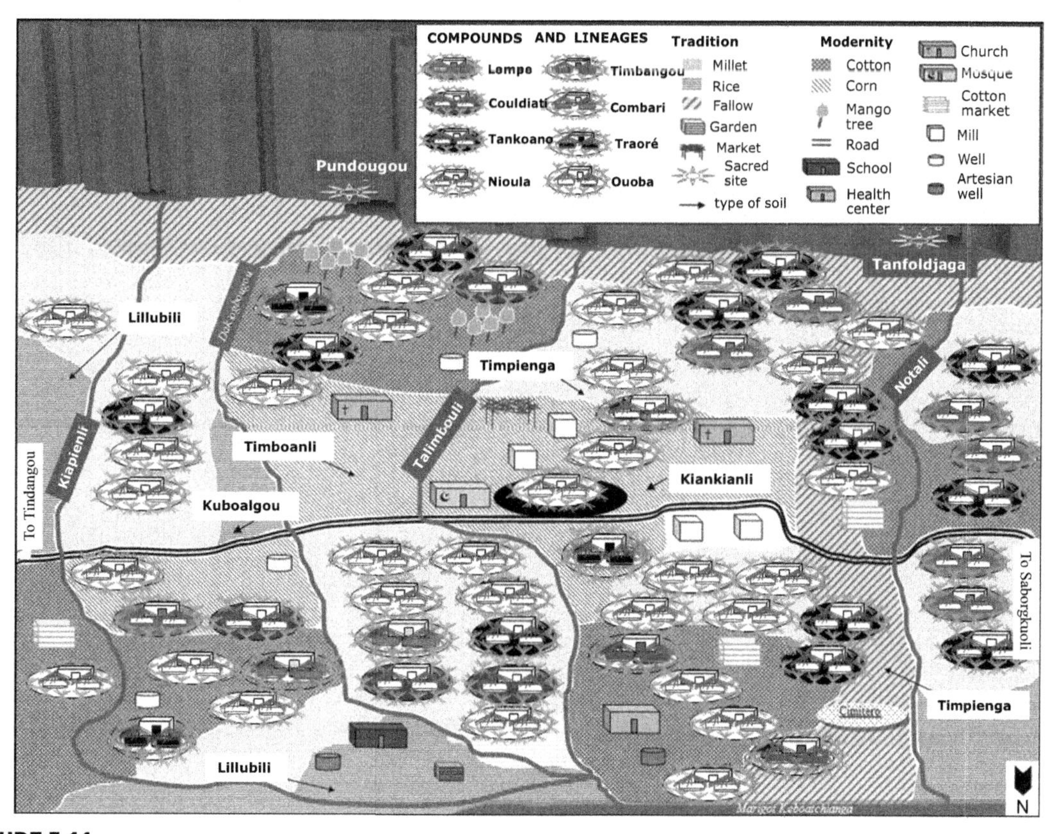

FIGURE 5.11

Layout of the Yirini village. (See a color reproduction at the address: booksite.elsevier.com/9780128035092.)

to soil composition and hygroscopic capacity: sandy clay soils (timboanli) on low slopes, silt-clay (kuboalgu) and clay (lilubili) soils in lowlands. The localization of fields intended for other crops recalls the status of the land and the role played by each member within the group: the fields of the village chief (kiankianli) as a distinctive symbol of his authority, occupy the front of the entrance to his concession. In addition to vegetable gardens within each concession, the fields of women (manloli) are located in the back area (dapuoli) along with those used by other members for private purposes. Fields located far off, be they either communal (kwa-kiamo) or private (sual-kwanu), are subject to crop rotation and their usage rights are established each time by the chief of the lands. Also, the map provides information on recently introduced modern facilities such as schools, places of worship, dispensaries, mills, the cotton market, and drinking water wells.

Nonetheless, if we intend to represent the landscape features of such settlements, we need to restore the visual shape of the village and its relationship

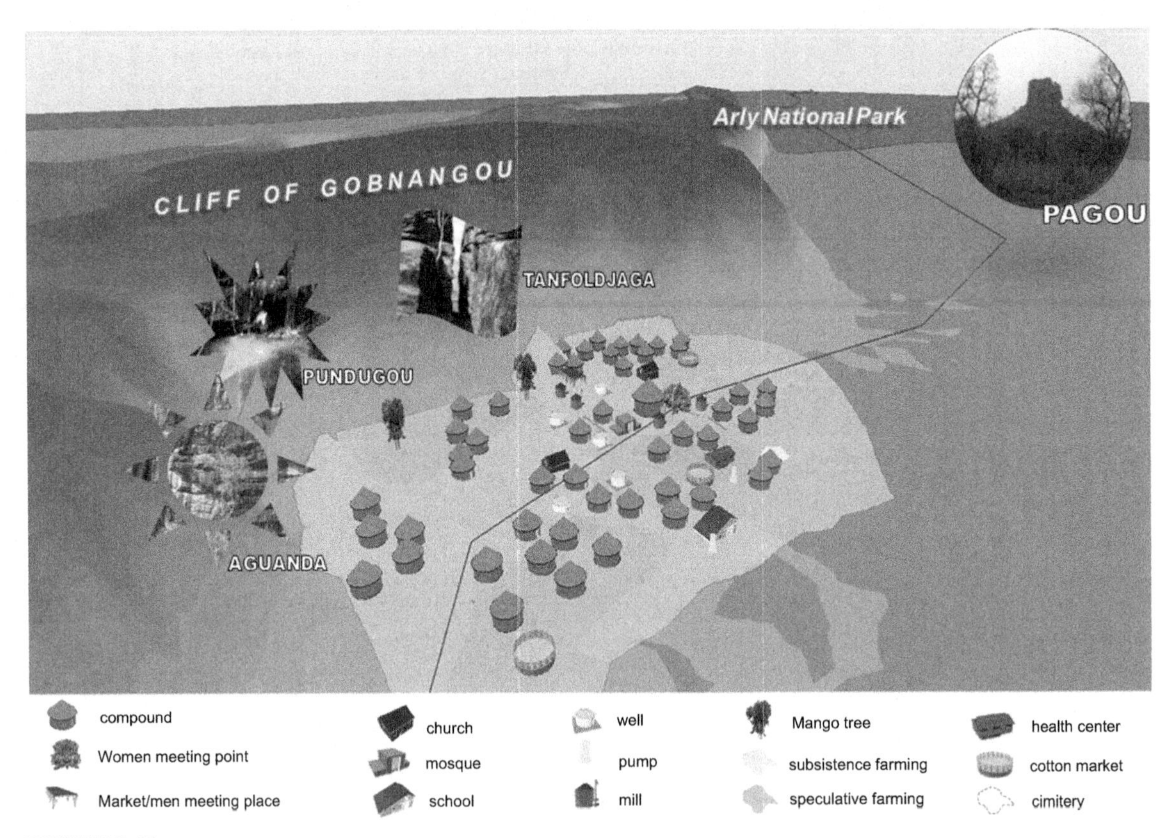

FIGURE 5.12
Landscape modeling in the Yirini village. (See a color reproduction at the address: booksite.elsevier.com/9780128035092.)

with the cliff (Fig. 5.12). In order to avoid photographic reproduction, which would quite indistinctly exhibit all its visual features, we advocate a modeling of landscape rooted in the iconemes which give us a hierarchy of cultural values. As cultural values are transformed into icons, landscape is communicated cartographically. Such modeling responds to the relevance given to landscape by the inhabitants of the village and thus highlights their cultural values which center around two iconemes: the cliff and the village. The technical expedient behind this consists of using three-dimensional (3d) rendering for both to recall their visual form and refer to the centrality on which identity discourse is formed.

The word *Yirini*, which in the Gourmantché language originally meant "peak," alludes to the provenance of its first founders, belonging to the Gnoalba and Nomoumba clans. It dates back to the 19th century. The narration handed down through the *griot* has the founders originally coming from the Cliff, led by a panther (*gambo*) which showed them the way and rescued them from enemy attacks. The group was followed by a cat who hid all their tracks and prevented enemies from chasing them.[45]

Concessions provide a sort of pattern which gives us a fixed record of clan history for the village and of the different relevance each of its eight founding genealogies has. Over time, in fact, to the two founding lineages were added six more which play specific roles based on the assignment of duties and responsibilities related to particular production activities within the community. The concessions, grouped into neighborhoods, gather the extended family belonging to the same lineage and express, within them, the different functional roles of its members. What is even more relevant is that these concessions mark the places from which the links to the Cliff originate. Near its entrance, each concession has a place for propitiatory rites meant to protect the family. Yet it does not cover rites of divination, which are instead practiced in sacred sites jealously guarded on the slopes of the Gobnangou cliff: *Pundougou* (the sacred Falls), *Tanfoldjaga* (the sacred Cave) and *Aguanda* (the Sacred Stone hidden in the bush).

I already dwelt on the significance that these iconemes have in the understanding of landscape and the transmission of its social values, which the community attached to the cliff. Such sacred places do not play the same role and are not subject to the same interdictions: some are sacred places anyone can approach to ask for personal or family-related favors; others are accessible only by the village chief being invested with political or religious power; some are places where interdiction applies only to specific sacrificial areas, where the *parkiamo* invokes divine intervention to solve issues that threaten the whole village community. In the area around these sites, members of the village are allowed at certain junctures of their individual life, such as the transition from one age group to another. In these places all the practices of resource use are forbidden: one cannot fish, cannot use spontaneous vegetation, tame animals are prevented access, and hunting is prohibited. The set of prohibitions as a whole is designed to promote respect for the site where the deity manifests itself, in an effort to preserve the site's original shape. It is therefore a landscape which must be respected and guarded jealously, not only as a way of honoring the local people, but also in order to protect a culture whose ways of relating to nature are unlike those of the west.

A figurative rendering of sacredness must of necessity leave behind the visual shape of landscape and use abstract icons to convey the connotative value of such values. It is the kind of abstract iconization found in Figure 5.13, where the cliff is stylized, concessions are rendered as spheres that are scaled in relation to their peculiar social role, and abstract forms identify sacred places. Everything

---

[45] In sign of gratitude, the members of the clan totemized the cat, the leopard and all the other animals of the cat family.

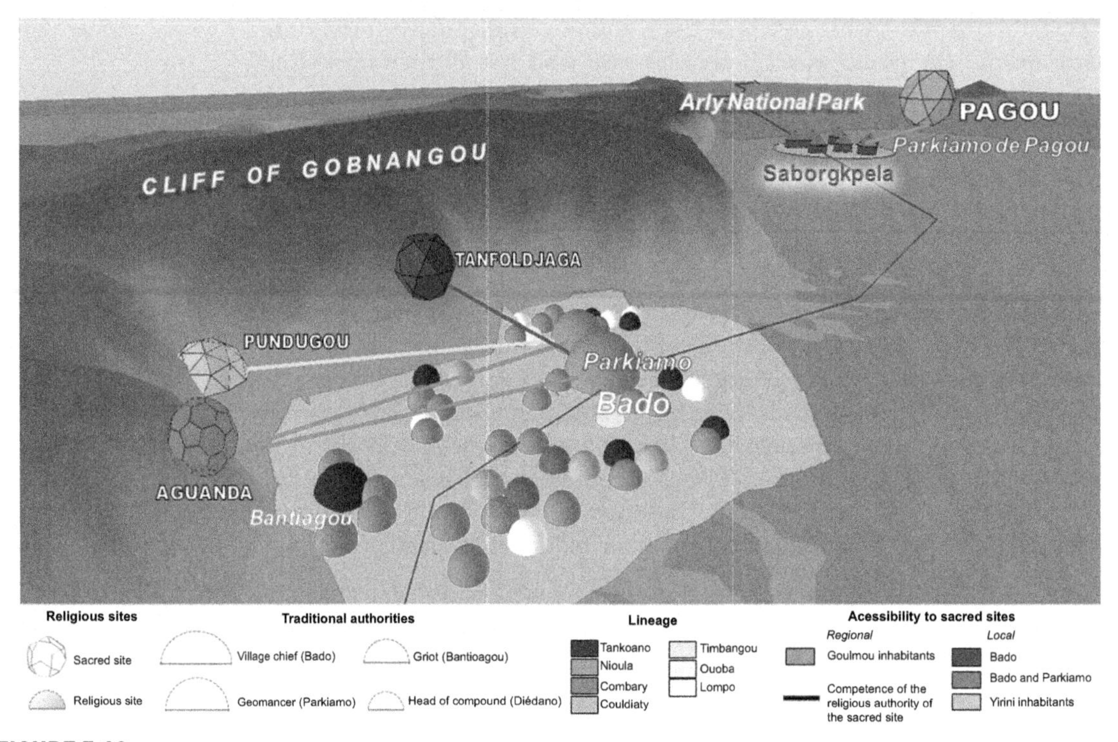

FIGURE 5.13

The sacredness of landscape in the Yirini village. (See a color reproduction at the address: booksite.elsevier.com/9780128035092.)

underlines a sacral dimension of nature, seen as *médiance.* In short, the mapping of sacred landscape in the Yirini village expresses a semantics of landscape that yields a set of syntactic rules able to show the links between the cliff and the village. Hence, the role the cliff plays in hypostatizing the powers that be, thereby sanctioning the social order, is enhanced. By considering such places as iconemes which direct interpretation and endow landscape with meaning, we cannot fail to notice that, as they exhibit the link between the village and the cliff, they also inform our reading of the settlement, because they provide a hypostatic ranking of concessions. In particular, the concession of the village chief will be the most important socially, since it is inhabited by the person who has exclusive access to *Tanfoldjaga* and who also partakes of the religious chief's right to access *Pundougou*. The dwelling of the religious chief is itself relevant for different reasons. It is actually the dwelling of the man who – albeit deprived of powers over land or the community – acts as an intermediary of divine will and therefore ensures a constant exchange with the cosmogonic world. The religious chief also has exclusive access to *Aguanda*. The concession of the *Bantiagou*, the *griot* in charge of handing down myth in the form of public narration, also plays a key social role in the village, although certainly quite subordinate to the roles of the *bado* and of the *parkiamo* by virtue of their link to sacred sites. If we stop to consider the landscape features that mark these sacred sites, it becomes clear that their meaning is closely tied to connotation and as such it is enhanced by an additive relationship. The *cave of Tanfoldjaga*, where the village chief receives the investiture of mystical powers and, therefore, where his social role is legitimized, alludes to the depth of the relationship he has with the divinity. It takes place inside the bowels of the earth, and therefore, inside the god's womb. The *Pundougou* waterfall, for initiation rites, to which both the *bado* and the *parkiamo* have access, recalls aspects of life associated with water, both in its purifying function and in its regenerative role of a cathartic rush, given by the abrupt break of the water's smooth flow. The waterfall, then, is the place where divine benevolence and the responsibility of the individual are evoked in the sanction of a vital pact. *Aguanda*, the rock hidden by vegetation, the access to which is reserved exclusively to the *parkiamo* for purposes of divination, refers to the inaccessibility and the secrecy with which the deity manifests itself. Finally, the map's backdrop outlines the site of *Pagou* and its link with the ancient village of *Saborgkpela* where the sacrificer dwelt. His role, as we explained, does not affect the status or organization of individual villages, but serves to evoke the sense of belonging to the ethnic group as a whole at a different scale, the regional one.

The map conveys the sacredness of the Gourmantché landscape not so much and not only by virtue of its tie with nature, but as an ordering principle of the *médiance* Burque considers vital for giving substance to places. Visually, landscape will not betray any reference to its social dimension. It is only by making sense of the mythical values attached to the cliff that one can build a plot or a story that presents what we have discussed. To be sure, we are dealing with experiments in the making. Yet, such tests pave the way to new modes of cartographic representation.

Landscape and Territorial Planning

The map we produced made it possible to bring out the cultural and social dynamics of territory that promote the recovery of landscape as UP and include it in protection strategies. In the area we considered, a solid and well-structured system of traditions indicates a potential for starting some type of exogenous development planning, since the representatives of that system could play the role of reliable interlocutors in the assessment of the proposal and of local agents, able to translate it operationally. At the same time, our analysis made it clear that landscape is not only a cultural resource but represents the *totality* whereby the community expresses itself and shows the territorial setup of the villages. In other words, the sacred sites should not be considered as isolated and isolable phenomena, but as poles of meaning around which landscape as a whole revolves. They are the iconemes which can endow the whole cliff and the whole cluster of villages with semantic and syntactic meaning. Landscape, therefore, must be bound to measures of protection and valorization able to place it within the context of a sustainable development strategy. Indeed, the quantity and quality of the area's cultural and natural resources leave no doubt as to the potential for eco-friendly projects, such as those which, in accordance with the values mentioned above, ensure local development and resource conservation. The ways to achieve this can be many, but each will have to take account of the social and functional articulation of the area, assigning to each village community a complementary role based on the degree of tradition or modernity that it presents and on its propensities for openness or closure towards visitors. It is imperative, however, that any development project be carried out with the involvement of the local populations and using instruments of participation and consultation similar to those that our research produced.

To conclude, the testing of landscape as UP outside the European context shows that the policy reasserted by the Council of Europe for its landscape is applicable to external contexts, where the intertwining between society and nature is marked by sacred sites. It also shows that, when supported by cartographic communication tools, such testing yields valuable information on which to reflect in order to seek *médiance* also within the European landscape which the Convention, first and foremost, addresses.[46]

THE LANDSCAPE-BASED DIMENSION OF ICONS

Although they offered glimpses of a possible chorographic horizon, the cartographic experiments presented so far also betrayed the limits of their graphic rendering, tied to the degree of empiricism involved in their formulation. Consequently, what seems binding at this point is to pause and reflect on the way in which icons may convey the various values expressed by iconemes. To do this we need to use as a backdrop the semiotic process of figurativization* which, I noted, takes over the designator and turns it into an icon. We will consider its phases: *spatialization*, *figuration*, and *iconization*. That will enable us to find a mode of intervention in the construction of the chorographic icon: spatialization, which in topography used the zenith-based point of observation and assumed a uniform scale, will now provide instead multiple points of observation and interaction of multiple scales for enhancing the topological dimension of space. Figuration will use surrogates either able to recall the symbolic meaning of iconemes in order to recover *médiance* or, conversely, capable of showing

---

[46] The European Landscape Convention is a document endorsed by the Committee of Ministers of Culture and the Environment of the Council of Europe on 19 July 2000. Besides giving a definition of *landscape*, the Convention lays out provisions related to recognition and *protection*, which member states undertake to enforce. It also defines the policies, objectives, the preservation and management related to the landscape heritage, upon recognition of his cultural, environmental, social, and historical relevance as a component of European heritage and a key element in ensuring the quality of life for people.

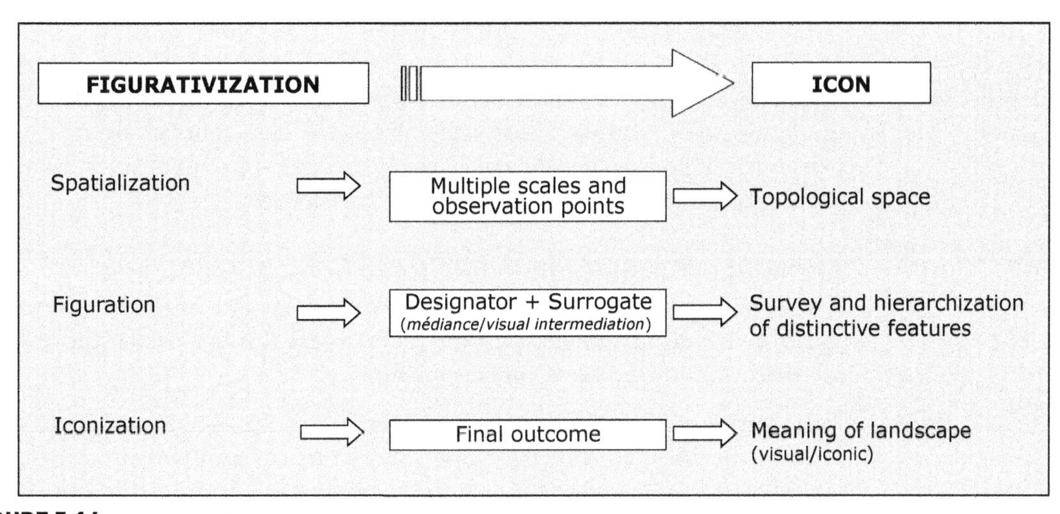

FIGURE 5.14
The landscape dimension of icons.

their visual irrelevance. The outcome will be to assign distinctive valences and prioritize the role of icons in communication. Finally, by jointly embracing the outcome of spatialization and figuration, iconization will convey in different modes either the iconic or the visual meaning of landscape (Fig. 5.14).

We should start by stating that iconemes play different communicative roles: some are located at the very end of a communication chain which informs us about the overall significance of landscape, while others intervene in the transmission expressing a specific social value landscape contains.[47] Their graphical transposition must therefore take into account their different communicative role. In the same way, the adoption of iconemes as perceptual units of landscape, to which icons should conform, is not intended to translate cartographically the meaning of the landscape taken as universal and unique. It aims, rather, to declare the futility of graphically identifying and classifying units of meaning suitable for any social sphere. We already found that the values of the landscape transmitted by iconemes vary according to the society that interprets them. Icons will thus have to try to convey their mutability. Our ultimate goal, free from any enumerative claim, is to envisage a cartography that, by positing a particular conjecture and acknowledging its informational variability, may use a changeable language

[47] For instance, the iconeme of a castle in an alpine setting consolidates the cultural sense of the entire landscape and directs the reading of other mountain iconemes. Conversely, the one given by electric pylons, which connect the Radisson power station placed in the far north of the Quebec region to the south, marking the transition from the taiga to the Canadian tundra, iconizes the technological overcoming of the climatic frontier; or, again, the iconeme of the woods which, in its perpetuation, makes it possible to specify the extension of vegetation resources.

to translate the diverse social values attributed to the different landscapes. And it is more consistent with our goal not to take iconemes as elementary units,[48] placed at the base of an order of landscape, but to focus on the different communicative role they play, in an effort to render them cartographically. In this perspective, it is worth recalling the archetype/cliché opposition, around which a number of theoretical possibilities may be envisioned.

Archetype/Cliché Opposition in the Selection of Icons

Philosophy defines the *archetype* as a preexisting and primitive form of a thought, a kind of universal prototype for ideas through which individuals interpret what they observe and experience.

In a narrower analytical context, that of linguistic *deconstruction*, Jacques Derrida specifies the role archetypes play in the dissemination of meaning within a text. He asserts that archetypes are to be found in what one writes (in one's unconscious) before they are *uttered*, with the features of permanence and deferral of the encoding/decoding process which writing typically has.[49] Assuming this textual perspective and bending it to our semiotic ends, we could claim that the archetypal nature of an iconeme is comparable to the primary dualism that obtains in its relationship with nature, as described by Berque.[50] Consequently, the icons that translate it in a cartographic setting will have to convey either the imbalance or the symbiosis of such relationship. For instance, a sacred site will be represented differently according to the role it has in the reference culture. In Gourmantché culture such a role crystallizes into iconemes which recall the symbiotic relationship and envisage *médiance* as the outcome of the interpretation of landscape through mythical thinking. Because of that, our map will record it as having an all-encompassing reach and ordaining the whole landscape. The case of Italian culture, characterized by an uneven relationship with nature, is different. Italy presents markedly religious iconemes, to be conveyed by icons which

---

[48] Among the units of elementary territorial meaning three were put forth: the choreme, the chorotype and genotype able to reproduce, in their combination, any geographic reality. The first two, developed by R. Brunet, differ by the fact that the chorotype accentuates the recovery of the social aspects of the choreme (R. Brunet, "La composition des modèles dans l'analyse spatiale," in: *L'Espace géographique*, 4, 1980, pp. 253–265; Id., *La carte, mode d'emploi*, Fayard/Reclus, Paris, 1987; Id., *Le défrichement du Monde. Théorie et pratique de la géographie*, Editions Belin, Paris, 2000). The third, on the contrary, is proposed by J. Lévy, who gives it the meaning of a "set of chorotypes that, in their interactivity bring out the importance of the situation considered as a spatial metaphor able to recover the complexity of geographical realities" (J . Lévy, "At-on encore (vraiment) besoin du territoire?," in: *EspaceTemps Les Cahiers*, 51–52, 1993, pp. 102–142).

[49] In Derrida the archetype is said to coincide with "archewriting": the ideal form of writing preexisting in humans before the creation of the language which engendered them. Although the French-Algerian theorist eludes any attempt at defining deconstruction, one may interpret it as a "listening" strategy to be activated each time, in order to reach the deeper meaning of the text and of the culture (J. Derrida, *L'Ecriture et la différence*, Seuil, Paris, 1967).

[50] See footnote 19 of this chapter.

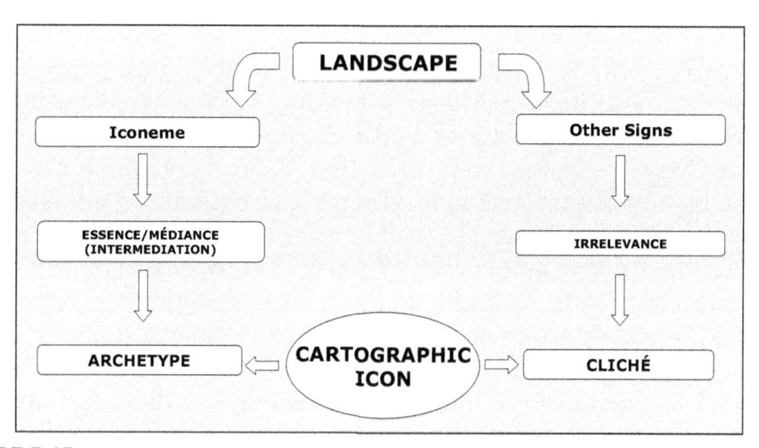

FIGURE 5.15
Landscape communication: from iconemes to cartographic icons.

evoke the spiritual values socially produced in *médiance*. Yet such icons will also expose the rift of values between nature and society. It is evident, however, that in either case figuration of the iconeme will be implemented by means of icons that relate to the value expressed in the specific social context.

With regard to this transposition we should recall both the evolutionary dynamism of the archetype and its link to the *symbol*, highlighted by Carl Gustav Jung when he claims that archetypes are endlessly reshaped by societies and represented via symbolic images.[51] On the basis of this authoritative statement, we may interpret the icon to be a symbolic figure which takes over the archetypal meaning of the iconeme produced by *médiance*. As demonstrated in the Gobnangou example, iconemes actually produce units of varying syntactic weight and render the value of landscape in accord with the role that, over time, topomorphosis* ascribed to them. Icons, then, use surrogates that underwent specific formal processing in order to express such a value cluster as well as the outcome of their evolution[52] (Fig. 5.15).

---

[51] Jung argues that, in modern times, the survival of archetypes is tied up with the outcomes of mass communication. A blockbuster film, a book, a widely popular television show can play a role in reviving them or weakening them (C.G. Jung, *Man and His Symbols*, Dell, New York, 1968).

[52] If we follow Arnheim's lead about the thought processes at work in visual perception, then we could represent iconemes which lie at the cusp of a communicative chain and shape the primeval significance of landscape by way of abstract icons, which refer to a concept more effectively: their shape, volume, and color will be aimed to underline that feature. Conversely, iconemes that intervene in transmission, yielding some cultural value sedimented over time, will take the form of stylized icons which, along with volume and chromatism, will promote understanding of their social role. In fact, the author argues that thought and hence knowledge are fundamentally perceptual, and examines how the abstraction of form and its intermediate stages affect communication in different manners (R. Arnheim, *Visual Thinking*, *op. cit.*, 1969).

Whether and how other landscape features should be represented, even if they fail to recover archetypal meaning, remains to be seen. Indeed, we already submitted that what matters is less the building of thorough maps than the production of maps expressive of a specific landscape feature which recalls *médiance*. Useful indications come to us from the concept opposed to archetype, that is, the concept cliché, defined as the way to produce a simplified and widely shared view. Cliché was originally a synonym for stereotype, but in time the term shifted, taking on the sense of a metaphor for any set of ideas reiterated identically, *en masse*, and with minor changes. A *cliché* is the antithesis of the archetype and by extension even of the symbol. Strictly speaking, it may not even be equated with the iconeme. We could say that it is present in landscape, yet fails to convey the landscape's social value. In short, the archetype and the cliché refer to separate planes in the transmission of knowledge (connotation and denotation, respectively) and differ substantially even as regards the symbolization of icons.

In this regard, exploring the problematic nature of symbolization and its repercussions on communication, Christian Jacob points out that, as it highlights the relationship between icons, the very analogical setup of maps develops a communication system which is in itself symbolic. In his view *miniaturization*, a procedure that brings about a condensation of meaning not only by reducing scale but also by imposing its own referential definition, entails the transmission of particular symbolic messages of the icon.[53] On the other hand, symbolization has crucial repercussions even in an intercultural milieu, since symbols in themselves cannot ensure mutual understanding between different societies. The transformation of a sign into a symbol occurs at a higher level than the one we are consciously embracing here. Symbolic communication must then refer to a horizon of understanding of the world that may enable each to appreciate different facets of the whole.[54]

However, it should be noted that these findings, while crucial for acknowledging the deeply speculative nature of cartography, become binding if the underlying aim is to build an objective and universal representation. The limited scope of our research leads us to consider symbolization as an opportunity for reinforcing the goal we pursue. In short, as we endeavor to isolate landscape iconemes that recall the archetypal quality of the relationship with nature and the evolution that occurred over time and is expressed by *médiance*, we are bound to take on a hermeneutic perspective, in which the symbolization of

---

[53] Although the term *miniaturization* is borrowed from computer science, especially nanotechnology, where it has a very specific technical sense, its original meaning pertains to the practices of symbolization. See: C. Jacob, *L'empires des cartes...*, *op. cit.*, pp. 146–147 and 414–427.

[54] On symbolization practices that pertain to the relationship between man and nature, Barthes' work on Japanese culture is still seminal: R. Barthes, *Empire of Signs*, Hill and Wang, New York, 1983.

the map is consciously endorsed and kept in mind during its construction, by bending it to specific ends. Indeed, rather than consider the symbolic nature of maps as a move away from reality, we ought to see such potential for abstraction as consistent with our goals. We should ask ourselves: how can cartographic icons, drawn as they are from culturally shared procedures of selection, simplification, classification, synthesis, and finally symbolization, possibly allude to the primordial meaning of iconemes and to the ongoing dynamism expressed by *médiance*? And again: are icons capable of rendering features of landscape that are nonrelevant, and therefore comparable to cliché, through signs that communicate exclusively via the denotative level without involving connotation, namely the symbolic level? The time has come for us to turn to the codification of icons and to examine their symbolization practices.

Icon Symbolization Practices

Before addressing cartographic transposition, we need to clear the field of misunderstandings about the usage of the term *symbolism*, because in the nonspecialist language of geographic maps it is taken to mean the encrypted set that substantiates information: shapes, colors, numbers, and names. Here, by contrast, we wish to bring out the different meaning that this encrypted set assumes according to the treatment it undergoes and the communicative outcome it produces. This process is clearly not homogenous across all surrogates that compose it. While numbers and names elude radical treatment, because they are inspired by the conventions adopted in the language of origin – numerical or lexical – more elaborate manipulations are to be expected in the case of colors and shapes. It is advisable therefore to relate icon symbolization to cartographic semiosis. Cartographic semiosis makes it clear that the icon is based on the pairing of a name to a surrogate word (shapes, colors, numbers, and positions) designed to recall the form or content of the phenomenon it represents. These surrogates may be expressed by signs derived from a visual semantics and transmitted via the denotative level. Alternatively, they may require the use of symbols that are derived from socially constructed values and transmitted at the level of connotation. To address icon semantics in order to ensure the preservation or loss of the symbolic content of iconemes means to find procedures able to recall the visual or conceptual meaning of landscape in the graphic treatment of surrogates. However, it should be noted at once that semiosis approaches icon symbolization in very specific terms: the cluster of signs and symbols the map may use is intended as a technical kit for communicating the meaning embodied in the name which identifies the phenomenon it refers to. This shows that, while not underestimating the impact of the ideological or social aspect of surrogates, semiosis regards the name or designator as the actual symbolic structure capable of directing the cluster of signs that refer to it. More precisely, as I pointed out in the first chapter of this book, even before becoming signs or symbols, colors, shapes, and numbers are surrogates of a verbal projection, whose purpose it is to communicate the intellectual appropriation territory achieved by denomination: the assignment of a name to a territorial phenomenon establishes its *referential, symbolic,* and *performative* meaning. Moreover, it should be stated that, in landscape transposition, referential meaning, transmitted at the denotative level, recalls one of its specific visual features. The other two meanings, derived from the metaphysical reservoir of a given society, refer to socially constructed values, which are conveyed at the level of connotation. We need to stress this to avoid the impression that the identification of the different role surrogates play in conveying the form or content of landscape contradicts the role that cartographic semiosis assigns to designators within icons. If we were to disregard this, we would risk reducing the icon to a symbolization practice divorced from the symbiotic relationship with territory, thus denying its iconizing outcome and the possibility of using it to our ends. Iconization, like the symbolization just mentioned, should not be considered a limit to the realization of a landscape cartography, since it requires an interpretation based on self-referential information. Rather, it is precisely the self-referential information feed that makes it possible to achieve it. In the first part of this book I repeatedly addressed the question of the semiotics of vision and of the cartographic use of perspective, color, and shape which either support or discount the iconization of territory. Now, to find directions in the graphic treatment of the landscape icon surrogates I will focus on the perceptual practices of images discussed by Arnheim.

The Perceptual Practices of the Image

Rudolf Arnheim, author of leading studies on the psychology of art, argues that there is no gap between thought and perception, because the latter is shaped over images of the world we live in and the second consists in mentally grasping relevant generic features of the object perceived: features of thought in perception and perceptual features in thought are complementary. The essential trait of this joint cognitive process is that it involves abstraction at every level. Abstraction also underlies the finding of the various communicative functions performed by images, which Arnheim sees in *representation* as well as in those of the *sign* and the symbol. He notes, however, that these functions are not structurally traceable within the image, but are defined in accordance with the degree of understanding they offer to the observer. For example, if an *image* provides no guidelines on how to perform abstraction, which would allow the viewer to reach understanding, it confines him or her to the level of visual perception. By means of *representation,* instead, which is always characterized by a degree of abstraction, the observer is helped to pinpoint salient features in which to ground its interpretation. It is when the author addresses the definition of abstraction that his line of argument becomes particularly relevant to our purposes. He claims that abstraction is not a form of incompleteness, but rather the means by which representation is achieved and the thesis it conveys is completed at any level of abstraction. However, he warns of the existence of different forms of abstraction in the image, which must be used to distinguish its other constituent functions, namely those of the *sign* or of the *symbol*. He argues that an image serves as a sign to the extent that it represents a particular content, without figuratively reflecting the object's features. Alphabet letters used in algebra are very close to acting as pure signs.[55] They do not in any way resemble the things that they represent nor could they be specified any further, under penalty of losing the generality of the proposition they convey. In this perspective, even the letters and the words of written language are signs that operate as pure references to the things they represent. They cannot be used in themselves to express the thought. Conversely, abstraction acts as a symbol when viewers are asked to make decisions on their own about the nature of what they see. Therefore, the image becomes representation to the extent that depicts things situated at a level of abstraction lower than that of a pure sign, considered above. As we grasp or render some qualities–shape, color, movement–of the objects or activities we portray, we interpret what is being portrayed across multiple levels of abstraction.[56] Arnheim claims that, according to the degree of abstraction, we build *figures* corresponding to the functions indicated previously, which pertain to three perceptual categories: *realistic images, highly stylized forms* and *extremely abstract drawings*. The first reproduce cognitive acts of the lowest order and do not in themselves guide understanding. Instead, they can even hamper identification because to identify an object is to recognize some of its salient structural traits. So, if the goal is to recall deeper meanings of the object, realistic images ought to be used with caution, because a faithful resemblance may paradoxically preclude the interpreter from the essential features of the object represented. Indeed, it is through the structural properties of lines and colors that the transmission of even some of its "material" persistences is best accomplished. At the other end of the abstraction scale there are very stylized forms, often purely geometric. They have the advantage of accurately isolating particular features whose meaning is decoded according to the situation. Finally, a really abstract drawing, which has little resemblance to its referent, will have to be limited to a single application or to refer in an explicit and binding way to the context it is in, so that such context may explain it. In fact, the situation will determine whether a cross plays the role of a sign, whether it stands for a religious or arithmetic symbol, or whether it refers to the crossing bars of a window, in case no semantic function is aimed at.[57]

---

[55] Although accurate, these signs insinuate a portrait or representation when placed in relation to each other (A and B).

[56] R. Arnheim, *Visual Thinking...*, *op. cit.*, pp. 173 ff.

[57] The progressive scale of figural abstraction is at the core of Arnheim's research and was addressed in these pages: *Ibidem*, pp. 192 ff.

Treatment of Icon Surrogates

The claims Arnheim made as to the issue of icon surrogates lead us to conclude that only limited action is possible on numbers and names because they both belong to the category of signs. It is possible, however, to prioritize their importance by sizing and the use of different typefaces, as topographic cartography already requires. However, in our perspective, their use is not so much intended to meet current topographic conventions, namely reinstating subdivisions into areas (continents, regions, etc.) or political entities (state, region, province, municipality, etc.). Rather, it is meant to recover the various relevance of icons, depending on the cultural value of the landscape that one intends to convey. The question is more articulate in the case of figural icons, because the progressive scale of abstraction is closely tied to the possibility of conveying concepts which become more and more complex. They express a process of abstraction whose degree rises in accordance with the different role they play in communication. They may present themselves as *figurative drawings* that reproduce reality mimetically. Alternatively, they may take on *stylized forms* which allude to the iconeme's content by virtue of their formal starkness. Yet again, they may refer to the conceptual essence of the iconeme and present themselves as *abstract figures* which, without recalling the profile of the phenomenon they indicate, represent it symbolically. It is well-known, in fact, that abstract figures portray the inner nature of the things and events of the world and as such preserve their relevance for the social aspects of territory[58] – hence the importance of choosing to recall alternatively either the denotative or the connotative level of communication. What is concretely at stake here is the opportunity to convey the archetypal or stereotypical meaning of icons. Archetypal meaning may be conveyed by underlining its conceptual density. Conversely, those who wish to recover stereotypical meaning will insist on its marginal value to the point of making it irrelevant.

That may be pursued by deploying a set of assembly techniques which in the case of the archetypical icon are aimed to:

- Make the icon more independent from the perspectival setup in which it appears: size, shape, and volume are divorced from the proportional criteria suggested by the basemap, in the hope that such alienation may force recipients to interpret the icon in exclusively connotative terms;

---

[58] With regard to a typology of signs, G. Anceschi argues that those used in cartography cover a very wide spectrum, ranging from fiction (orographic cartography, shaded cartography, color cartography), hatching (Euclidean cartographic base), to the iconic (thematic maps). (G. Anceschi, *L'oggetto della raffigurazione…, op. cit.*, pp. 45–51). However, if we consider the contribution of J. Bertin, who worked in a milieu very close to that of linguistic semiotics, we identify three categories of representations: the diagram, the network (réseau) and the map. In the discussion of symbolization issues, the latter is the only reference to the figurative/abstract question (J. Bertin, *Semiology of Graphics: Diagrams, Networks, Maps*, University of Wisconsin Press, Madison, 1983).

- Make sure that surrogates refer less to the formal or material qualities of iconemes than to the features of content: cultural aspects, social knowledge, and the landscape-based propensity of the recipient are guidelines the icon should conform to. Here, too, we deem it essential that interpretation be carried out in connotative terms by using a miniaturization capable of making a concept intelligible through various degrees of abstraction. Icons that recall the essence of the relationship with nature will be rendered via symbolic abstractions (sphere, prism, pyramid and so on) already circulating in the culture they belong to, while icons that refer to social values will recover the abstract/figurative pair in a stylization of the iconeme. The objective, in this case, is to activate an iconic junction* at the level of syntax such as may promote, via a system of figural interference, the formation of syntagms, that is, relational sets able to envisage the different social role played by icons. Icons will then differ also from clichés, which will be shaped in terms quite opposite to the ones in the previous discussion.

We can complete our overview by touching upon volumetric or plane figures, marked with the 3d or 2d acronym in order to differentiate archetypical icons from cliché icons. A two-dimensional object lacks depth and develops only along a flat surface. Photographs are examples of two-dimensional representations. Conversely, what goes under the name of "volumetric space" is characterized by the rendering of all the three dimensions and differs from planar space, equipped with only two dimensions and generally called Cartesian. Using three-dimensional computer graphics,[59] namely techniques such as perspective and shading, we simulate object perception by the human eye. Widely used in film to reproduce a perceptual experience, it is an option now available to computerized cartography, where it features as an icon assembly technique for diversifying icons.

Here we propose the use of 3d figures for archetypal icons and of 2d instead for those related to clichés. The underlying aim is to overshadow the Cartesian model to the benefit of the landscape-based view. Their inclusion in a perspectival or zenithal framework will be a subsequent mark of distinction. Icons that convey the essence of landscape will appear in 3d and be independent from the system in which they are placed: their size will reflect neither perspective nor scalability. Those meant to convey *médiance* will hold an intermediate position: their form will be volumetric, but they will abide by proportions imposed by the basemap. Finally, besides being rendered in 2d, icons related to the cliché will depend on the framework in which they are placed.

---

[59] The science, study and projection method for the mathematical representation of three-dimensional objects using a two-dimensional image.

ICON TYPE	COMMUNICATION LEVEL	ICONEME MEANING	DIMENSIONS (with regard to the structure of the basemap)	PERSPECTIVE	COLOR (reiterated in syntactical relationships)	SHAPE
ARCHETYPE	Connotative	Essence	Autonomous	3D	Abstract Chromatism	Abstract Figures
	Connotative	Médiance (intermediation)	Dependent	3D	Analogical Chromatism	Stylized Figures
CLICHÉ	Denotative	Irrelevant	Independent	2D	Black and White	Figurative

FIGURE 5.16
Cartographic icons.

In short, icons that convey a social value of landscape will take a volumetric form because they recall the landscape perceived by the subject; those related to the cliché will have a two-dimensional form on account of their irrelevance for conveying social values (Fig. 5.16). It should also be added that the choice of the level of figuration and of the two- or three-dimensional form of icons affects the iconization process with a view to favoring the rendition of a conceptually meaningful landscape.[60]

We should not forget that this "instruction manual" provides only tentative guidelines, since the representation of landscape is plural and ever-changing, in keeping with the intellectual constructs that a given society produces. These guidelines are not arbitrary, however. They are anchored in studies on the semiotics of vision which, applied to chorography, help us to find within cartographic semantics the most effective ways to communicate the different values contained in iconemes. Thus, such anchoring avoids the temptation of submitting arbitrary codifications, like the taxonomic ones I illustrated in the first part of this book, based on catalogs of a positivist kind. I already addressed the issue of the problematic nature of color when I mentioned maps whose goal is at variance with the one pursued here, and aim at compiling a thorough taxonomic list by using a very large number of variables.[61] Suffice it for now to stipulate that the use of color in landscape cartography will be reserved for archetypal icons, using a limited range of variables corresponding to primary colors, as required by the theory of color in cartography.[62] Black and white will be used instead for icons that render clichés. Finally, icons that refer to *médiance* will be characterized

[60] The zenith-based and volume-based dimension will be addressed in the next chapter.
[61] Chapter 2 of the present volume.
[62] J. Bertin, *Semiology of Graphics, op. cit.*, pp. 85–91.

by a color that reproduced the one visually perceived in real life. As I recalled earlier, color ensures immediate perceptual attention because class progression is made to correspond to that of the perceived color spectrum and thus to what is generally and conventionally recognized as distinctive of such spectrum. In this perspective, even the cartographic codification of color is matched to the coding of reality, on the assumption that color has the power to draw attention. Yet, to avoid possible interference in symbolization, its use must comply with the dynamics involved in sight and perception.[63] If color is used as an alternative to black and white and, in keeping with perceptual rules, is attached to the icons that belong to the same communicative syntagm/portion, then it will work in communication as any other surrogate. At the same time, the fact that its use is reserved exclusively to archetypal icons necessarily places color at the hierarchic top of the information chain, thus preventing interference with the socially relevant meaning of icons. The pairing with other icon surrogates that can convey elaborate concepts and complex intellectual overviews will ensure that color focuses attention without altering their meaning. As we deploy color by the same standards as any other surrogate, we reduce its iconizing possibilities over the transmission of landscape value.

In sum, cartographic icons are meant to hierarchize the semiotic role of iconemes, triggering procedures that can ensure the expression of the social meaning of landscape. Such procedures, which pertain both to the layout of the basemap and the assembly technique of its surrogates, differ in accordance with the type of icon.

In the next chapter we shall see how the emancipation granted by the multiple points of observation together with anamorphosis* – both features that chorography provides for – may ensure a flexibility of the basemap which can actively affect icons. Now, with regard to the technique for building icons, I should stress that sizes and angles by which the archetypal icons were submitted differ from the restrictions imposed by the layout of the basemap. Those that recover the nature-society relationship should be bound to such relationship; and those that are supposed to transmit *médiance* may instead depend on it. In sum, all the icons that recover iconemes will be rendered using 3d, but will have a different relationship to the basemap. With reference to shape and color, those referred to its archetypal value will have abstract forms and colors and be independent from the basemap setup. Those referred to social values will have stylized forms and analogical colors and will rely on the setup of the basemap. The features that instead are

---

[63] Their intensity and their combination will take account of what was masterfully discussed by: J. Bertin, *Semiology of Graphics, op. cit.*

irrelevant with regard to the transmission of landscape values – the clichés – may be left out or if present they will have flat shapes, with sizes dependent on the basemap. They will be rendered in black and white and via two-dimensional perspective. The objective is to activate, by means of an iconic junction, a system of interference of communication that favors the formation of syntagms able to prioritize information and forcefully bring out the cultural value from the landscape.

To conclude, since landscape cartography is less interested in showing visual analogy than in recovering the values of landscape, it may use a visual form to facilitate identification, and exhibit its value hierarchy on the basis of the recipient's interpretation.

For the sake of simplicity, in this chapter I addressed the construction of the icon separately from the structure in which the icon stands – the set of assumptions, rules, and graphics codifications of the basemap. My objective is, however, one and the same. My approach depends on the fact that, as we start to examine landscape cartography, it is essential to clear the field of any topographic legacy that heavily curtailed our ability to focus on the icon and on its semantic implications. In the next chapter I will look at the basemap as the final step in the construction of a chorography.

CHAPTER 6

Coming Full Circle: Towards a Chorography

The shape of the heaven is of necessity spherical; for that is the shape most appropriate to its substance and also by nature primary.

Aristotele, *De caelo*, Liber II

This chapter proposes the implementation of a chorography that intervenes on the base map and on the set of assumptions, rules, and graphic conventions that govern spatialization. *In the previous chapter we considered icons as outcomes of topological figuration. Here we will focus instead on spatialization and the possibilities offered by IT for producing icons that better reflect the notion of a reticular, or network-like, spatiality. While a topographic view is zenith-based and relies on a single scale, proportional to the extension of the territory it represents, the topological dimension of space, which we propose to outline and assess here, centers on the use of multiple points of observation and the interaction of multiple scales. I will propose using* anamorphosis *as a technique able to undermine the basic assumptions of the topographic basemap by injecting topological features into it: namely* extension/contraction *and, going beyond the concepts of "continuity" and "limit," the concept of "vicinity," which is typical of reticularity. That will reopen the issue of map scalarity arising from the need to take account of the* multiscalarity/transcalarity *of planned interventions on the part of agents. The route we take will offer a wealth of examples: pre-modern cartography, a rare instance of prehistoric petroglyph, web-based electronic maps and the latest technological resources. And we will do so without following a strict chronological order, but rather an exploratory trajectory that favors dynamic associations of thought. We thus come full circle: leaving topography behind, we finally show other ways of dealing with the representation of a chorographic world.*

CONTENTS

CARTOGRAPHIC RENDERING OF SPATIALITY THROUGH THE CENTURIES

The history of cartography presents us with a relevant fact: topographic maps, based on Cartesian logic, are neither the only nor the best maps possible. Cartography changed over time as to its forms, means, and languages, taking on very different configurations. Its use also changed: an instrument of appropriation, of propagandistic praise, of mnemonic application or a military, political, administrative tool. Yet, the one recurring assumption that

Reflexive Cartography. ISSN 1363-0814, http://dx.doi.org/10.1016/B978-0-12-803509-2.00006-9

the history of cartography seems to have preserved over time is a reliance on metrics, or the succession of various metrics which structured cartographic representation. Such metrics came from the various concepts of space each culture developed. Ancient Greeks used *ecumene* to designate the world as an inhabited area. The Roman period sanctioned *reticular, or network-like space* based on the roadway system that ensured imperial control. *Odologic space* underlay nautical maps, which called for a route-based, linear space. The Middle Ages introduced *creationist space* to explain the world as God's design. In the Renaissance, the success of Signories equated domain over territory with extensive knowledge. Finally, during the Enlightenment, when accurate measures prevailed as tokens of national identity, space was conceptualized as a *topography.*

All these representations share a reliance on the concept of spatiality, of the way in which humans relate to the world, as they elect either one of the primary functions of a map: describing or conceptualizing.[1] Maps *describe* the world when they endeavor to render its features as seen through a direct observation of reality: in other words they explain to us how the world is made up. Alternatively, maps *recount* the world, tell us how it works according to categories of representation which arise from one possible interpretation of it. We are then in a position to distinguish between maps that favor the communicative mode of description and maps that rely instead on conceptualization, and thus propose a worldview only partially modeled after canons of real-world mimesis. The medieval Ebstorf Globe (Fig. 6.1)[2] envisions the Earth as an extension of Christ's body. Effigies of his face, hands and feet outline the band of water that surrounds the Earth.

It is a creationist conceptualization that refers to Genesis and neglects the description of the earth's shape by overriding it. It is based on the interweaving of information from the Scriptures and the mystical literature circulating at that time, which the Ebstorf globe embodies. Fra Mauro's map has a different aim. He intends to offer for the first time a description of the world based on the information of those who experienced it (Fig. 6.2).

Dating a few centuries later, this map was built in Venice, on the island of Murano, by a Camaldolese monk who culled information from Marco Polo's voyages of exploration and from other voyagers/travelers. The map gave Renaissance society a representation of the world that faithfully depicted the

[1] These functions coexist in real life, but what I would underline is that the prevalent function determines the conception of space of the society in which the map was produced.

[2] The document itself is quite large: before its partial destruction during the Second World War the Ebstorf map had a diameter of 3.5 meters. See: D. Woodward, "Medieval Mappaemundi," in: J.B. Harley, D. Woodward, eds., *Cartography in Prehistoric, Ancient, and Medieval Europe and the Mediterranean*, Vol. 1, *The History of Cartography*, University of Chicago Press, Chicago, 1987, pp. 286–370, esp. 290–291; 307–310.

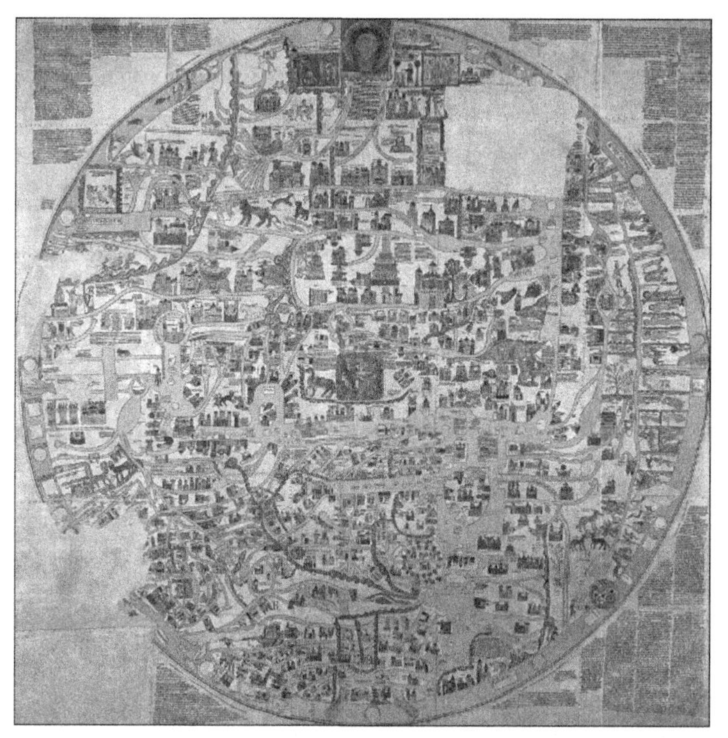

FIGURE 6.1
The Ebstorf Globe, 13th century. (See a color reproduction at the address: booksite.elsevier.com/9780128035092.)

data reported in that odeporic literature.[3] Recognized as the cartographic work that marks the birth of modern cartography, Fra Mauro's map of the earth is not without references to the medieval heritage, developed on variously intertwined narrative levels which allow for very elaborate interpretations.[4]

---

[3] The large circular planisphere lies within circumference of nearly 2 meters in diameter, which is in turn framed in a square that bears graphs and inscriptions. The map drawing, on parchment glued onto a wooden support, bears nearly 3000 inscriptions, which include not only designators, but also historical and geographical notes. The set is an extraordinary *compendium* of Renaissance geographical knowledge. It is of great importance not only because some of its content is highly innovative (for instance the data concerning African geography), but also for its particular setup, which in many ways marks both a point of connection with and of departure from medieval culture. Moreover, this document bears witness to the circulation of ideas in the Mediterranean, since its layout strongly alludes to Arab cartographic production, especially the one of the Moroccan cartographer Al-Idrisi.

[4] Among these see: P. Zurla, *Il Mappamondo di Fra Mauro camaldolese descritto e illustrato da D. Placido Zurla dello stess'ordine*, Venice, no publisher, 1806; G.L. Bertolini, "I quattro angoli del mondo e la forma della terra nel passo di Rabano Mauro," in: *Bollettino della Società Geografica Italiana*, n. 47, 1910, pp. 1433–1441; T. Gasparrini Leporace, ed., *Il Mappamondo di Fra Mauro*, presentazione di Roberto Almagià, Istituto Poligrafico dello Stato, Rome, 1956 (2001 reprint); D. Woodward, "Medieval Mappaemundi," *op. cit.*, especially pp. 314–318 and 324–328; P. Falchetta, *Fra Mauro's World Map*, Brepols, Turnhout, 2006 (with CD-ROM containing a high-resolution reproduction of the world map).

FIGURE 6.2
The Fra Mauro Globe, 15th century. (See a color reproduction at the address: booksite.elsevier.com/9780128035092.)

It should be noted, however, that in both representations the shape of the Earth is reduced to that of a disc surrounded by water, in accordance with the Scriptures. On the other hand it could not have been otherwise, since both were designed in the sphere of religion. The Ebstorf world map is the manifesto of creationism and the Venetian map is the work of a monk who had to follow the doctrine of the Church: all theories purporting the spherical shape of the globe were banned as heretical. Conversely, it should be noted that the spatial concept underlying both documents is area-based. Both adopt a metric that depicts the world as based on its extension. Whatever its form, round or spherical, the world is represented as a set of seas and lands scattered with figural and Scriptural information that qualify its features. This area-based feature, often neglected in textual analysis, is not present in all cartographic documents. In maps aimed to describe the world for practical purposes – those that take the experience of the world as their guiding principle – space is conceived in different terms. It metamorphoses into a linear or reticular, network-like spatiality. Consider, in this case, two important documents built to envisage mobility. The Tabula Peutingeriana, a map drawn in the Roman period, and the Renaissance map of Battista Agnese, depicting the new American territories as a transoceanic destination. The former is a parchment whose skins, brought together into a long strip (680 × 33 cm), depict hundreds of thousands of kilometers of roads, about 555 cities and 3500 other geographical features[5] (Fig. 6.3).

The map was drawn as a representation of the *cursus publicus*, that is, the road network reorganized by Augustus, with post stations and services at regular distances which supported Imperial traffic and trade. The delineation of such a network covers the entire Roman Empire but is not confined to its borders. It stretches, albeit in general outlines, to the Near East and India, all the way to the Ganges, Sri Lanka (*Insula Trapobane*) and China. Distances, drawn in red lines and marked by travel times, define a mesh whose nodes consist of the imperial cities Rome relied on for the smooth functioning of its administration. The whole political structure of the Empire depended on those connections. Even though the world is depicted in its analogical set of relations, it is distorted by stretching. Italy, stretched horizontally and surrounded by the Mediterranean, is bordered on the left by the northern European countries, at the bottom by the coast of Africa, at the top by Mediterranean countries and Eastern Europe and to the right by the territory that extends to the Asian regions. Yet, despite the fact that the representation of surfaces is completely misleading for assessing their extension, their layout and their contiguity relations are respected and allow an interpreter to pinpoint the territory in which they are inserted. Mistakenly, the map is considered as "a symbolic representation, a kind of diagram similar to a subway map, which allowed one to easily move from one point to another and know the distances between the stops, but would not yield a true representation of reality."[6] This assessment, based on the Euclidean model, demeans the spatial conceptualization that underlies the map, because, even as it disregards the topographic metrics, the road network is syntactically placed within the territory. What changes, instead, is the metrics, which aimed to show relevant features (road networks that define the social space) and was reticular. It did not aim to depict the extensive features of the continents, but took mobility as its priority in order to underline information on distances between the cities. The *Tabula Peutingeriana*, therefore, is not comparable to a subway map. In other words, it

[5] The cities are represented by two houses; the host cities of the Empire – Rome, Constantinople, Antioch – are marked by a medallion. The original, now lost, was dated between 109 BC and 328 AD. Even the dating of the manuscript copy in our possession is uncertain but could be placed in the 13th century. It is supposed to be the work of an anonymous copyist monk from Colmar, who presumably composed the oldest document in 1265. The *Tabula* was finally printed in 1591 in Antwerp as the *Fragmenta tabula antiqua* by the famous publisher Johannes Moretus. The manuscript copy of the *Tabula* is currently kept at the Vienna Hofbibliothek. An in-depth analysis of the issue is found in: O.A.W. Dilke, "Itineraries and geographical maps in the early and late Roman Empire," in: J.B. Harley, D. Woodward, eds., *Cartography in Prehistoric, Ancient, and Medieval Europe and the Mediterranean*, Vol. 1, *The History of Cartography*, University of Chicago Press, Chicago, 1987, pp. 234–257.

[6] A naive cartographic notion of "faithful representation of reality" leads to the adoption of topography as the only possible model. See: http://en.wikipedia.org/wiki/Tabula_Peutingeriana.

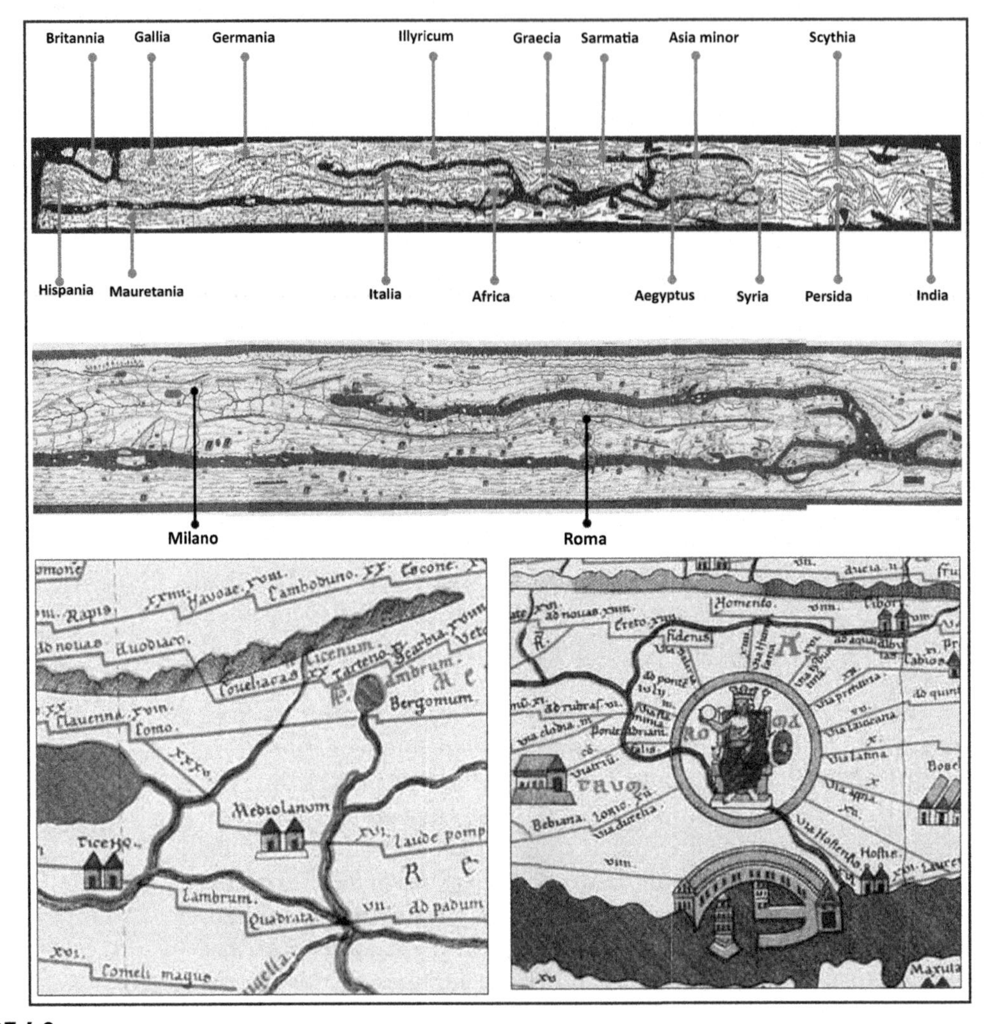

FIGURE 6.3
Tabula Peutingeriana (Peutingerian Table), first century BC–4th century AD (13th century copy). (See a color reproduction at the address: booksite.elsevier.com/9780128035092.)

should not be considered as a representation divorced from the form of territory, but as one that embraces network-like spatiality, based on relationships and connections and meant to showcase mobility as the phenomenon that distinguishes the Roman world.

A subsequent change of spatial perspective is recorded by the second document in question, a map of the Atlantic world designed by Battista Agnese, a leading Genoese cartographer who worked in Venice centuries later, namely

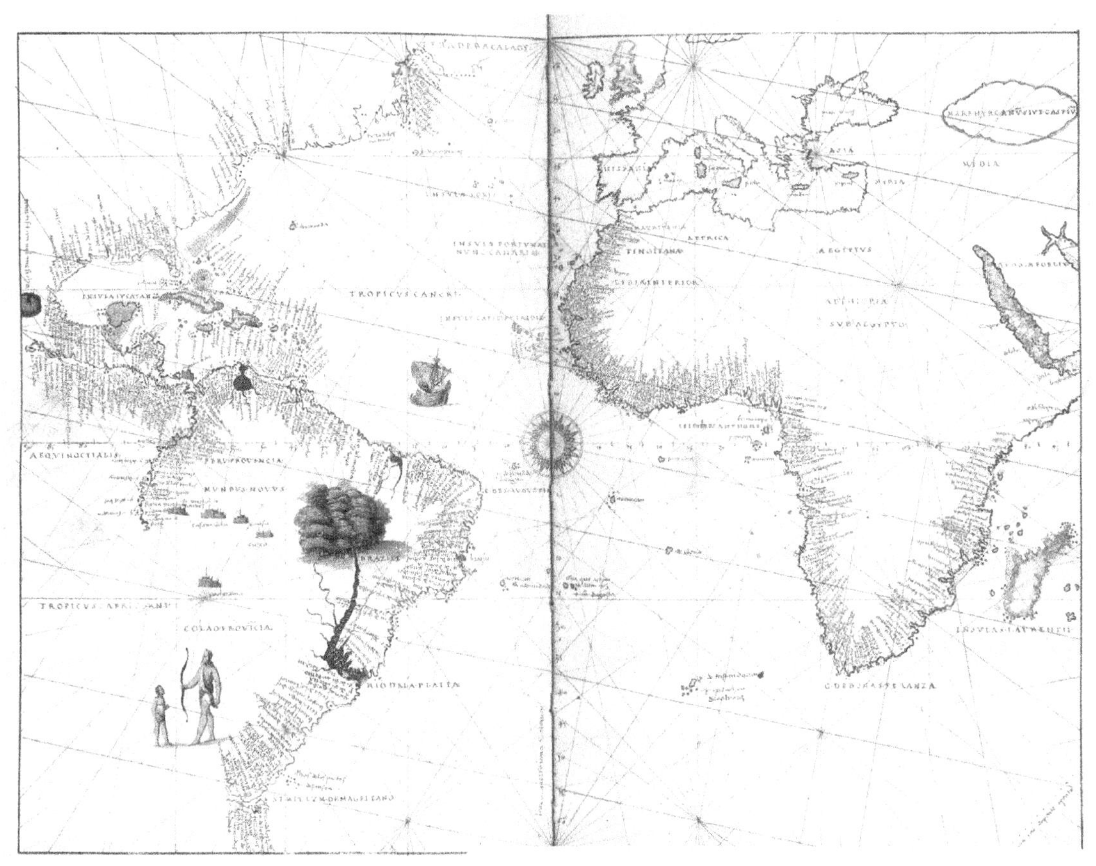

FIGURE 6.4
Battista Agnese, Map of the World, Nautical Atlas, 1553. (See a color reproduction at the address: booksite.elsevier.com/9780128035092.)

during the Renaissance. Agnese's map adopted a new spatiality called linear or, quite aptly, "odologic"[7] (Fig. 6.4).

While the *Tabula* we saw previously was intended to show road routes that, in their intertwining, formed the mesh of an inhabited and known territory, Agnese's nautical map traced sea routes for reaching a recent world, only partially explored. It used "wind diamonds"[8] to mark the beginning and the end of the journey in a specific coastal area. The coast, which could be named and referenced, was the

[7] On odologic space and the different world view of the ancients see: P. Janni, *La mappa e il periplo. Cartografia antica e spazio odologico*, Bretschneider, Roma, 1984.
[8] This term commonly refers to the set of directional lines derived from the compass rose. In their intersections, such lines allowed seafarers to calculate the angular direction for sailing the high seas.

sole basis for orientation. The map shows that by using land-sea designators which mark gulfs and headlands. Although there is no area distortion, the total lack of information on inner and continental areas bears witness to the map's linear spatiality. The outline that surrounds the Atlantic is shown as a dividing line between two empty surfaces: land and sea. The only exception in this communication perspective are the figures placed in correspondence of South America (an armed giant, a lush forest accessible by river navigation) to recall symbolically travel stories which described it as a wonderland inhabited by men and giants, a land rich in impressive natural resources.[9] These figures are also linked to mobility, because they project expectations about the journey's end, depicted via a figurativism that leaves room for imagination.[10]

It must be emphasized, however, that the two documents have a mainly descriptive function and both differ from an area-based conception of space. This should be seen in relation to what emerged in recent studies that examined the role of spatiality as a social category related to daily and practical life. For beyond the realm of pure speculation, spatiality has been shown to inform our own experience, our own daily life.[11] The pioneering work of Michel de Certeau demonstrates how the space we practice is continually re-encoded against the general pattern. It is the results of operations that are multiple and often conflicting and yet somehow manage to coexist.[12] Travelers who walk across a city rebuild the perception of its urban layout in their imagination and in the social imagination (through the observation of others). They ascribe meaning to the names of the streets, choose alternative routes, and do so regardless of the constraints imposed. In the author's view, the experience of the city compels an individual to discard *topos*, denying the presumed thoroughness of its material assumptions in order to go "beyond," and recover *chora*.[13]

---

[9] Battista Agnese (c. 1500–1564), of Genoese origin, operated mainly in Venice. Between 1534 and 1564 his workshop produced at least 71 lavishly decorated atlases of nautical charts. The one considered here is part of an atlas of the world comprising 29 cartographic maps (426 × 576 mm) built in 1553 and preserved at the Museo Correr in Venice, Port. 1. It is a representation of the new world centered on a compass rose in the center of the Atlantic. See: S. Biadene, ed., *Carte da navigar, portolani e carte nautiche del museo Correr 1318–1732*, Marsilio Editori, Venice, 1990, pp. 65–70; G. Romanelli, M. Milanesi, eds., *Atlante nautico di Battista Agnese 1553*, Marsilio, Venice, 1990.

[10] This is confirmed by other maps such as the one included in the 1544 pilot's Atlas n. 5 (see reproduction on the Library of Congress website: http://memory.loc.gov/am-mem/gmdhtml/gnrlagn.html). In that case, however, figuration is limited to the indication of a large forest washed by the Rio de la Plata. The insertion of these figures is bound up with the essential conceptualizing function of maps and to what Eric Dardel claims when he says that a naively scientistic vision of the world should be seen as man's temptation to abdicate his critical capacity. E. Dardel, *L'homme et la terre*, Colin, Paris, 1952, p. 91.

[11] Among the most significant: M. de Certeau, *L'écriture de l'histoire*, Gallimard, Paris, 1975; Id., *L'invention du quotidien. 1. Arts de faire*, UGE, Paris, 1980 (new edition by Gallimard in 1990).

[12] As keenly noted by Giorgio Mangani, Michel de Certeau reverses the place/space model, attributing a strategic function to the latter which offers a possibility for redemption: G. Mangani, "Intercettare la chora," in: E. Casti, ed., *Cartografia e progettazione territoriale. Dalle carte coloniali alle carte di piano*, UTET, Turin, 2007, pp. 31–41, esp. 37–39.

[13] M. De Certeau, *L'invention du quotidien…, op. cit.*, 1980, pp. 186–187.

In short, the erratic movement that characterizes contemporary cities is actually the generalized social experience of a "deprivation of the place," a denial to which the experience of the *chora* is profoundly related. To refer to spatiality as the practice of everyday life also emphasizes its connection with landscape, because both arise from the social context and both shift the plan of analysis from *topos* to *chora*. The point at issue lies rather in how to mark cartographically the meaning of spatiality shown in the everyday practice of its actors.[14]

Spatialization and the Analogical System

In the cartographic documents we just addressed, quite different in terms of the concepts and spatialities they uphold, the analogical setup of the base map is never questioned. That is linked to what we said in the first chapter of this book, namely that maps structurally appear as *sorting systems* that endeavor to master a complex cluster of data by homing in on the most relevant geographical phenomena and arranging them in the same sequence in which they are perceived in reality. That line of research enables us to shed light on the foundational aspects of cartographic representation. Maps reproduce space according to the principle of analogy, which aims to arrange objects using a topological order which relies on a *cognitive spatialization*: a procedure whereby objects are invested with spatial properties in relation to an observer who is in turn "spatialized" with regard to them.[15] Such feature is essential for building a cognitive network of territorial referents without which no territorial action would be possible.

All this is crucial, despite my earlier statements to the effect that maps are hypertext systems based on a dual communication system, analogical and digital and that, contrary to common perception, maps are not analogical models of reality but models that use both the analogical and the digital systems in a most particular combination. The analogical system aims to depict objects as they are in real life, intended as a *continuum* that is based on actual physical laws arrived at through *differentiation* (one object differs from another because it is located at a given point, and because its features are unlike those of other objects). For its part, the digital system uses a wide range of icons that isolate and convey a few features of the designator's meaning by setting up *distinctions*, that is to say by underlining its characteristics. Thus, the analogical features of maps provide the "context" for the digital systems and the two work together to produce a third system – the *iconic*. This latter system organizes information in a new, distinctly cultural perspective and by doing that it iconizes the world. Clearly, if we wish

---

[14] For a tentative analysis of this concept see also the "Spatialité" entry edited by Michel Lussault in: J. Lévy, M. Lussault, eds., *Dictionnaire de la Géographie..., op. cit.*, pp. 866–868.

[15] A.J. Greimas, J. Courtes, eds., *Sémiotique. Dictionnaire raisonné de la théorie de langage*, Hachette, Paris, 1993. On cognitive spatialization and on the formation of a network of spatial references, see: A. Turco, "Dire la terra: la costituzione referenziale del territorio in Costa d'Avorio," in: *Terra d'Africa*, Unicopli, Milan, 1994, pp. 15–58.

to recover the actual social meaning of territory by using a chorography, we need to maintain both communicative systems of maps. If anything, we need to act with a renewed awareness of their relevance in order to use them to our ends.

This premise is meant to address criticism against the use of analogy for representing a reticular spatiality. Some researchers claim that the constraints imposed by analogy force us to conceive of a spatiality that cannot be but topographical. To be free of such constraints, they argue, means to embrace representations that are completely detached from any analogical framework. What they advocate is thus maps which represent social relations and the dynamics of phenomena in terms removed from any analogical basemap, claiming that that is what topology calls for. To these authors, conceiving a network-like spatiality means building self-extensive maps that develop on a neutral plane and envision new spaces relying precisely on the fact that they are medium independent.[16] Aesthetically pleasing and intriguing in their elaborate informational tension,[17] such drawings explore the new paths trailed by the language of art as an escape route from topographic hegemony, especially when the underlying aim is to represent motion[18](Fig. 6.5).

It should be noted, however, that such representations do not discard the analogical framework, but "warp" linear distance, which, as I have shown, is an altogether different matter. Ultimately, we are not supposed to underestimate the relevance of the analogical framework, namely of spatialization, with its topographical rendering of distance. And of course we should clearly have in mind what a denial of spatialization entails. Spatialization ensures that the designators, the substance of icons, enhance localization when placed within an analogical system, thereby recalling the cultural valences expressed by icons. If spatialization is discarded, a world devoid of referentiality will be envisaged. Thus we will end up reproducing the very world chorography is meant to oppose: an abstract, placeless world, a falsification and useless reduction of the complex world of phenomena. The risk we run into if we reject the

---

[16] We should clarify that the issue falls in fact within a wider context, where reticularity is addressed in all its complexity and explored both in its theoretical and its representative features. J. Lévy, "Una svolta cartografica? Colmare il divario tra le scienze e le tecnologie dello spazio abitato," in: E. Casti, J. Lévy, eds., *Le sfide cartografiche..., op. cit.*, pp. 23–32. Maps 4 and 5 may be consulted at www.unibg.it/sfidecartografiche.

[17] Cross-referencing data on travelling times, individual wishes, actual extent or virtual quality of movement and others.

[18] See: J. Lévy, ed., *Échelles de l'habiter*, PUCA, Paris, 2008; A. Ourednik, "Cartografare in due dimensioni la realtà diacronica dello spazio abitato," in: E. Casti, J. Lévy, eds., *Le sfide cartografiche..., op. cit.*, 2010, pp. 65–77; and in the same volume, J. Lévy, "Una svolta cartografica?," *op. cit.*, pp. 23–32. The need for a new representation of the world is felt by a large array of artists who start from cartographic suggestions to experiment with elaborate forms of spatiality. Along these lines, on the representation of an elsewhere, see the interesting production of Patrice Cujo: www.patrice-cujo.net.

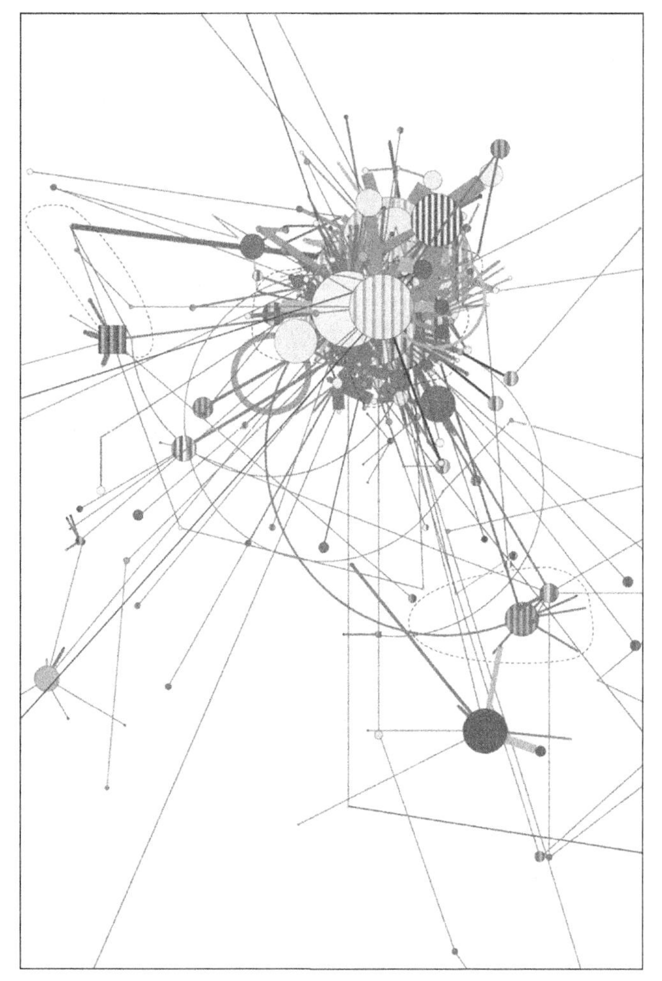

FIGURE 6.5
Mobility among inhabitants of Paris (in: J. Lévy, *Échelles de l'habiter*, PUCA, Paris, 2008). (See a color reproduction at the address: booksite.elsevier.com/9780128035092.)

analogical basemap is to convey the notion that the "ruggedness" of the world, and thus the form of its landscape, are irrelevant. Supporters of self-extensive maps counter this objection with the claim that spatiality cannot be severed from geography and that analogical spatialization, instead, encourages such false belief. It is therefore necessary to think of other forms of representation that no longer rely on it.[19]

---

[19] J. Lévy, "Una svolta cartografica?," *op. cit.*, pp. 23–32. See also the "Spatialité" entry edited by Michel Lussault in: J. Lévy, M. Lussault, eds., *Dictionnaire de la Géographie...*, *op. cit.*, pp. 866–868.

FIGURE 6.6

The world: from a flat to a spherical dimension. (See a color reproduction at the address: booksite.elsevier.com/9780128035092.)

The issue, in my opinion, is misplaced. The challenge does not consist in discarding the analogical framework of the base map, but in overcoming Cartesian logic by conceptualizing at the same time both the spherical shape of the world and thus its irreproducibility on a plane and its landscape features, as illustrated in Fig. 6.6.[20]

But, before tracing the cartographic process of such a perspective, we should address the fundamental limit of topography, which consists in flattening the world and reducing it to two dimensions, discarding its tridimensionality. We will do that on a prehistoric example of momentous interest.

Bedolina: Map or Tridimensional Representation?

Anthropic activity in the Alps dates back to thousands of years before Christ, and records exist as regards hunting, herding and agriculture. On the Italian side, we have quite important, sprawling valleys that show tangible signs of their ancient social setup: Val d'Aosta, whose inhabitants were long opposed to Roman rule; Valtellina, possibly the wealthiest section of Lombardy in the Middle Ages; Val d'Adige with its two major branches of Val Pusteria and Val Venosta, where Ötzi the iceman – whose mummified body was exhumed from the ice a few years

[20] The map is rendered via stereographic projection from a panoramic photograph of the Parc Interdépartemental des Sports of Paris. It was made by "Gadl," a user of the Flickr social network, and is part of a wider collection named "Wee planets": http://www.flickr.com/photos/gadl/sets/72157594279945875/

ago near the Similaun mountain on the border between Austria and Italy[21] – is supposed to come from, and finally Valcamonica, which straddles the provinces of Bergamo and Brescia. Valcamonica, which extends from the shores of Lake Iseo in the south for approximately 50 miles to the north,[22] is arguably the most important among the valleys we mentioned from a historical and archaeological point of view, due to the presence of a large number of Neolithic petroglyphs made by the ancient prehistoric people known as Camuni. Petroglyphs are of course also found elsewhere, in neighboring valleys at an altitude ranging from 300 to 2,500 meters. It is here, however, that they appear in significant numbers. Their discovery goes far back in time, but research on the petroglyphs has gathered momentum only over the last few decades, thanks to rigorous scientific efforts on the part of many researchers, including archaeologist Emmanuel Anati, founder of the *Camuno Centre for Prehistoric Studies* in Capo di Ponte.[23]

The number of engraved stones found so far is well over 2,500, and cataloguing is still in progress. Several dozens of these petroglyphs have been identified as resembling maps and one of those mappiform or map-like petroglyphs in particular drew the attention of researchers worldwide for its outstanding state of preservation: the "Bedolina map."[24] The name comes from the small town where the engraved boulder is located on the west side of the middle valley, within the area of Pescarzo, a hamlet of Capo di Ponte. On a large, smooth to

---

[21] Radiocarbon dating sets an age of between 3,300 and 3,200 BC, which corresponds to the Chalcolithic or Copper Age, a time of transition between the Neolithic and the Bronze Age.

[22] The river Oglio runs through Valcamonica, which at Edolo branches out into a small valley leading to Valtellina via the Aprica Pass, while to the east it leads to Val di Sole and then Val d'Adige via the Tonale Pass.

[23] Since their discovery dating back to the 19th century, these rock engravings were classified as planimetric representations of agricultural areas and were the subject of numerous studies, including: E. Anati, *The Civilisation du Val Camonica*, Arthaud, Paris, 1960; W. Blumer, "The oldest known plan of an inhabited site dating from Bronze Age," in: *Imago Mundi*, 18, 1964, pp. 9–11.

[24] Described in 1932 by Raffaello Battaglia, who presented it at the first International Congress of Prehistoric and Protohistoric Sciences as "a very neat representation of fields and fences" (R. Battaglia, "Ricerche etnografiche sui petroglifi della cerchia alpina," in: *Studi etruschi*, 1934, vol. 8, pp. 11–48), will be the subject of studies, from the 1960s (E. Anati, *Rock Engravings in the Italian Alps*, 1958, 11, pp. 31–59; W. Blumer, "The oldest known plan of inhabited site dating from the Bronze Age," in: *Imago Mundi*, 18, 1964, pp. 9–11. In the course of the 1980s the engraving was recognized as a map (P.D.A. Harvey, *The history of Topographical Maps: Symbols, Pictures and Surveys*, Thames and Hudson, London, 1980) and towards the end of the same it was officially included among the prototypes of cartographic history, on the occasion of the publication of the first volume of *History of Cartography* (J.B. Harley, D. Woodward, eds., *The History of Cartography*, University of Chicago Press, Chicago, Vol. 1, 1987. In that volume, scholar Catherine Delano Smith included the Bedolina petroglyph in her list of "complex prehistoric maps" (C. Delano Smith, "Cartography in the prehistoric period in the old world: Europe, the Middle East, and North Africa," in: J.B. Harley, D. Woodward, *op. cit.*, pp. 54–101), along with other petroglyphs found mainly on the same western side of the town of Capo di Ponte. Delano Smith's list endorsed Sansoni's hypothesis as to the existence of a "map route" (U. Sansoni, *L'arte rupestre di Sellero*, Ediz. del Centro, Capo di Ponte, 1987; U. Sansoni, S. Gavaldo, eds., *L'arte rupestre del Pià d'Ort, la vicenda di un santuario preistorico alpino*, Ediz. del Centro, Capo di Ponte, 1995).

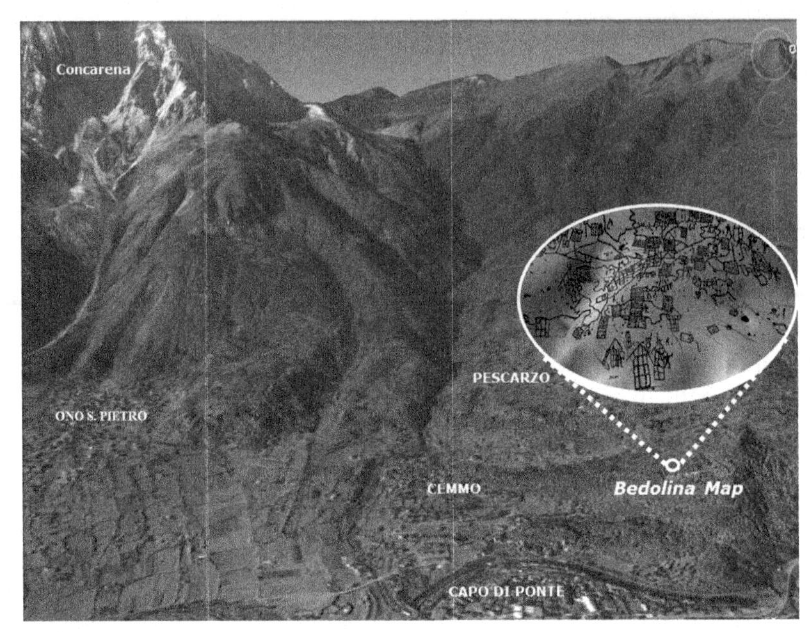

FIGURE 6.7
Valcamonica with the localization of the Bedolina map. (See a color reproduction at the address: booksite.elsevier.com/9780128035092.)

the touch but tridimensional, indented rock with curvatures that create bumps and hollows there are engravings of figures, lines, symbols, virtually a set of signs which, being spatialized and syntactically related, may be read as an instance of figurativization (Fig. 6.7).

The pattern is formed by long, winding lines which connect closed rectangular shapes, within which several small "cups" (small circular shapes engraved in the rock) suggest some kind of arcane inner articulation. Elsewhere, stylized figures reproduce men and animals engaged in various activities. There is no doubt, therefore, that it is some sort of spatialized figuration, though all researchers confidently classified it as a "map" on account of its resemblance to topography. Its prevailing layout is admittedly geometric: the long lines that run through a large portion of the engraving were identified as a road system. The rectangular shapes, which stand out as sharp figurative elements, were identified as "fields," "dwellings" or, more generally, as enclosures. The other circular forms were made to correspond to functional places, probably of community relevance. Such interpretative assumptions were made without special reflection: rather they were derived from the geometrics of the design, presumably indicating a precursor to topography. Unwittingly, that unreflective attribution was also upheld by the work of archaeologists, who reproduced it on a sheet of paper, on a flat

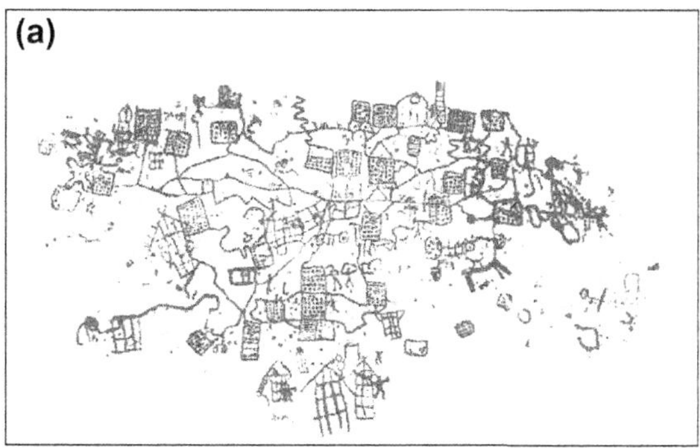

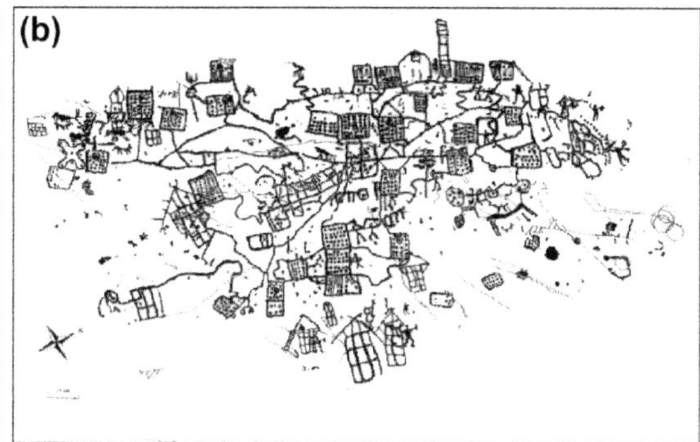

FIGURE 6.8
Archaeological survey of the Bedolina map, according to: a) Miguel Beltran Lloris, 1969 and b) Cristina Turconi, 1997.

surface in order to make the drawing more readable, thereby obfuscating the tridimensional structure of the natural "base map"[25] (Fig. 6.8).

The next step in this interposition was to consider the map as a topographical "land register" or cadastral map, namely a drawing meant to record various properties within the Camuni community.[26]

Scholars seem to have focused on different features of the engraving, on the basis of personal experience and their respective fields of interest. Ausilio Priuli (1991) and Andrea Arca (1997, 2004), for instance, approached the petroglyphs as art historians, in line with a method that derives the presumed meaning of a given picture from comparison and contrast with

[25] Ultimately, findings have led to suggesting two conjectural and quite different datings for the Bedolina map. The first hypothesis, supported by Emmanuel Anati, is based on observations made in 1969 by Miguel Beltran Lloris, who dated the map between the second millennium BC and the third century BC. The second hypothesis is based instead on a survey by Cristina Turconi, 25 years later, who dated the map between the eighth century BC and the first century AD. See: M. Beltran Lloris, "Los grabados rupestres de Bedolina (Valcamonica)," in: *Bollettino del Centro Camuno di Studi Preistorici*, 8, Edizioni del Centro, Capo di Ponte, 1972, pp. 121–158; C. Turconi, "La mappa di Bedolina nel quadro dell'arte rupestre della Valcamonica," in: *Notizie Archeologiche Bergomensi*, 5, 1997, Bergamo, pp. 85–114.
[26] More precisely, researchers sought to identify representations of agricultural spaces comparable to land register records. See: E. Anati, *La préhistoire des Alpes. Les Camuniens, aux racines de la civilisation européenne*, Jaca Book, Milan, 1982; A. Arcà, "Fields and Settlements in Topographic Engravings of Copper Age in Valcamonica and Mount. Bégo Rock Art," in: P. Della Casa, ed., *International Colloquium PAESE '97*, Zurich, 1997, pp. 71–79; C. Bicknell, *Guida alle incisioni rupestri preistoriche nelle Alpi Marittime Italiane*, Istituto Internazionale di Studi Liguri, Bordighera, 1972; A. Priuli, *La cultura figurativa preistorica e di tradizione italiana*, Giotto, Bologna, vol. 1, 1991; C. Turconi, "La mappa di Bedolina...," *op. cit.*, pp. 85–114. Unfortunately, archaeologists seem to employ geographical terminology with a certain measure of approximation: terms such as map, land, settlement, topography, parcel or landed estate are often used quite freely without rigorous semantic denotation. That engenders descriptive approximation and prevents a genuine interdisciplinary exchange of views on the complex drawing that the Bedolina petroglyph presents to us.

similar figurations; archaeologists Geoffroy de Saulieu (2004) and Raffaele de Marinis (1994), on the contrary, set out to date and study the stone itself, once again neglecting its specific figurative quality.[27] In short, the petroglyph was considered by all as yet another form of artistic figuration, not to be measured against cartographic criteria which instead would have provided an interpretation capable of highlighting the map's iconizing propensity.[28] Around the same time, iconologists did start to advance the idea that the simultaneous presence of pictorial and abstract figures on the map might imply a double communicative system of landscape and territory, but their suppositions were met with very little consensus within the scientific community.[29] The results achieved by geographers specializing in prehistoric agricultural systems used to be held in low esteem. Yet, over the years, they have developed concepts and descriptors now quite essential for modelling ancient agricultural systems. So, their contribution is invaluable today for interpreting petroglyphs, though they obviously do not directly engage the issues raised by the area those petroglyphs represent. What those specialists do offer are useful interpretative clues for making sense of the agricultural shapes found on the engravings, although, in the case of Bedolina, they merely record the high degree of formalization of the geometric agrarian space reproduced by engravers, without going any further.[30]

Ultimately, all the hypotheses advanced so far lack hard evidence or solid arguments: the Bedolina map has yet to be thoroughly deciphered. We are groping in the dark. Archaeological studies seem to have reached a stalemate, as the analytical method they used is based mostly on stylistic comparison with other petroglyph findings elsewhere, which were also classified with little documentary evidence and are not in fact nearly as complex as the Bedolina engraving. Arguably, one feature that was overlooked or rather written off as irrelevant is precisely what is

[27] See: A. Priuli, *La cultura figurativa preistorica,..., op. cit.*, 1991; A. Arcà, "Fields and Settlements in Topographic Engravings," *op. cit.*, pp. 71–79; A. Arcà, "The Topographic Engravings of Alpine Rock-Art: Fields, Settlements and Agricultural Landscapes," in: C. Chippindale, G. Nash, eds., *The Figured Landscapes of Rock-Art Looking Pictures in Place*, Cambridge University Press, Cambridge, pp. 318–349; G. de Saulieu, *Art rupestre et statues-menhirs dans les Alpes. Des pierres et des pouvoirs, 3000-2000 avant J.C.*, Errance, Paris, 2004; R. De Marinis, "Problème de chronologie de l'art rupestre du Val Camonica," in: *Notizie Archeologiche Bergamensi*, 1994, n. 3, pp. 99–120.
[28] As pointed out by Marlène Brocard, however, archaeologists did carry out interesting research on these engravings by studying the plan of settlement remains and ascertaining that the weapons represented in the maps refer to the typologies found in excavations. M. Brocard, "Les gravures rupestres à parcellaire," in: *Études rurales*, 2005/3, n. 175–176, pp. 9–28.
[29] As the studies of: P.D.A. Harvey, *The History of Topographical Maps,..., op.cit.*; N. Thrower, *Maps and Civilization: Cartography in Culture and Society*, University of Chicago Press, Chicago, 1996.
[30] On prehistoric agricultural systems and their influence in the Roman fragmentation, see, among all: G. Chouquer, "L'émergence de la planimétrie agraire à l'age du fer," in: *Etudes rurales*, 2005/3, n. 175–176, pp. 29–52.

signally needed in order to derive semiotic clues that go to the heart of the matter and establish whether the map is in fact a cartographic representation or, vice versa, a sort of symbolic representation whose meaning eludes us. I am referring to the tridimensional shape of the rock and its role as basemap. Recently, Joseph Brunod and others supposed, in that regard, that the drawing was not in fact adjusted to the roughness of the stone but, rather, that the stone was specifically chosen as a base for the engraving by virtue of its tridimensionality, presumably fit for reproducing the same geomorphism that the Camuni wanted to represent.[31] Unfortunately, that insightful idea was only aimed at identifying which territory the map represented: it looked for similar morphological features in the surrounding territory, with rather disappointing results.

Yet, an interpretative analysis carried out with adequate theoretical tools could go further. Brunod's thesis finds strong corroboration by the application of cartographic semiosis. Pulling apart the two layers of the map (i.e., the one of the basemap and the symbolic one, given by the set of overlapping marks), we are in fact given the opportunity to highlight possible reasons for the design choices and to focus on the relationship that exists between them. What we need first is to abandon all attempts to tie the map to a representation of some territory existing in the valley today. Then we need to consider the two cartographic layers separately in order to identify within figurativization the two phases of spatialization and figuration.[32]

As for *spatialization*, if we consider the tridimensional quality of the base map with regard to the engraving, we notice significant elements that confirm the conjecture whereby the lines running along the rock correspond to paths. The meshes of this network, in fact, are quite consistent with the shape of the rock: they run straight in flat areas, and become winding over convex or concave areas, to the point of resembling hairpin bends when the stone is especially indented. The most notable instance occurs in the middle portion of the rock, along a concave section, where lines start taking a twisting course. And the same meandering line is visible in the top right portion of the rock,[33] where lines tend to wind up as they connect some rectangular shapes. In some cases, as in the upper margin of the rock, where the stone inclination is steep, we find a winding course that recalls a proper hairpin bend (Fig. 6.9). From an analysis of this first figurative feature, it seems reasonable to suppose that this is in fact a road network in a mountainous area. Such hypothesis is in line with the basic assumptions advanced so far, and serves to clear the field from

[31] G. Brunod, A. Ramorino, A. Gaspani, *Bedolina, la città ritrovata: 5000 anni di vita in Val Camonica incisi sulla roccia*, Noster Mond edizioni, Brescia, 2004.

[32] On this semiotic aspect, see Chapter 1 of this book.

[33] The reading direction of the engraving can be inferred from the anthropomorphic figures it represents.

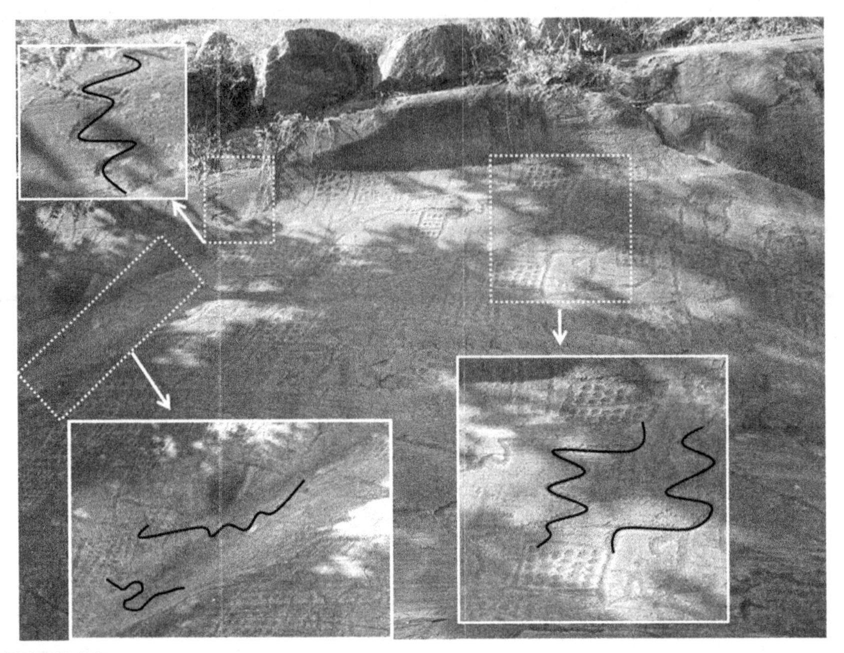

FIGURE 6.9
The curviform or winding shape of linear signs. (See a color reproduction at the address: booksite.elsevier.com/9780128035092.)

esoteric or cosmogonic readings of the engraving, quite alien to the representation of territory.[34]

If we turn to consider the placement of the rectangular shapes (called fields or built-up areas), we notice that it also seems to be affected by the various inclinations present on the surface of the rock. Useful information in that respect comes from the analysis carried out by a French researcher, Marlène Brocard, who studied the layout of "such fields" by superimposing a grid square that revealed a double orientation she called isoclines/anisocline (Fig. 6.10).

More specifically, we find, quite surprisingly, that the shapes corresponding to convex or concave areas belong to the second type: anisocline. And, if we are willing to grant that the articulation of the rock represents geomorphism, we should then conclude that the orientation of those shapes depends on the fact that they are on

[34] Even though such cosmogonic dimension is to be traced back to ancient communities founded on a sacred link with nature, as highlighted by Levy-Strauss in the wake of Durkheim. Rather, attention should be directed toward the specifically linguistic systems created with the aim to order such a worldview (J. Goody, *The Domestication of the Savage Mind*, Cambridge University Press, Cambridge, 1977) or toward studies on the foundation of archaic world religions, such as Mircea Eliade's groundbreaking research, which sees in the category of the sacred the anthropological primacy of the experience of the world (M. Eliade, *Cosmos and History: The Myth of the Eternal Return*, translated by W.R. Trask, Princeton University Press, Princeton, NJ, 1954).

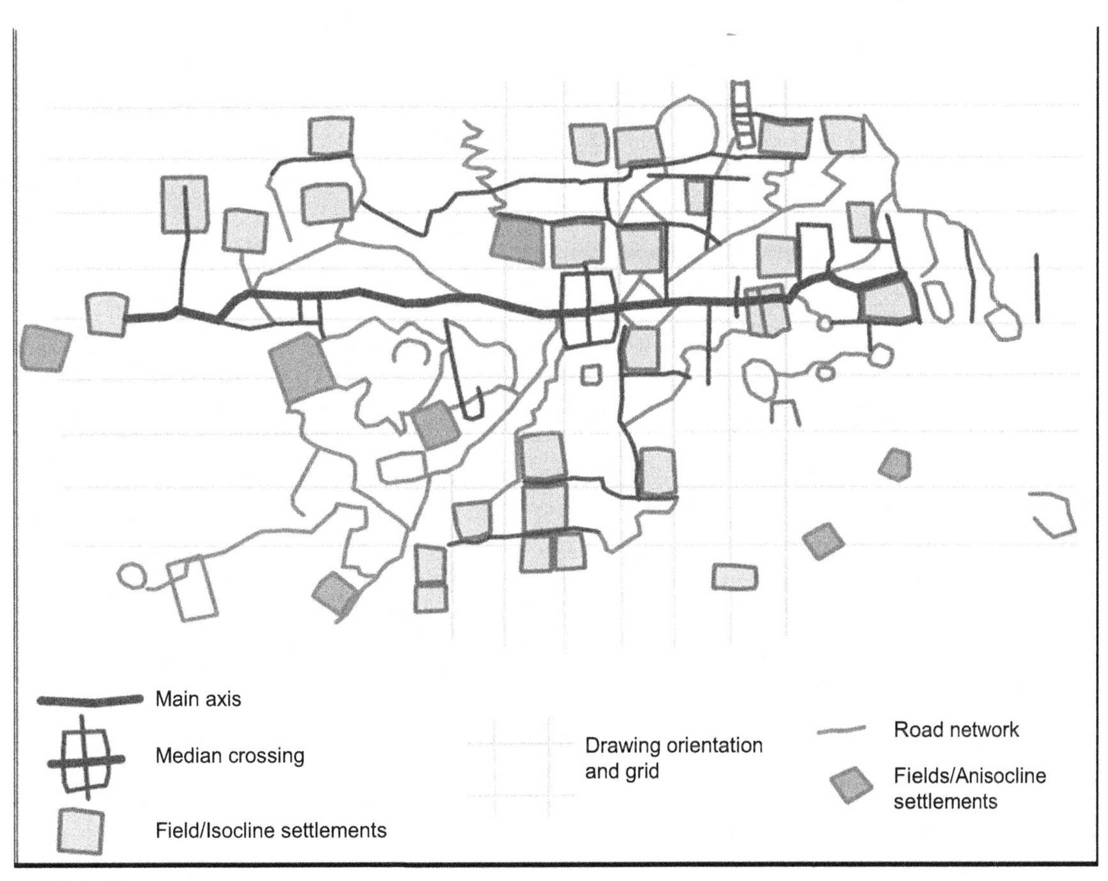

FIGURE 6.10
Interpretation of the particle-like pattern of the Bedolina map (Brocard, 2005, p.12). (See a color reproduction at the address: booksite.elsevier.com/9780128035092.)

steep terrain, which prevents isocline alignment and requires adjustment to inclination. That seems to validate our argument as to the importance of the base map.

That is evident in the digital model obtained through three-dimensional survey of the boulder made with a Total Station, high definition photo survey. Trails can easily be seen winding up or down in the folds of the rock. Also, the anisocline layout of the *fields* quite clearly matches the inclination of some sections of the rock[35](Fig. 6.11). Ultimately, without relying unduly and erroneously on

---

[35] 3D and photographic surveys were made by Studio Ambiente e Tecnologia s.r.l in collaboration with Giovanni Ginouliach, and later assembled by Stefano Giavazzi of the Studio f+g Architettura e Ingegneria. The whole experimental procedure was performed under the guidance of Federica Burini from the *Diathesis* Cartographic Lab. 3D reproduction is based on a survey of rock morphology obtained by measuring the elevations at which Cristina Turconi's archaeological survey was carried out. The procedure was completed with an on-site survey and a photographic reproduction made by piecing together single HD shots of the various sections of the stone.

FIGURE 6.11
3d rendering of the Bedolina map with superimposed survey.

geometrics or on a third-party, mediated survey of the drawing and after a thorough land inspection carried out in subsequent months, I feel I can submit that the rock base map on which the design is engraved is an integral element of the representation. It conveys a spatialization reproduced on a three-dimensional medium that was chosen, in all probability, because it matched the shape of the mountain landscape meant for representation.

To move on in our analysis and address the issue of *figuration*, we need to focus on the sequence of steps that led to the creation of the petroglyph. All archaeologists agree that the engraving was done at different stages, which match different locations on the stone: the oldest design is placed in the upper left corner of the boulder, although some scattered small engravings are found to the right. That design is followed by the more specifically "cartographic" pattern, that is, the one that radiates from the central part of the boulder, refers to paths and fences, and continues with the so-called "constructions" at the bottom. Anthropomorphic and zoomorphic figures were presumably added at a later stage.[36]

[36] Miguel Beltran Lloris identified four engraving stages spanning between the second millennium BC and the third century BC. In his view, the "topographic" phase is the second one, dating back to the middle of the second millennium BC. Conversely, Cristina Turconi identifies six engraving phases and possibly a seventh stage based on a thread-like technique, between the eighth century BC and the first century AD. For Turconi, the topographic survey must have been carried out in the third phase, in the period between the seventh and fourth centuries BC. Both scholars in fact agree on the sequence of the petroglyph. See respectively: M. Beltran Lloris, "Los grabados rupestres...," *op. cit.*, pp. 121–158 and C. Turconi, "La mappa di Bedolina," *op. cit.*, pp. 85–114.

The most intriguing feature for our purpose is the perimetric collocation of the initial engraving phase, which cannot be reasonably explained without assuming a specific desire to locate the drawing on the outer margins with regard to the whole stone surface. It would otherwise be hard to explain the choice of a marginal area that bears no relevant features in terms of smoothness or otherwise, when the whole stone surface was available. Such conclusion would seem to be borne out by the fact that the icons used for this primitive design are analog representations to do with the activity of hunting (there are recognizable animal figures) – and thus identify a cynegetic area necessarily removed from the main settlement, which will be reproduced only at a later stage. All this reinforces the idea that we are dealing with a composite representation of a vast mountainous territory that the natural tridimensionality of the rock matched with a good degree of approximation.

The analysis made so far substantiates the claim that the configuration of the rock closely matched a geomorphism. Therefore, its shape presumably provided less an obstacle than an expressive opportunity for the people who carried out the drawing. This new approach forces us to review all the studies conducted so far. What is more, it revolutionizes the approach of all those who will follow. First, we will have to underline that it is in fact a cartographic representation; secondly we will have to specify its type: if a picture was engraved on a "base map" that volumetrically reproduced some area of Valcamonica, an apt denomination is not "map" but rather "relief model." To be sure, such a claim matters not much in notional terms but, rather, because it disrupts the wrong assimilation of the map to a topography and includes it among tridimensional representations, that is, tridimensional maps. These are presumably more suited to rendering visual data because they match the shapes of terrain. Whatever the message it was meant to convey, the Bedolina map/relief model tells us that its engravers thought it crucial and indispensable to be rooted in the morphology of the valley and its landscape.

Striking suggestions in this regard come to us from other maps found in Valcamonica and precisely by observing the landscape of the sites in which they are located.[37] If we look beyond the engraving themselves to the features of the landscape and to the outlines of the stones on which they are impressed, we

[37] As indeed shown in Casey's keen discussion of landscape cartography, when he mentions the landscape surrounding the Bedolina map to conclude that the localization of the map, on the mountainside, provides for a cartographic reassertion of the adjacent landscape. The boulder-map of Bedolina, he maintains, relocates the landscape and the region in a perennial dimension: the landscape in the edges of the map and the region in the engravings that make up the map. E.S. Casey, *Representing Place Landscape Painting and Maps*, University of Minnesota Press, Minneapolis-London, 2002, pp. 137–38.

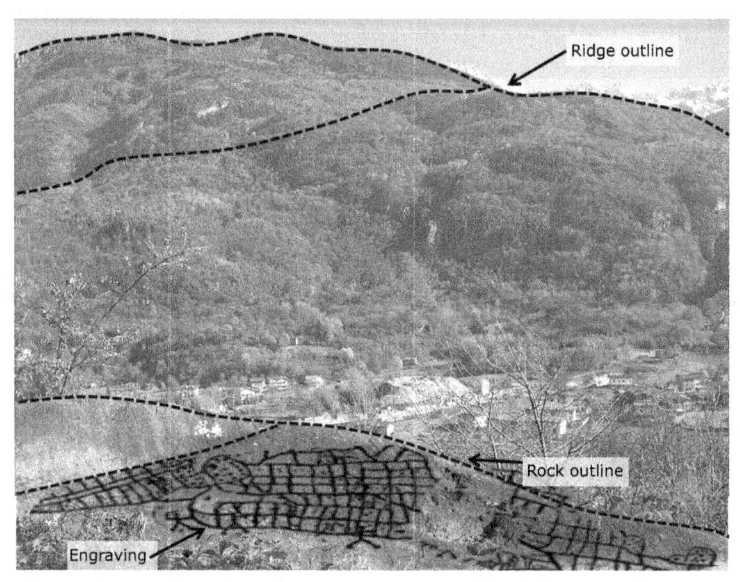

FIGURE 6.12
Comparison between the ridge profile and the engraved outline from Seradina I n. 57 (superimposed enlargement of the engraving). The relief engraving, made by Alberto Marretta in 2012, is available at: http://www.parcoseradinabedolina.it/parco.html. (See a color reproduction at the address: booksite.elsevier.com/9780128035092.)

can easily notice the similarity between the contours of the mountains and the shape of the engraved stones[38] (Fig. 6.12).

The Bedolina "map/relief model" presents us with a territory/landscape duality and reminds us of the importance of a three-dimensional basemap for reproducing the latter.[39] The importance of that finding for our purposes is obvious because it asserts the relevance of three-dimensionality for rendering landscape, which I discussed in the previous chapter of this book.

In short, even the fascinating Bedolina petroglyph leads us not to exclude, *a priori*, from our experimental horizon the possibility of a 3d basemap, since IT advances make it possible. However, before addressing technical issues, let us pause to consider how cartographic spatialization occurs in such techniques.

---

[38] It should be pointed out, however, that the Bedolina map makes such comparison impossible. Even Brunod's analysis, which pinpoints the area around the hamlet of Ono S. Pietro and the hamlets of Pescarzo and Cemmo Capo di Ponte arguably lacks comparative evidence (Cfr. G. Brunod, A. Ramorino, A. Gaspani, *Bedolina, la città ritrovata…*, *op. cit.*, 2004).
[39] At the *Diathesis* Cartographic Lab of the University of Bergamo research is being conducted to complete the semiotic interpretation of the map and to reconstruct the traditional socio-territorial layout of that mountain. Our objective is to advance a hypothesis on the typical features of Camuni *spatiality*, also by using comparative analysis of other maps from the Valley.

CARTOGRAPHIC SPATIALIZATION OF GLOBALIZATION

Issues related to the treatment of the base map also include the relevance of using projections to translate the spatialization of a globalized world,[40] for projections are nothing more than the support grid used to envisage geographical phenomena on a plane in the same arrangement in which they are found in reality. The radical criticism leveled against their use comes from the fact that, by distorting space, projections irreparably affect the transmission of the cartographic message, since they show as flat what is in fact curved. Jacques Lévy argues that "assuming that a map is not reality itself but a specific cognitive project of a given reality, the question of projections can and must be reformulated. In many cases we have no need for them, and the question is: how else could we replace them to ensure an accurate relationship between the site and the map? The harder we strive to give a realistic picture of the world, consisting of a nexus of interweaving sites, territories and networks, the more we notice that the role of the projections fades and vanishes. Will we be able to adapt our mapping of the globe to the encompassing phenomenon of globalization?"[41]

It should be noted at this point that what is being questioned is less the use of the analogical basemap than the role played by projections in ensuring the relationship between site and map, achieved by reducing the earth's spherical shape on the basis of mathematical principles.[42] Disclaiming the analogical base map as part of a general rejection of topography, which is based on projections could have the disastrous outcome of "throwing the baby out with the bathwater" – that is, to dispose of what must be instead preserved. Following Lévy's lead, I will try to reformulate the question of projections by revealing the gap between them and the analogical setup of the base map.

Circle or Sphere? Projections and Google Earth

I already stated in the second chapter of this book that projections appear as a set of conventions that regulate the topographical basemap and aim to adapt the spherical surface of the world, multivariate and tridimensional, to a planar and

---

[40] Also on the technical implications this entails. On cartography's problem areas see: E. Casti, J. Lévy, "Svolta prossemica: dalla rappresentazione alla spazialità cartografica," in: E. Casti, J. Lévy, eds., *Le sfide cartografiche..., op. cit.*, pp. 7–12.

[41] J. Lévy, "Una svolta cartografica. Colmare il divario tra le scienze e le tecnologie dello spazio abitato," in: E. Casti, J. Lévy, eds., *Le sfide cartografiche..., op. cit.*, pp. 29–30.

[42] It bears repeating that that is where the specific role of cartography and the basis of its difference from other forms of visual representation of the world lie. Semiotic analysis shows, in fact, that cartographic language is unlike other forms of visual communication, since it is based on two communication systems: analogical and digital, and that their combination produces a much more complex message, which is the harbinger of self-referential messages and is capable of producing iconization. See ch. 1, paragraph 1.5.

homogenous representation. This is because projections, as is well known, should respect or at least approximate three analogical correspondences with the properties of Euclidean space: *equivalence* (the extension of the areas is preserved), *equidistance* (linear distance is guaranteed), *isogony* (the virtual geographic grid that envelops the earth is reproduced with intersecting angles to preserve its spherical shape).[43] Compliance with these correspondences is the essential condition for the world to be considered as faithfully reproducible on a plane.[44]

No projection can avoid distortion, given the impossibility of developing the geoid surface onto a plane. Yet the smaller the area to be represented and the larger the cartographic scale, the smaller will such distortion be. Only globes can preserve with mathematical accuracy the lengths, angles and areas corresponding to reality, through equidistance, isogony and equivalence.[45] In the case of geographic maps, instead, any expedient can thoroughly fulfil but one of the last two conditions (isogony or equivalence), because equidistance is confined to set limits and in certain directions. Hence, distortions are not of equal magnitude within a single map (they normally grow from the center toward the edge), and in some map scale varies from section to section, at times so noticeably that it is worth marking it with latitudes and longitudes, as in the case of the Mercator map.

The choice of a projection depends not only on the content and/or the purpose of the map but, first of all, on the width of the region to be represented and on its position with respect to the values of geographical coordinates. More temperate or polar regions appear more distorted than equatorial ones. The projection type of map is not, therefore, chosen case by case, but takes into account the size of the surface to be represented, its shape, its latitudinal position and extent, and the objective to be achieved or the phenomenon to be highlighted. Numerous projection methods have been investigated, some based on the laws of geometrical projection, others obtained by the use of mathematical ratios.[46]

---

[43] Such features are designed to maintain unchanged on the map the metric properties of distance or at least to refer to their quantitative sizing. According to their technical characteristics, they are broken down into conventional projections (made on an empirical basis) and actual projections obtained by geometric and mathematical rules.

[44] Approximation is yet another misleading system, because, while it endeavors to respect real-world measurements, it actually generates a deviation, an unavoidable error. At the same time, by stressing approximation as a geometric prerequisite, cartographers convey the idea that, as a building system, it is capable of ensuring accuracy. Much was said on this topic even from the viewpoint of ideology. See among others Jacob's *L'empire des cartes..., op. cit.*, pp. 153–161.

[45] A number of scholars have recently addressed this issue, albeit on different levels: F. Farinelli, *Geografia. Un'introduzione ai modelli del mondo*, Einaudi, Turin, 2003; Id., *La crisi della ragione cartografica*, Einaudi, Turin, 2009; C. Grataloup, *Représenter le monde*, La Documentation Française, Paris, 2011.

[46] For an analysis of the multiple projections see the website of the Institute of geometry and discrete mathematics at the Technische Universität in Vienna, which lists 36 projections http://www.geometrie.tuwien.ac.at/karto/index.html and a site created by the Center for Spatially Integrated Social Science at the University of Santa Barbara, California, dedicated to information about map projections, which also includes a comprehensive bibliography: http://www.csiss.org/map-projections/Reference/Bull1856.pdf.

While projections do work for redressing some of the abuses in plane representations of a curved space, some distortions inevitably impinge on geographical coordinates. We know that coordinates make it possible to accurately communicate the position of a point on the Earth's surface, since they rely on a fixed reference point, located outside the earth, referred to that of a star which ensures its relative immutability. Thus, the checkered pattern of coordinates appears self-evident because it is the projection of the framework of the cosmos onto our microcosm. Until quite recently, people generally relied on maps or the calculations that could be made on them in order to find out earth coordinates. In any case, they relied on adjustments implemented by the cartographer, on distortions already allowed for from the outset in the graphical transposition by way of projections.

Today, things have changed and computer technology gives us the opportunity to go beyond projections. Before addressing such techniques, however, we need to assess the repercussions that the use of projections has over one final and certainly not minor aspect: namely the vectorization of space.

I have repeatedly claimed that maps reproduce space according to the principle of analogy with reality and aim to arrange objects using a topological order which relies on a *cognitive spatialization.* I should now state that the latter is the result of a *vectoralization* of space implemented by a subject. He or she sets up the vectors on the basis of which surrounding space can be composed. Vectors are therefore a human product, not a natural phenomenon. Vectors taken may vary according to their reference culture. In the West, a powerful vectoralization is enshrined in the east/west and north/south axes, which mark the rising and setting of the sun: its culmination or, conversely, its absence. This vectoralization is indispensable for expressing the world which has its roots in visual perception, a world which, over time, is enriched with socially produced values and becomes precisely a cognitive spatialization. Even the localization of objects on the map abides by vectoralization rules, so it is also a human product. We should recall that *spatial localization* is one of the spatialization procedures. It can be defined as the construction of a system of references that enables researchers to situate objects based on their horizontality, verticality and perspective (front/rear). The role played by maps in the semanticization of the world is therefore to propose an organizational structure based on topological categories that localize objects. In this respect, the map is a *device for representing the earth's surface in which objects are arranged in a topological plane governed by a principle of cardinality.*

In short, I wish to point out that every constructive choice – and the use of projections *in primis* – with all its attendant editing – scale, orientation, observation point, the centrality of representation – has not only the technical value of producing spatiality. It is also dictated by cultural choices, since it adopts the spatialization typical of a given society. And above all such choices are plural and changeable, along with the social interests in which they are anchored.

Topographic metrics, for instance, which is based on an objective view of the world, adopted the principle of cardinality yet ruled out the presence of the subject nonetheless, producing an intangible vectoralization that weakened spatialization and its cultural values. The geographic coordinates that virtually encompass the round shape of the globe, via projections, are placed onto a flat surface that distorts them and turns them into an abstract system of reference.

It should be noted, however, that this is not invariably present in cartographic production, not even in the West, but pertains exclusively to topography. If, in fact, we assess how vectoralization was addressed over time, we find that the subject was not always excluded. Medieval nautical maps, for example, which used neither coordinates nor projections, had directional lines for rotating the document according to the position of the sailor and the direction that he intended to follow to reach his destination. Because of that, designators themselves were vectoralized.[47] We should therefore consider aspects regarding vectoralization separately from those related to projections. Vectoralization starts from the point where the subject is placed and is made to derive either from directional lines or directly from geographic coordinates, taken as a conventional grid ("graticule") which envelops the planet without the interposition of projections.

Today, satellite detection, namely the global positioning system (GPS), potentially makes this operation available to everyone. GPS accurately pinpoints each location of the earth's surface by means of numerical data (three, if altimetry is included as well).[48] Now, if in the phase that follows, that is cartographic visualization, we adopt a sphere instead of a flat base map – something GIS technology enables us to do today – in theory we no longer need to interpose any projection. The advantage of that lies in the fact that the analogical setup of the globe's rendering would be matched to a set of points projected directly onto a spherical shape, without the intermediation of any projection. What I am pointing out is that, although satellite detection and GPS exclude the interposition of maps, data still continue to be placed on maps built in accordance with projections. And I want to stress that such graphic transposition need no longer be made on maps. It could instead be made onto a spherical surface or better on an

---

[47] Cartographers provided for the need of sailors who, having to follow a route on the basis of data gathered from the pairing of compass and map, were facilitated in their task if they could rotate the map. That produced a multiple radial vectoralization, given by the multiple ways of reading the map. The parchment depicting Lake Garda is a great example of this. See: E. Casti, "State, Cartography and Territory in the Venetian and Lombard Renaissance," in: D. Woodward, G.M. Lewis, eds., *The History of Cartography*, Vol. 3, The University of Chicago Press, Chicago, pp. 874–908. http://www.press.uchicago.edu/books/HOC/HOC_V3_Pt1/HOC_VOLUME3_Part1_chapter35.pdf

[48] In practice, GPS also requires a system of conventional reference, namely WGS84 datum which, however, is not a projection but a mathematical model, as we will see shortly.

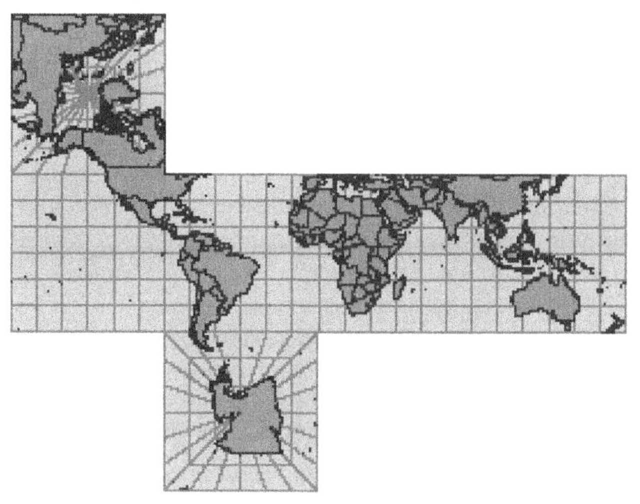

FIGURE 6.13
Cubic projection as employed in the ArcGlobe application. (See a color reproduction at the address: booksite.elsevier.com/9780128035092.)

ellipsoid solid as the Earth is, on which most GIS software is based.[49] The latter uses two projection systems: the *spherical geographic coordinate system* that uses a spherical three-dimensional surface to locate points on the earth's surface; and *the Cartesian or projected coordinate system,* defined on a flat and two-dimensional surface that has constant lengths, angles and areas and is still based on a geographic coordinate system of reference. There are hundreds of systems of either type that users may select when creating their maps.

ArcGlobe is an application of the "3D Analyst" system, part of the ArcGIS software package produced in the US by ESRI. It allows for the visualization of data on a globe made via a particular coordinate system called Projection Cube, in which the world is projected onto a figure consisting of six squares, with meridians and parallels as straight lines (Fig. 6.13).

As its reference ellipsoid, ArcGlobe uses WGS-84 (*World Geodetic System 1984*), the most widespread international geodetic system. It is a mathematical model of the Earth from a geometrical point of view, geodetic and gravitational, built on the basis of measurements and of scientific and technological know-how available in 1984. Unlike classical geodesy systems, which may be defined on a local or regional basis and provide accurate geoid approximation only around the point of emanation, the WGS84 global system is valid for the whole world

---

[49] ArcGIS Desktop, for instance, features a 3D Analyst plugin which contains the ArcGlobe application and allows users to visualize data on a globe, in their actual geodesic position. For further details see: http://help.arcgis.com/en/arcgisde-sktop/10.0/help/index.html#//00q800000053000000.

because it uses the EGM96 standard, which approximates the geoid as a whole, thus providing a considerably more effective tool,[50] in terms of the constructive and semantic perspective adopted here.

If we turn to analyze the outcome this has on the recipient or, better, if we consider the pragmatics of cartography and, therefore, how communication takes place, we discover that the whole process revolves around interactivity and that communication of the earth's sphere is accomplished at the highest level with WebGIS. We already addressed the interactivity of such systems in Chapter 4. Now I will only mention that in an interactive environment the subject, who builds or orientates the map, is a user who wants to know the world and does that via the resources that the new geographic systems provide online. The user acts on the image first by establishing scale: flying over the world, zooming in on a specific region, looking up details provided by the system. The process is structurally dynamic: users may zoom in or out to find their bearings, to understand what the system has to offer. As they do that, they input names, evidence that the cartographic message starts there. Only then, in fact, the image they have before their eyes changes from an image into a map, even though it is not a map.

Let us consider the most widespread and widely available WebGIS system: Google Earth. Even though, as we shall see, Google Earth is still influenced by topographic metrics, it seems the right context for future innovations. The experience of using the system is exhilarating. One looks at the globe that spins and stops at a mouse click which selects an area for viewing. That area is automatically zoomed in until one issues a second click, as territory takes on increasingly recognizable features, to the point of showing buildings, streets, gardens and anything else in ever sharper and somewhat worrying detail, since any privacy boundary is presumably erased.[51] Google Earth browsing provides a host of other options. One can "drag" to connect the area to adjacent ones, thus expanding the horizon, or restricting it to zoom in on another area that also will be displayed as a visual rendering of the world based on our selections. This system, apparently a "global" photographic image, is, in practice, a sham, since it submits as a picture what is in fact a complex manipulation of data and projections that fully embrace topographic metrics. In fact, Google Earth uses cylindrical projection with WGS84 datum*. Through interactivity, however, which leads us to believe we can fly over

[50] WGS84 was adopted for air navigation in 2000. In Italy, both Aeronautica Militare and ENAV, l'Ente Nazionale di Assistenza al Volo (National Air Traffic Control) completed data conversion from the old ED50 and MM40 formats to WGS84, in compliance with new international standards issued by ICAO.
[51] To the point that the governments of various countries, including Italy, are demanding guarantees of confidentiality and privacy in the publication of material or personal data. Google Italy is already working to ensure users of the various services offered, such as Google Earth, Google Maps, StreetView, the right to privacy of personal documents or on the location of the user. See: http://www.google.it/intl/it/policies/privacy/.

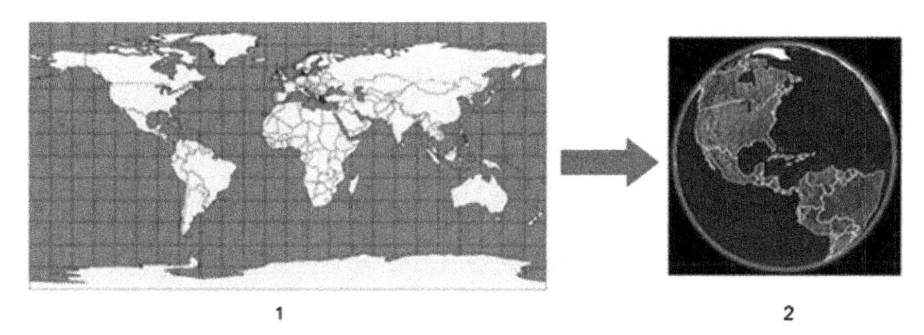

FIGURE 6.14
Equirectangular cylindrical projection (Plate Carrée) and a base of Google Earth images. (See a color reproduction at the address: booksite.elsevier.com/9780128035092.)

the world and master its "natural" layout in complete freedom, Google Earth becomes a sophisticated instance of cartographic iconization (Fig. 6.14).

Strictly speaking, it is a projection in which meridians and parallels are represented by equidistant straight lines that intersect at right angles. However, since this projection is aimed at representing the earth on a flat surface, WGS84 datum is used to reduce that cartographically obtained surface to a form that reproduces, in mathematical terms, the ellipsoid shape of the planet. A *datum* is essentially an algorithm that defines the matching of geographical coordinates to points on the spherical surface of the Earth. *Datum* also forms the basis for the measurement of altitude. This interposition, and the inevitable distortion that follows, cannot be avoided, since the Google Earth system is the result of a collation of many satellite images that, in order to match, must be adjusted to the angle of observation of the sensor that recorded the data. These images will of course vary, depending on the orbit of the satellite and the time when the survey was conducted. The use of projection is necessary because this adjustment is more likely to be achieved on a flat surface like a sheet which, as I already stated, by the use of the *datum* is referred back to the ellipsoid geometric shape of the planet.[52] The interposition of *datum* does not *per se* represent an issue for our purposes, because it entails no distortion of the planet areas. If anything, it produces a deviation in the metric accuracy at the oblate poles and the equatorial bulge, that is to say it is an adapter of the planet's ellipsoidal shape. However, as this does not affect the representation of the curved surface of the Earth, this interposition can be taken, at least for

[52] For the sake of thoroughness we should add that WGS84 datum is one of many possible mathematical interpretations of the shape of the earth, but it has the advantage of being usable without cartographic intermediations. The data it refers to and their various mapping options are listed in the official documents published by the US National Geospatial Intelligence Agency: http://earth-mfo.nga.mil/GandG/publications/tr8350.2Ar8350_2.htmL

FIGURE 6.15
3d model of Mount Etna elevation (Source: www.astrium-geo.com). (See a color reproduction at the address: booksite.elsevier.com/9780128035092.)

now, as a provisional requirement of satellite detection, given that innovation in this field is accelerated and consistently progressive.

Since 1991, when data was collected by radar from satellites ERS-1 and its successors ERS-2 and Envisat, all from the European Space Agency (ESA), geodesy has made great strides and the calculation of the geoid has achieved great accuracy. Moreover, there have been recent reports to the effect that satellite detection is entering a new era. It will no longer rely on photographs or images, but on radar detection, which in addition to allowing for the plotting of altimetry will no longer require the intermediation of projections. The news comes from the German Space Agency Deutsches Zentrum für Luft-und Raumfahrt (DLR), owner of the Tandem-X and TerraSAR-X satellites manufactured by EADS Astrium. These satellites are able to acquire data of the same area simultaneously from different angles and can achieve a detail resolution of up to one meter.[53] In addition to ensuring high quality features (vertical accuracy, horizontal raster image and global homogeneity even in overlapping between image strips), the system aims to create a three-dimensional digital elevation model of the earth surface in high definition (Fig. 6.15). This allows us to envision the disposal of projections, and the consequent adoption of direct geographic coordinates and the three-dimensional rendering of the world with a high degree of detail.

[53] In this case it is a "bistatic" radar, a radar in which a satellite transmits the radar signal to earth and both receive backscatter, or return echo signal, simultaneously. In this regard, see the website of the GEO-Information division of Astrium Services, who own the marketing rights for such data: http://www.geo-airbusds.com/terrasar-x/.

This innovative satellite detection system shows, then, paths that lead to a resolution of the cartographic representation of the globe. A map created with new technologies which, as we saw, offer new possibilities over printed maps, will eventually be able to render the world in its actual configuration. And through interactivity, such configuration will change, becoming plural in accordance with the observer who interprets it. In short, once the world has been depicted according to its actual shape, a base map will be offered on which the data coming from the social action can be placed. This will open the opportunity to reflect on what information to enter in relation to the multiscalarity of the chosen phenomenon and through the translation of the transcalarity* of the actors' operational plans. To be sure, multiscalarity cannot be settled by entrusting it to zooming, which progressively increases the detail level of the phenomenon being viewed. Multiscalarity must instead be ensured by giving users different options for viewing a phenomenon and letting it take on different semantic senses in accordance with scale. We understand, at this point, that technical potentials will have to combine with analytical and territorial skills, properly geographical know-how, the only one able to give substance to the recognition of phenomena. In this way, mapping will become the ground where geographic knowledge can test its potentials for speculation and also for inclusion, on the basis of participatory methodologies that take into account the variety of actors involved in the construction and management of territory.

Cartographic innovations are abundant in this direction and the possibilities offered by Geoweb grow wider by the day. The web promises cartographic capabilities that allow dynamic, interactive and multimedia access to a huge number of territorial data as well as the effective and operational sharing of information with users. In April 2010, the city of Québec received an award from ESRI-Canada for the publication of its new web cartography system which provides access to a large number of geographic data related to urban territory.[54] Basic zoom, pan and print functions allow users to visualize and overlay information layers with administrative boundaries, orthophotos and buildings perimeters.[55] Users can also search, make comments, and access links. The cartographic representation simultaneously retains an overview of the area and a detailed view, but also the type of information on the phenomenon that different scalarity reveals. It is a powerful tool that encourages the use of cartography in participatory practices. This and other innovative tools allow citizens to express themselves, to voice their opinion on several projects and issues and to mobilize citizen knowledge with

---

[54] An in-depth analysis of this product was conducted by: M. Ricci, *Cartografia e politiche della partecipazione: il caso del Québec*, Master thesis, University of Bergamo, academic year 2010–11.

[55] A video format obtained by zooming the central portion of a map and adapting it to the required format, for instance by cropping off its sides or the black bars that would have been produced in the adjustment phase.

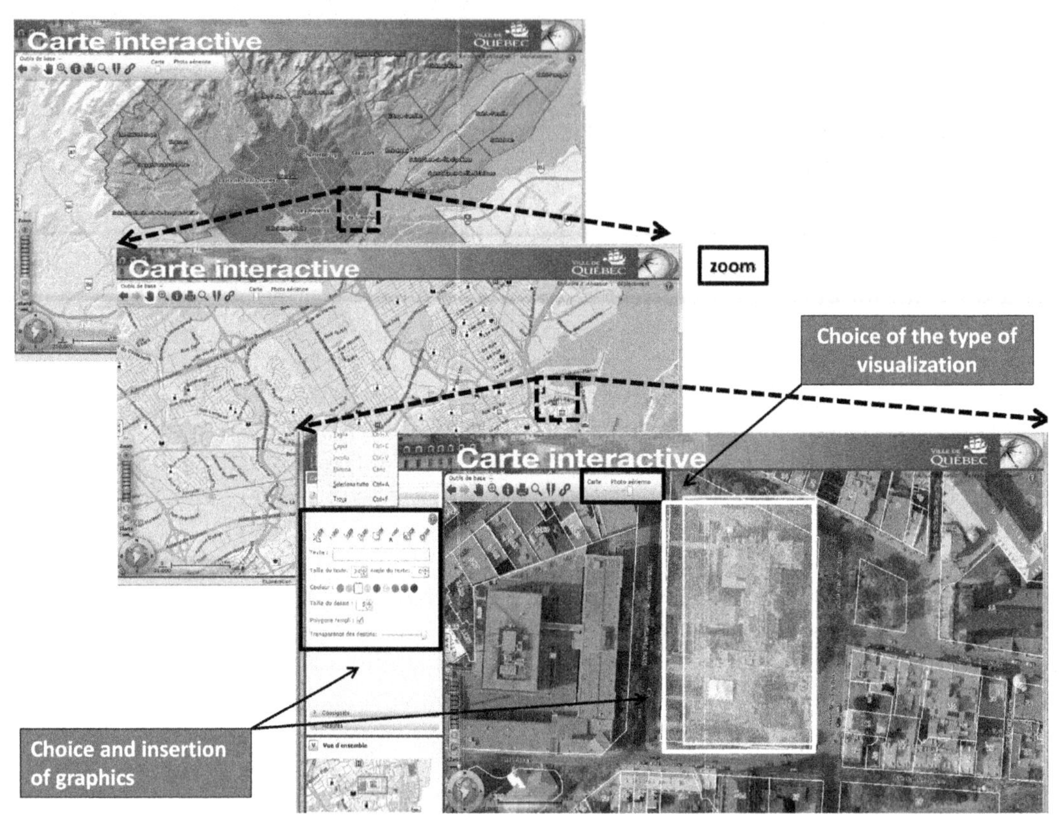

FIGURE 6.16
Interactive map of the city of Québec (www.ville.quebec.qc.ca/carte_interactive). (See a color reproduction at the address: booksite.elsevier.com/9780128035092.)

a view to co-management. And all this is possible because the GIS system envisages a semantic transcalarity which shows different features of the phenomenon according to the chosen scale and the interests of the actors involved (Fig. 6.16).

This corroborates our view that the representation of the globe cannot be conceived of as a small-scale image, for that would only be a hasty outline of territorial features which overlooks the data from the local contexts that make them up, just as it would be unrealistic to think of a local representation juxtaposing other places in order to depict the complexity of globalization. Globalization is to be conveyed by embracing the new scalar perspectives of reticular space and the ways in which such space can be rendered cartographically.

Multiscalarity/Transcalarity and Distance

I mentioned this issue earlier. Now, if anything, we need to ask ourselves whether, in order to render a globalized world, cognitive spatialization should be extended to a double scaling, local and regional. As we do so, we bring back into play the *multiscalarity* and *transcalarity* of the contemporary world, that is to say the properties whereby the same phenomenon may occur and be grasped at multiple territorial levels pertaining to the global and the local. These are not only scalar geographic levels, but two different logical dimensions governed by scalar mobility, that is, the transition from one scale to another that allows for the exchangeability of points of view. Even though in most cases reticular spatiality is represented as a homogenous, isotropic space, conceived as a seamless continuum of constant features in the distribution of phenomenical qualities, it actually involves actors who – albeit with a certain degree of asymmetry – are able to shift from one scale to another without losing sight of their local rootedness. In other words, when these actors operate on a local scale, although they evoke a world that comes to life in its diversity as culturally and historically shaped over time, they also take into account regional dynamics or local economics. Both these scalar levels, therefore, marked as they are by expressions of territoriality, stand up against the kind of nondescript localization inherent in the distortion that topographic imaging entails.

This leads us to think that the representation of a topological space that can invoke these properties must simultaneously rely on the same base map. The challenge is to show a phenomenon by recovering both *multiscalarity*, namely its specificity to occur at multiple scales, and its *transcalarity*, or its ability to be grasped from exchangeable points of view. In short, the normal relationship between geographic scale* and cartographic scale should be overturned and the former should be given priority over the latter. To recall social complexity both at the local and at a regional level is to abandon the idea that it is the map scale to determine the size of the area to be represented and thereby to decide which and how many of its aspects must be shown. Rather, we should cartographically highlight the multiscalar relations that such complexity produces. On the other hand, we should not forget that digital cartographic techniques offer satisfactory solutions in this regard, sparing us from the need to exclude features. A WebGIS system allows us not only to pass from one scale to another, increasing or decreasing the level of detail, but also to ascribe the ratio between one scale and another to the scale set that is being dynamically created.[56] Of course, the cartographer should ensure rendition of this complex phenomenon semantically, by using the outcomes of geographical analysis. And it is up to the web user to become aware of the options at his disposal for highlighting it.

---

[56] To keep regarding the medium of paper as the limit of cartographic potentials is anachronistic. The reference scenario must be digital, where, as we showed in Chapter 4, maps lose their static format and with it most of their visualization restrictions.

In short, what we just said about the new scalar perspective of globalized reticular space does not invalidate the use of a single base map. If anything, a single basemap establishes a reference set and paves the way to a reflection on how to render the last inescapable data of the contemporary world, that is, *distance*.

In a spatialization centered on a reticular system instead of an area-based system, distance is the primary connective datum. We need to ask whether it is possible to show the distance between phenomena in accordance with social variables. I am not here referring only to a way of assessing distance and, therefore, the parameters used to measure it (traveling time, cost, attractiveness of the means, fatigue, etc.). Rather, I am arguing that distance assessment relies on a convention, which in the case of topographic metrics is restricted to linear measurement (metric or kilometric) between two points on the earth's surface. Conversely, the topological perspective discards such strict topological convention to consider the relationship between subjects who form networks where distance depends on the intensity of these relationships. In this perspective, the *réseaux* (networks) are seen as the spatial framework on which the calculation of distance is made. That challenges the assumption that contact between the two areas occurs exclusively by contiguity. There are, in fact, *vicinity* relations that transcend it. As Jacques Lévy noted, the political significance of this position, based on topological space, depends on the fact that the criteria underlying this metrics will be the same as the one adopted to manage territory and, therefore, will be those that affect intervention decisions.[57]

Going back to consider the parameter for assessing distance, we see that the outcome varies considerably. Mileage between Venice and Paris is about 620 miles, but in terms of time it is an hour and a half flight, the same time it takes to drive to Belluno from Venice. This means that if the mile-based parameter is replaced with the time-based one, introducing the variable of the type of means of transport used, distance not only changes, but takes on a social value, as other factors are involved in the choice of the means. It should be noted though that in either case the subject is only marginally accounted for. That is not so in all cultures: the direct experience of the Elsewhere is a valuable awareness tool in this respect. In the territory of Mozambique, for instance, among the Matzua people distance is measured in relation to the traveler and to the fatigue he or she endures in a day's walk. In this case, the metrics is dictated by the fatigue of the subject, who is thus at the center of the calculation. When asked how many miles we still have to go or how long it takes to reach our destination, the village of Chikelene, on the bank of the River Save, Joao, my guide, gave no response or said he did not know, because he did not understand which assessment parameter for distance I was referring to. That suggests

[57] J. Lévy, "Métrique," in: J. Lévy, M. Lussault, eds., *Dictionnaire de la Géographie…*, *op. cit.*, p. 608.

that distance is a cultural concept which excludes or includes the central role of the subject. The Matzua concept of distance demonstrates what is meant by topological distance though, of course, it is unthinkable to adopt it as a parameter in complex societies like ours. What, however, can be adopted is the notion that distance is an abstract concept and it is modulated in practice according to the cultural values used within a given society.

In the current world the importance of distance increased significantly, following the creation of ever larger urban clusters. These call for new parameters in transport planning, for the number of residents is not very indicative if taken in isolation. It must instead be assessed in relation to the number of commuters and to all that relates to generalized mobility toward production areas or service centers. The very design and functional management of residential areas would be invalidated if we were to neglect the dynamism brought about by accessibility over the reduction or the increase of their distance against a metropolitan world structurally designed to be experienced online.

Returning to chorography and to the rendering of the distance, we need to explicitly declare the parameter used to measure it. Internet users must be informed of the reasons underlying our choice. So, assuming that the concepts of space, metrics, projections, distance are cultural variables, we can overcome the idea that topographic metrics is ineludible and turn unbiasedly to techniques that dispose of it, like anamorphosis.

Such technique, which I will address in the next section, can display distance calculated on metrics other than topography. It does so by distorting both the areas and the linear distances. It was recognized that in the past exclusive adoption of topographic metrics and its attendant mile-based parameter to measure distance was brought about precisely by topographic maps and not because it met the real needs of society.[58] At the end of the 19th century, Italian emigrants chose their destination less on its distance in kilometers than on the cost of reaching it, that is, on the cheapest fare. So, between the railway network that led to the rest of Europe and the steamboats that served the transoceanic route for the Americas, people chose the means of transport they could afford, regardless of where and how far it would lead.[59]

Now, urban commuting that characterizes metropolitan areas is based on speed and fare costs, and not so much on distance in kilometers. To represent that reality is not simple, although anamorphosis allows for an assessment of the relative positions of points by interconnecting them with regard

[58] Karl Schlogel underlined the influence of topography in the determination of distance in his: *Leggere il tempo nello spazio*, Bruno Mondadori, Milan, 2009, esp. pp. 64–74.
[59] A. Lamberti, "Geografia, politica e migrazioni: una lettura dell'emigrazione italiana," in: E. Casti, ed., *Il mondo a Bergamo…*, *op. cit.*, pp. 41–64.

to distance-time, cost, preference, proximity or other. Such data pertain to the multidimensional analysis of proximity or preference, although alternative methods may be used, such as Waldo Tobler's "trilateration" or Jean Claude Muller's elastic cartography.[60] In short, showing the alterations caused by the construction of an expressway or the activation of a new air route requires current data, preliminary analyses, and more or less elaborate processing. One must take into account, for instance, that cities do not enjoy the same conditions of access to rail transport, but depend on the shape of the network and the organization of the lines. In Italy, long distances were greatly shortened with the introduction of the "Eurostar" and TAV rail systems. By contrast, we witnessed an expansion of short distances. These are covered by "through trains" that are overturning the experience of space. Represented by anamorphosis, such space now appears as dilated, contracted, into unusual and disorienting shapes. Similarly, if anamorphosis is used to plot the accessibility of the capitals of northern Italy, what emerges is the "ruggedness" of space determined by the various policies in terms of mobility.[61]

Clearly, the analytical approach I am illustrating here for rendering distance and, more generally, reticular spatiality, is the same I took for landscape icons. While maintaining the importance of localization, this approach aims to establish new links that affirm the social sense of territory online. It should not be forgotten, in fact, that the effectiveness of a topological representation does not come from envisaging the phenomena of territoriality taken individually but, rather, as ranked into syntagms, able to yield a territorial syntax. Before we conclude, we need to consider what anamorphosis is.

Anamorphosis and Chorographic Features

The word anamorphosis (from the Greek *anamorphoun*, "to transform") was used during the Renaissance to refer to a painting technique whereby an image was deformed deliberately so that it was not easily detected when observed from a frontal viewpoint. The viewer is asked to "optically correct" this distortion by changing position and looking at the picture almost in profile. Hans Holbein's *The Ambassadors* is arguably the most famous example of anamorphosis in painting (Fig. 6.17). At the foot of the figures of the ambassadors, the artist added an element not recognizable at first sight but which, when viewed in profile, takes shape and becomes a macabre human skull. In this context, the use of anamorphosis

[60] C. Cauvin, "Au sujet des transformations cartographiques de position," *Cybergeo: European Journal of Geography*, Cartographie, Imagerie, SIG, article 15, posted online 14 January 1997, modified 27 April 2007: http://cybergeo.revues.org/index5385.html.

[61] P. Langlois, J.C. Denain, "Cartographie en anamorphose," in: *Cybergeo. European Journal of Geography*, Cartographie, Imagerie, SIG, article 1, published 14 January 1996, modified 14 March 2007: http://cybergeo.revues.org/index129.html.

FIGURE 6.17
Hans Holbein, *The Ambassadors*, 1533 (National Gallery, London) (with anamorphic object rendering). (See a color reproduction at the address: booksite.elsevier.com/9780128035092.)

was intended to emphasize the inadequacy of the image to render the meaning of things and, consequently, to warn not to rely on appearance unthinkingly.[62]

It took cartography longer to give evidence of a similar reflexivity,* able to cast doubt on the topographic model. Topography was only discarded of late. That occurred in the second half of the 20th century thanks to the work of Waldo Tobler, who assigned a solid mathematical model to the construction of anamorphic cartography.[63] So anamorphosis became an IT-based technique (obtained by interpolating a vector field with a gravity model), which undermines the assumptions of the topographic base map by distorting them to fit the social data being represented.[64] Treating the base map as a sensible phenomenon of the reality to be depicted, anamorphosis released us from the dictatorship of topography. The base map is no longer the rigid container on which data are

---

[62] This artistic technique still in use today has taken on other meanings, including that of a disorienting optical perspective and a manipulative play on representation. Among the many sites dedicated to anamorphosis see: www.anamorphosis.com.
[63] W. Tobler, "A Continuous Transformation Useful for Districting," in: *Annals of the New York Academy of Sciences*, vol. 219, n. 9, 1973, pp. 215–220.
[64] For a thorough bibliography on this issue see: C. Cauvin, "Bibliographie sur les transformations cartographiques," in: *Cybergeo: European Journal of Geography*, Cartographie, Imagerie, SIG, article 9, 2007: http://cybergeo.revues.org/index146.html.

inserted but a flexible, adjustable space that is forced, expanded, or contracted in relation to the relevance of the data. Territory thus ceases to be an intangible feature of the map and allows for engagement with the themes represented. The communicative advantage of this way of treating the basemap rests with the fact that the relationship between Euclidean space and social phenomenon is reversed: it is the intrinsic qualities of the second that trump the extensive ones.

There are various algorithms for making the surface of the basemap proportional to a given quantitative variable. Without working out their details or assessing their results[65] I wish to recall the one developed within ArcGIS.[66] The transformation principle consists of wielding forces that start from the center of the polygon (corresponding to each identified sector of territory) towards the points that define its outline. These forces represent the deviation between the initial surface of the polygon and the final one that is proportional to the quantitative variable of the social data inserted. What, however, is imperative in order to prevent anamorphosis from disrupting information is that graphic proportionality be made explicit in the legend, so that recipients may be alerted to the meaning they are expected to attach to it instead of relying on the outcomes of self-reference mapping. In fact, topography has convinced us that area proportionality corresponds to a homogeneous and uniform scale; in anamorphic maps, on the contrary, the statistical method of linear regression disrupts such correspondence and allows us to appreciate the proportionality of the data through an extension of surfaces, as exemplified in Fig. 6.18. It should be noted, however, that even though distances are considerably distorted, such transformation preserves area contiguity. As such, it integrates with cartographic semiotics both when it produces area information and when it shows linear or punctual data.[67] Of course, such metamorphosis could have negative effects over some undersurveyed areas, like those where visualized data is especially thin. Possible distortions may however be corrected technically.[68] The proportionality of the chosen model is substantially correct, even though it may seem arduous to the general public and may require some time

---

[65] See for instance: C.J. Kocmoud, *Constructing continuous cartograms: A constraint-based approach*, Master thesis, Texas A&M Visualization Laboratory, Texas A&M University, 1997 (www.viz.tamu.edu/faculty/house/cartograms/Thesis.html.

[66] This program, developed by C. Jackel and later enhanced by A. Agena, may be downloaded from http://www.arcscript.esri.com See: D. Andrieu, "L'intérêt de l'usage des cartogrammes: l'exemple de la cartographie de l'élection présidentielle française de 2002," in: *Mappemonde*, n. 77, 2005, http://mappemonde.mgm.fr/num5/articles/art05105.html.

[67] Area anamorphosis must be based on an equivalent projection, and linear anamorphosis on an equidistant projection. See: P. Poncet, "Lire le monde par la carte," in: J. Lévy, ed., *L'invention du Monde*, *op. cit.*, pp. 19–36.

[68] Such distortions undergo further corrective adjustments which change in accordance with the system used. See: J.A. Dougenik, N.R. Chrisman, D.R. Niemeyer, "Creating continuous cartograms in ArcGIS9," in: *Professional Geographer*, n. 37, 1985, pp. 75–81.

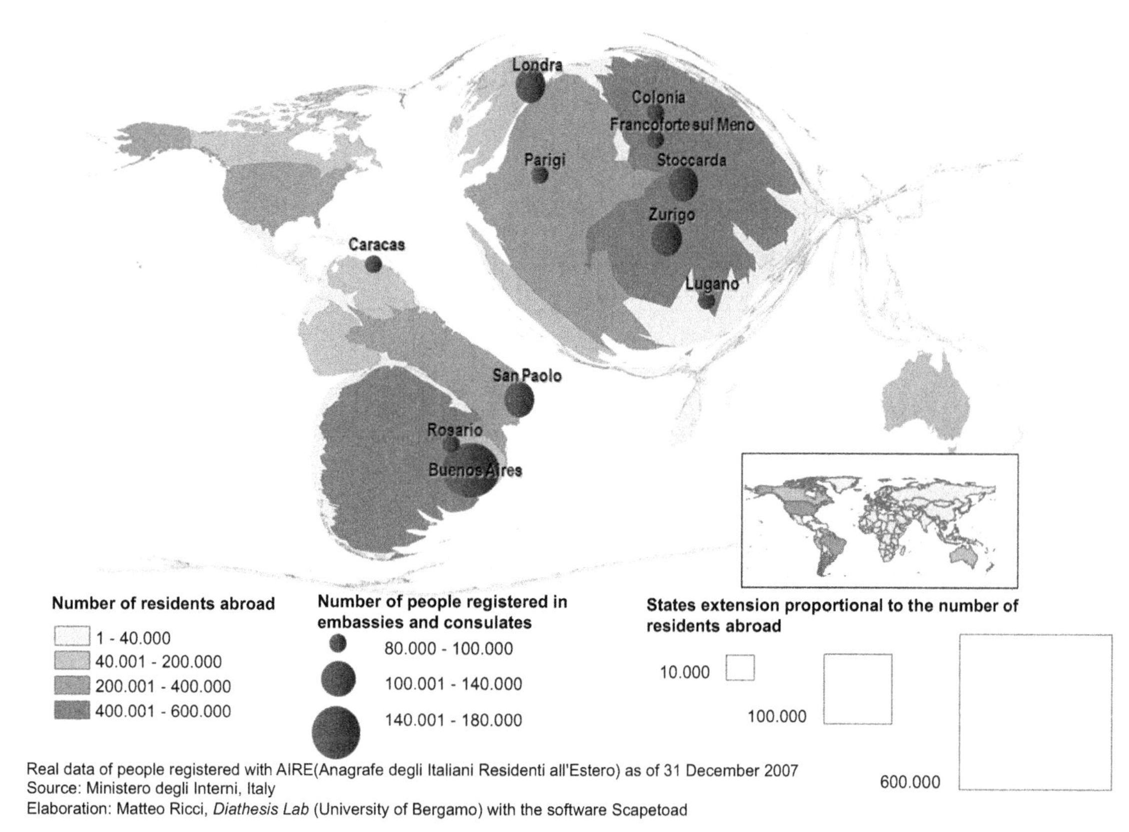

FIGURE 6.18

Anamorphosis: Italian emigration abroad to 2007. (See a color reproduction at the address: booksite.elsevier.com/9780128035092.)

to become part of popular knowledge. Similarly, the current processing time for building a database, which is today particularly taxing, will no longer be an issue in the future as the Statistics Agencies grant online access to their data banks. For the map to become relevant, it needs to offer within its representation area many sectors which can be matched to partial data of the represented phenomenon in order to define its distinctive internal features. That entails substantial processing power.

However, among the many techniques available to bind the surfaces in an indefinite series of variables, anamorphosis looks the most effective. It allows for the creation of maps based on a single social information set (thematic anamorphosis), or on the cross-referencing of several indicators (in which case we have a *cartogram*).[69] Cartograms are semiotically intriguing and deserve special attention on

[69] J. Lévy, M. Lussault, eds., *Dictionnaire...*, *op. cit.*, p. 74.

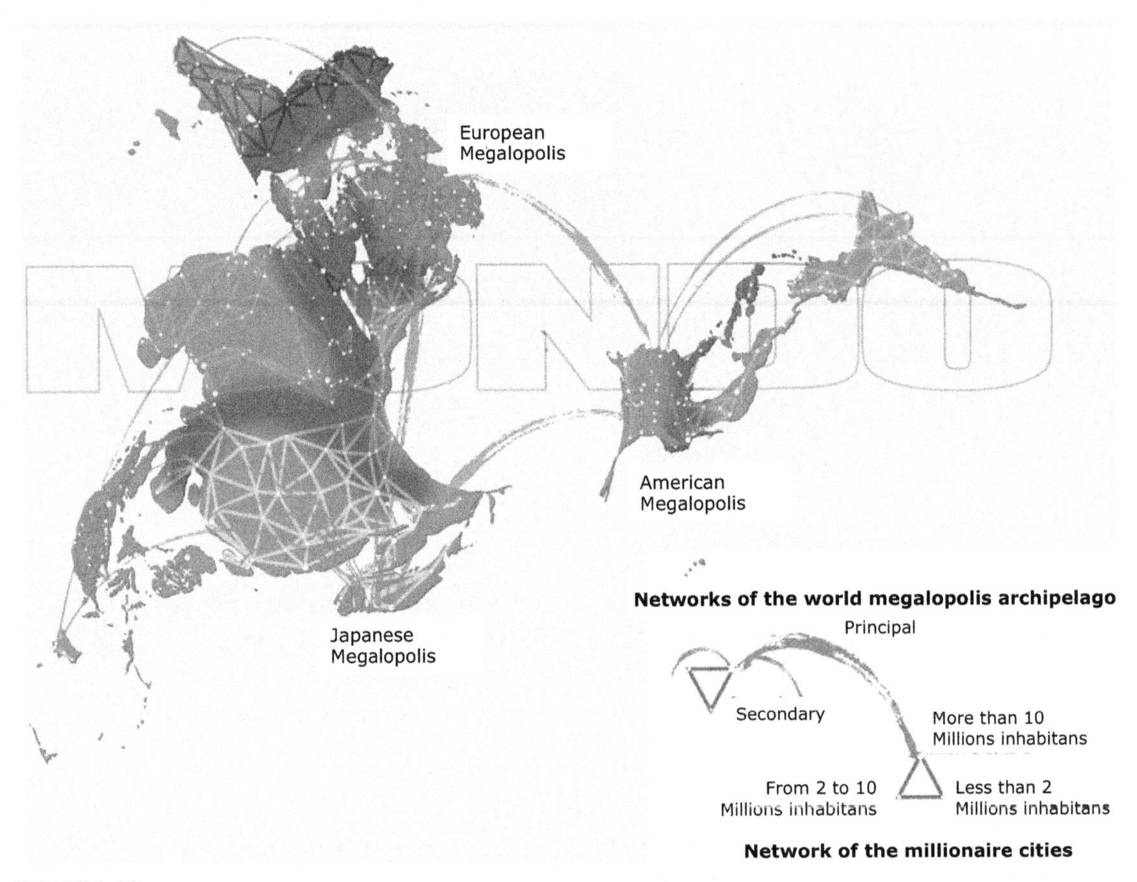

FIGURE 6.19

The co-spatiality of the contemporary world (from: Lévy, *L'invention du monde*, 2008). (See a color reproduction at the address: booksite.elsevier.com/9780128035092.)

account of their communicative propensities. They aim to create advanced, second-level knowledge and thus to render the complexity of current phenomena. The analytical categories that framed the world according to the parameters of center and periphery are no longer fit to promote the spatiality built by the network of exchanges and contacts. Cartograms advocate a new, more far-reaching category based on the megalopolis model: the category of functional nodes which innervate the mesh of structured exchanges along a horizontal interdependence. The cartogram shown in Fig. 6.19, built at the EPFL *Chôros Laboratory in Lausanne,* shows such spatiality: the continents are distorted according to the distribution of the population on which is superimposed information that recalls the network-like setup of companies based on the polarity of the megalopolis. We can thus quantify the phenomenon, and also qualify it by suggesting cultural assonances and dissonances (European, American, Japanese). What is envisaged here is a world united in its relations and yet subdivided into cultural areas. As it coordinates the various cities, this world in turn puts forward a model of urban co-spatiality.[70]

The representation thus obtained is opposed to the one traditionally based on the political subdivision of states. It has the advantage of underlining bands of fluid transition (*Horizon*) where marginality ensures the creation of new territories (*Confins*) which depend less on contiguity relations with their immediate surroundings than on relational flows.

Such an approach also leads to stealing the advantage topography ascribes to areas considered most important and placed in the center of the map: the whole area considered irrelevant, because virtually unaffected by social phenomena, is thus left empty or equated with the North Pole, as shown in the map of Fig. 6.18. To conclude, the potential of a cartogram lies in the fact that, by taking advantage of the structure underlying topography,[71] it offers other views of the world and helps us to counter the self-reference maps have acquired over time.

That does not mean we should indulge in an unreflecting application of anamorphosis, because even anamorphosis is the result of technical choices that interpret data and affect its visualization. At the same time, for the recipient to benefit from the data being conveyed we will need to promote awareness of the fact that anamorphosis is not a falsification, for that would be tantamount to claiming that only topographic spatialization is correct. Rather, we are in the presence of a different "construction," capable of favoring social data over topographic ones.

Cartograms are among the most effective tools currently available to us for inserting in maps the distinctive features of topological space, that is, extension and contraction. These go beyond the notions of "continuity" and "limit" and introduce that of "vicinity," which is peculiar to reticularity. Of course the reading of cartograms is not immediate and may even seem arduous at first because of the cartographic self-reference whereby topography prevailed. The deeper one's training in traditional cartography, the harder one finds it to accept cartograms. Also, we should not overlook the risk that the preservation of contiguity between areas may draw disproportionate attention to urban clusters at the expense of the rest. In general, anamorphosis has the immense advantage of having demonstrated that topographic metrics can be superseded. Once their lack of objectivity and accuracy are acknowledged, maps become one of the many ways to depict the world, finally released from conventions which, under the pretext of ensuring rigorous information, invalidate data irreparably.

---

[70] The map, designed by Patrick Poncet and assembled with Karine Hurel, may be consulted in: J. Lévy, ed., *Inventare il mondo*, *op. cit.*, p. 348.

[71] This uses the Mollweide and Buckminster Fuller polar projections, which have the advantage of preserving surface relations and shapes, albeit the representation of the world they yield is unusual. See: P. Poncet, "Leggere il mondo attraverso la carta," *op. cit.*

FROM REPRESENTATION TO CHOROGRAPHIC SPATIALITY

In conclusion, we can say that anamorphosis, in its various forms, is of crucial importance for the return of a topological space and must be included, with satellite technologies, which envisage the recovery of the "ruggedness" of the world, and with participatory technologies, acting for the recovery of the sense of the landscape, within a new panorama of experimentation. These are different lines of research which all aim to achieve a chorography. In the course of our argument, it became clear that the recovery of the visual aspect of landscape does not aim to assert visual reality but to promote a paradigm shift. Similarly, when we advocate the spatiality of globalization based on an analogical base map, we are not striving to perpetuate a reassuring topographic framework. On the contrary, we want to provide a performative rendering of the dynamism and of the process-like quality of phenomena through a preservation of their spatial distribution.

Undoubtedly, the recovery of the multiple models and plural languages which can convey *chora* puts forward a new way of conceiving a map: not merely a representation of spatiality but spatiality itself. Even though cartographic experimentation failed to exhaust all the issues and did not indicate with certainty the way in which the maps of globalization are to be built, the goal can be reached.

In the course of this book we have ascertained that this is not a matter of blind faith, but a scientific conviction, grounded in the theories and observations coming from the history of cartography and especially of cartographic interpretation. What triggered our research on the possibility of building new maps was not the structural features of cartography (be they cultural, technical or pragmatic). Rather, it was the evolution which cartographic interpretation has undergone. The object-based perspective which took the map as a self-consistent tool to be interpreted like other representations was discarded. As the focus of analysis shifted, maps appeared less a *mediation* of territory than *operators* able to determine courses of action over territory. To consider maps as "symbolic operators" expressing a very particular configuration of the world meant to acknowledge that maps are powerful metamorphic devices, achieving the *map = territory* equation not as something objectively definable, but as a potentiality by which and through which the society-space relationship is established. Such communicative outcome, which in the past gave maps authoritative status as reliable and objective systems of communication, may now be turned to our ends. That can be done by multiplying forms, languages, points of view, but especially by showing the figurativization procedures whereby the societal substance of the world is prospected.[72]

[72] As Louis Marin suggested in his comment on Jorge Luis Borges' thought on 1:1 scale "the perfect map was the one that preserved the mark of its difference [from the Empire] in the enunciation of the reduction rules that shaped it." (L. Marin, *Utopiques: jeux d'espaces*, Les éditions de minuit, Paris, 1973, p. 292).

Of course, the risk is that such fragmentation reduces the social role of maps. But the very opposite may also happen. If its iconizing outcome, as we have variously argued in the course of this book, can be kept in check, then the path is open to new communicative possibilities, which cartographic interpretation has been advocating for some time now, able to meet new social needs.

Today we can achieve a chorographic metrics that envisages the societal substance of the world as a salient feature in recording the dynamics of relationship between actors. We need to render cartographically a world made up of communities which metamorphose physical contingencies into a *lived space* and rely on reticular spatiality.

Such spatiality, we have seen, is a proxemic system, that is to say a communicative and anthropological domain that touches upon the perception, the organization and the use of space, of distance and of territory. The communication exchange does not merely concern the semantic level of the message, but also envisages the need for a shared view of the world, rendered by metrics that can grasp the essence of this exchange. It is the same vision shared to understand generalized mobility, the new forms of citizenship, and the emergence of problem areas in complex societies like our own. In sum, the static world is obsolete and out of touch with the current set of events, which needs to be interpreted through suitable representations. It is at this point that cartography comes on the scene as the only operator able to promote such interpretation based on a revisited symbiosis with geographical analysis. For this to happen, an epiphanic revolution is needed in the way we look at the power maps wield as they intervene in social dynamics in order to represent the kaleidoscopic features of the world. The multiple points of view, the many techniques, and the many languages that combine and intersect, all express the value of a societal world adopted in its relentless becoming. That is the true meaning of the chorography which I endeavored to illustrate here.

Glossary/Compass

Concepts and Definitions for Navigating the Text

Anamorphosis derived from the Greek word *anamorphōsis* ("transformation"). This term refers to a painting technique used during the Renaissance whereby an image was distorted deliberately so that it was not easily detected when observed from a frontal viewpoint. Now the term also refers to IT-based data viewing (obtained by interpolating a vector field with a gravity model), which undermines the assumptions of the topographic basemap by distorting them to fit the social data being represented. Territory thus ceases to be an intangible feature of the map and allows for engagement with the themes represented.

Cartesian logic moving away from the notion of place as a socially understood phenomenon, Cartesian logic uses cartography as a means to render territory in purely abstract terms, divorced from any interpretation. Phenomena are therefore shown through their material features within a geometric space.

Cartographic icon a semiotic figure based on the pairing of a designator with other signs in a map. This figure endows names with particular values and governs how names should be interpreted in territorial practice. As they neutralize any excess of signification and prescribe directions for interpreting what is depicted and what is excluded, cartographic icons draw attention to some of their features at the expense of others. Cartographic icons are at the core of the autopoietic process, whereby information is not only processed and communicated but also effectively produced.

Cartographic scale it is the reduction ratio between territory and its representation, pertaining to an analog principle between linear distances verifiable on the ground and those present on the map. This relation is rendered on the map through a mathematical and graphical ratio. Positivist geographers named maps on the basis of their scale, but scale denominations have over time taken on different meaning in different countries.

Cartographic semiosis a theoretical approach that views a map as a semiotic field in which information is produced and conveyed in the presence of an interpreter. Within the *semantic* domain, meanings are formed via sign encoding; in the *syntactic* domain new meanings develop from the relationships signs are made to establish; in the *pragmatic* domain a map is interpreted and becomes a matrix for social behavior. The theory of cartographic semiosis envisages the abandonment of topographic metrics, an essential step towards redeeming *chora*. As it shifts focus from what portion of reality the map reproduces to what it conveys with regard to the meaning of territory, cartographic semiosis paves the way to mastering iconization to its own ends.

Chora a territorial concept that gives voice to assets, values, and social interests formulated by local inhabitants and made explicit by landscape. Augustin Berque rereads the classical *chora/topos* dichotomy in terms of the relationship between the physical dimension of space (*topos*) and its narrative dimension, with a view to expressing cultural identity (*chora*).

Chorographic metrics it envisages a topological space and a landscape-based logic for use in cartography. Firstly, two-dimensional Euclidean space and its attendant criteria of a unique observation point, metrical accuracy and linear distance are abandoned. Chorographic metrics embraces three-dimensionality, thereby opening up to subjectivity, relativity, and topological distance. Secondly, icons are developed to render less a cluster of properties selected by virtue of their materiality than a set of landscape values that are socially produced.

Chorography this term used to refer to regional cartography. In this book, chorography designates a kind of cartography that endeavors to render the "ruggedness" of the world as well as the social and cultural value of territory. Chorography relies on transparency and aims to produce maps that enable recipients to make critical sense of what they find on maps, thus avoiding iconizing prescriptions.

Cognitive spatialization The process whereby objects are invested with spatial properties in relation to an observer who, by creating a network of territorial references, is in turn, "spatialized" with respect to them. In cartography, spatialization lies inside the communicative procedure activated by icons to "show" the enunciation of territorial semanticization.

Critical cartography an analytical method aimed at an ethical interpretation of the role which cartography plays or has played in the social realm. It exposes cartography's iconizing outcomes. Such method relies on *reflexivity*, because it implies the researcher's involvement both in the study and in the effective solution of socially relevant issues, such as the role digital technologies play in empowerment, or the potential cultural assimilation brought about by cartographic tools.

Datum An algorithm that defines the matching of geographical coordinates to points on the spherical surface of the Earth. Datum also forms the basis for the measurement of altitude.

Euclidean space an "absolute/positional" space, conventionally empty, abstract, in which territorial phenomena are inserted in accordance with certain properties, such as: i) *continuity*, i.e. space as a seamless continuum; ii) *contiguity*, the uniform representation of proximity and, therefore, of effects related to contact or distance; iii) *uniformity*, or the uniform measurement and treatment of distance. In cartography, Euclidean space has invariably been rendered through the criteria of uniqueness of the observation point, metric accuracy and compliance with linear distance.

Figurativization a communicative procedure set in motion by cartographic icons in order to "show" the enunciation of territorial semanticization. Its stages are *spatialization* which establishes and bolsters the referential features of the designator; *figuration* which highlights its distinctive features; and, finally, *iconization* which, embracing the outcome of spatialization and figuration, invests the designator with social values.

Geographic Information Technologies (GIT) a set of systems and technologies aimed at acquiring, composing and managing geographic data to be analyzed, formalized, and displayed in graphic or cartographic form.

Geographical map a symbolic operator which, released from its customary function of mere tool for recording reality, becomes a highly advanced *medium* for hypertextual communication. Such medium cannot only describe the world but also conceptualize it, that is to say indicate how it functions on the basis of an hypothesis. That is why maps play an active role in territorial praxis.

Geographic scale different dimensioning of territorial realities which results in the production of different discontinuity thresholds, in the measurement of distance and in the assessment of phenomena. A *place* or *area* may be identified as *large* or *small* depending on how it is measured or assessed. If the social dimension is defined in relation to topological distance, one must define the value of this distance by adjusting the use of scales, which will depend on the value ascribed to the phenomenon, rather than on the value of its topographical dimension. Thus, the specificity of social space leads us to distinguish between degrees of size and degrees of complexity, to state that the larger space encompasses smaller spaces but does not contain them, i.e. it is unable to express some of their specific features.

Iconeme a semiotic feature comprising emergent or salient features. It endows landscape with order and communicative cohesion. As defined by Eugenio Turri, iconemes generate the concept of landscape because they reflect the iconic quality of place. Iconemes rely on the coherent set of relations built around one outstanding feature or between several features and, therefore, between several iconemes. It follows that, in its semantic layout, landscape is a mental construct which demands the intervention of an interpreter.

Iconic junction a semiotic figure which envisages a mashup between landscape figuration and map figuration. It promotes the formation of syntagms, or in other words of relational clusters able to affect iconization. At the level of syntax, iconic junction is also involved in the creation of a landscape cartography more attuned to the multiple communicative role played by icons (archetypes/clichés) in hierarchizing information and bringing the cultural value of landscape to the fore.

Iconization The communicative process whereby, on the basis of a map's self-referential outcomes, highly conjectural facts are expressed as truths. Iconization suggests using a map as a theory to be relied on in order to evaluate all information being conveyed. It activates a system whereby information and concepts – which intersect on the double level of cartographic communication (description and conceptualization) – are free to circulate endlessly across a wide range of intermediations.

Identity discourse The process, defined by Angelo Turco, whereby the set of values that are issued from territory is dynamically shaped. Such values will sanction the territory's social identity as iconized and conveyed through landscape.

Landscape in its basic sense landscape is the visual shape of territory, but from the point of view of its semantic layout it is a mental construct which demands the presence of a subject. For landscape is not merely the result of sensory experience linked to the physiology of the eye, but depends on the processing of such sensory experience via the brain. The cartographic transposition of landscape focuses less on its perceptual features than on the rendering of concepts through a recovery of their social values and the use of perspective. As it organizes objects hierarchically, perspective causes observers to perceive what is "near" or "far" and invokes the presence of a subject.

Landscape-based logic assuming landscape in its two territorial functions (establishing its visual form and expressing its cultural value), landscape-based logic envisages chorography as a cartographic practice able to express the cultural substance of territory. While such logic takes into account the material properties of phenomena, it pairs them with their intangible, symbolic and functional features, in a ranking system that depends on the assessment made by the subject (or local inhabitant).

Médiance "a complex form of representation, at the same time objective and subjective, physical and phenomenological, ecological and symbolic," able to address both the scientific and the artistic issues ascribed to landscape (A. Berque).

Mediatization the role a map plays as it autonomously mediates communication. Maps do not play exclusively a role of interposition between reality and those who interpret it (mediation); rather, they take an active part in shaping reality. By envisaging an interface between reality and society, maps fulfill the equation map = territory not as something to be defined objectively, but as a potential through which and in which the society-space relationship occurs.

Metrics a method for measuring and managing distance. The set of technical skills that contributes to managing space measurement and the localization of geographical phenomena.

Miniaturization a figurative procedure that brings about a condensation of meaning not only by scaling down its object, but also by imposing its own referential definition, which entails the transmission of special symbolic messages via cartographic icons.

Multiscalarity the property whereby each phenomenon may be displayed on multiple territorial levels pertaining to the global and the local. Such levels appear less as scalar levels of geography than as two different logical dimensions governed by scalar mobility. In WebGIS systems, multiscalarity is relevant because interactivity enables users to choose different ways of viewing a phenomenon and may make it possible for that phenomenon to take on different semantic senses in accordance with scale.

Reflexivity in the study of socially relevant phenomena, it refers to an analytical perspective aimed at achieving practical goals by acting upon researchers. Researchers are made to reflect upon themselves and the actions they are doing or have done; actions which in turn sketch the scenario in which they are to operate.

Reticular spatiality based on relations and connections. Nowadays, as exchanges and contacts intensify, reticular spatiality involves actors who – albeit with a certain degree of asymmetry – are able to shift from one scale to another without losing sight of their local rootedness. Cartographers have been testing new ways to render such spatiality. While preserving the prominence of localization, such new forms are meant to envisage connections that affirm the social sense of territory within the network.

Self-reference refers to the map's ability both to be taken at face value and to intervene in the communication quite independently of the cartographer's intentions or the awareness of the recipient/user. By virtue of self-reference, a map becomes a sign-based system which speaks for itself once it has been established, remains relatively independent from all that preceded it, and goes beyond the uses for which it was initially intended.

Semiotic field a map may be defined as a semiotic field within which signs become actual sign vehicles, that is to say, they acquire depth once their meaning has been established and interpreted. Each *sign vehicle* – the designator attached to any given sign – activates three orders of relationship that pertain to: 1) the formation of meaning; 2) the interconnections between signs; 3) the recipient's interpretative act.

Spatial localization one of the spatialization procedures. It may be defined as the construction of a system of references that enables researchers to situate objects based on their horizontality, verticality and perspective (front/rear) in relation to a subject.

Spatiality The set of features belonging to the spatial dimension of a social reality. In cartography, spatiality identifies the way humans relate to the world, as they elect either one of the primary functions of a map: describing or conceptualizing. Maps *describe* the world when they endeavor to render its features as seen through a direct observation of reality: in other words, they explain to us how the world is made up. Alternatively, maps *recount* the world; tell us how it works according to categories of representation which arise from one possible interpretation of it.

Substance the third component of topological space (together with metrics and scale); it aims to reject the postulate whereby social phenomena may be considered "without" space. Unless it can conclusively be shown that the contribution of space to the description of reality is irrelevant, all phenomena must be seen to enclose a cluster of spatial qualities not only in their form but also in their content. Jacques Lévy defines substance as "the non-spatial component of spatialized objects," which does not correspond to the residue of spatiality but is an integral part of it.

Symbolic operator role played by the map as it adjusts the complexity of geographical space, prescribing how territory ought to be assessed and directing actions to be performed within it. A map becomes a symbolic operator when seen in its role as an elaborate communicative system, whose distinctive quality lies in the fact that it is a meta-geography.

Territoriality In addition to the physical and natural relations between objects within a social setting, territoriality refers to all the actions, techniques and tools that are mobilized in order to act territorially.

Topographic metrics it envisages an abstract space through a Cartesian codification of signs and the preservation of features dictated by Euclidean space (contiguity, continuity, and uniformity). These are brought together in a method of measuring distance whose main interest has less to do with the representation of object properties than with a preservation of their relationships and their size.

Topography A cartographic genre which involves the use of abstract and coded language derived from Cartesian logic. It relies on a survey and mapping system aimed towards the representation of a well-defined concept of space: i.e. Euclidean space.

Topological distance a measurement of the distance between phenomena which rejects the "rigidity" of territory as a Euclidean space (a continuous space, marked by strict boundaries and tending to self-containment) in favor of the "flexibility" of a network, a space built upon a system of relations, whose features are discontinuity and incompleteness.

Topological space it envisages the recovery of a Leibnizian space characterized by a "relative/relational" dimension between social actors. Based on the extension/contraction duality, Leibnizian space goes beyond the notions of "continuity" and "limit" and introduces that of "vicinity," which is typical of reticularity. Extension and contraction are closely linked to the layout of a topological space based on the complexity of relationships and the existence of multiple paths other than the linear paths typical of topographic metrics. In chorography, topological space ushers in three-dimensionality and embraces subjectivity, relativity and topological distance.

Topos a territorial concept that refers to a physical space and endows the perceptual aspect of phenomena with value by identifying their size and location, while *de facto* voiding their social meaning. In topography, such concept was translated into a system of signs that does not preserve the substance of objects, but charts their mutual links and relationships, via an analogical preservation of distance.

Transcalarity the actor-based property of the contemporary world that enables us to see the same phenomenon at various territorial levels pertaining to the global and the local, by changing our point of view. Together with multiscalarity, such property, marked as it is by expressions of territoriality, stands up against the kind of nondescript localization inherent in the distortion that topographic imaging entails.

WebGIS high-level online cartography that takes over the functions of GIS in a Web environment. Along with web-based applications, this database-driven mapping system is noticeably at odds with traditional cartography in terms of usability and data management. The unique success of Web Graphic Information Systems, brought within the fold of cartography with truly unprecedented outcomes, is also attested by the fact that it differs quite markedly from more traditional modes of online mapping.

WikiGIS online mapping technologies based on the principle of geo-collaboration. It is a WebGIS built online via collaborative actions that require interactions between participants. The merging of traceable contributions produces coherent geo-related representations that may be easily updated and enriched. Unlike many other map services that undertake the creation and volunteered update of geo-localized content – in which case users do not effectively alter the existing database – WikiGIS ensures recorded traceability at all stages of contribution.

Zenithal projection it is the most abstract way to render the world cartographically. It provides for each feature of the represented territory to be rendered from an observation point perpendicular to it. Zenithal projection does not require a single point of observation, but many viewpoints as there are objects represented. Such projection disposes of the human gaze and, in turn, the type of world ranking which a perspectival view preserves. What is lost in the process is iconic significance, to the advantage of description.

Author Index

Note: Page numbers followed by "f" indicate figures.

Subject Index

Note: Page numbers followed by "f" indicate figures.